DRIVEN BY NATURE

*A Personal Journey from Shanghai to Botany
and Global Sustainability*

DRIVEN BY NATURE

*A Personal Journey from
Shanghai to Botany and Global Sustainability*

PETER H. RAVEN

Edited by Eric Engles

MISSOURI BOTANICAL GARDEN PRESS

ISBN 978-1-935641-19-3
Library of Congress Control Number 2020939163

Publisher: Liz Fathman
Managing Editor: Allison M. Brock
Editor: Lisa J. Pepper
Press Coordinator: Amanda Koehler

Cover and book design by Katie Koschoff

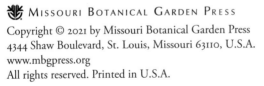 MISSOURI BOTANICAL GARDEN PRESS

MIX
Paper from
responsible sources
FSC® C008955
FSC
www.fsc.org

The rule of no realm is mine, neither of Gondor nor any other, great or small. But all worthy things that are in peril as the world now stands, those are my care. And for my part, I shall not wholly fail of my task, though Gondor should perish, if anything passes through this night that can still grow fair or bear fruit and flower again in days to come. For I also am a steward. Did you not know?

—J. R. R. Tolkien
The Return of the King

If we love our children, we must love our earth with tender care and pass it on, diverse and beautiful, so that people, on a warm spring day 10,000 years hence, can feel peace in a sea of grass, can watch a bee visit a flower, can hear a sandpiper call in the sky, and can find joy in being alive.

—Hugh H. Iltis
"Man First? Man Last? The Paradox of Human Ecology"

— Contents —

— Foreword —

A scientist of creative stature is metaphorically an explorer—an adventurer turned loose in unmapped terrain. Such is literally true in the case of ecologists and taxonomists. They discover and chart the little-known faunas and floras of the world. Their breed was exemplified by Carl Linnaeus, Alfred Russel Wallace, and Charles Darwin during the early centuries of biology. Peter Raven is one whose career has decisively continued the traditions into our own time.

Raven has graced this memoir with a vivid account of his formative years as a child naturalist, which alone would make *Driven by Nature* a valuable contribution. He describes vividly a pathway that has been followed by many scientific naturalists during their formative years. Because I have known Peter Raven only as the formidable botanist and scientific leader of his full maturity, I had at first no idea of the early events that led him into his career. I learned in this book that although he and I grew up in radically different parts of the United States in different kinds of families, our key experiences in scientific study were remarkably similar. At the age of eight or nine we became fascinated with butterflies and built up substantial collections. Raven lived close to the California Academy of Sciences in San Francisco; I spent two years only a few city blocks from the National Zoo in Washington, D.C., and often took a ten-cent trolley ride to the National Museum of Natural History. Later he turned to beetles and finally plants. Having moved to Alabama with my parents, I shifted to snakes, then ants. We both had supportive parents, and the fortunate assistance of sympathetic mentors. We both devoured stacks of *National Geographic*, filled with science and travel. We were strongly encouraged by advice from professionals to become scientists and given opportunities to travel widely. We both conducted research at an early age.

The difference, however, between Peter Raven and most of his same-age colleagues, myself emphatically included, was that he not only conducted a great deal of research early on and worldwide in scope, but also managed to find time to serve as a visionary leader in education and administration, comprising a professorship at Stanford University, authorship of the leading textbook in botany (*The Biology of Plants*), leadership roles in the U.S. National Academy of Sciences, and, above all, directorship of the Missouri Botanical Garden, which over decades he built into one of the most accomplished scientific institutions in the world.

In spite of his heavy responsibilities in research and administration, Peter was also a pioneer in biodiversity conservation. The following account is from my own memoir *Naturalist* (Island Press, 1994):

> I was finally tipped into active engagement by the example of my friend Peter Raven. . . . Peter was determined and fearless. He had no qualms about activism. By the late 1970s he was writing, lecturing, and debating those still skeptical about the evidences of mass extinction. In 1980 he chaired a National Research Council study of research priorities in tropical biology, putting stress on the urgent problems of deforestation and biological diversity. More than anyone else Raven made it clear that scientists in universities and other research-oriented institutions must get involved; the conservation professionals could not be expected to carry the burden alone. One day on impulse I crossed the line. I picked up the telephone and said, "Peter, I want you to know that I'm joining you in this effort. I'm going to do everything in my power to help." By this time a loose confederation of senior biologists that I jokingly called the "rain forest mafia" had formed. It included, besides Raven and myself, Jared Diamond, Paul Ehrlich, Thomas Eisner, Daniel Janzen, Thomas Lovejoy, and Norman Myers. We were to remain in frequent communication from then on.

To read *Driven by Nature* is to learn about a unique and distinguished life within a seldom-reported theater of the biological sciences.

Edward O. Wilson

E. O. Wilson and me on the Galapagos Islands, May 14, 2011, after a half century of cooperation in conservation efforts, enjoying the plants and animals of that magical archipelago.

— PREFACE —

Writing an autobiography was first suggested to me, in the waning years of my tenure as director of the Missouri Botanical Garden, by the remarkable Jonathan Kleinbard, the Garden's assistant director at the time, who had served for more than twenty years as a vice president at the University of Chicago. This highly reflective man, steeped in literature, considered my adventures at the Garden and in the broader arc of my life as stories worth telling.

Thinking over the matter, it became clear to me that an account of my life might be useful and interesting to others, offering them insights and perhaps even inspiration. And it could promote better understanding of the field of botany and the need for conservation of Earth's biodiversity, two of my deepest passions. So I decided it was a project worth pursuing. However, I knew myself well enough to realize that it would be difficult to write an autobiography in the normal way—that is, by actually writing all of it. As you will soon learn in reading the book, it is in my nature to take on more projects than I can reasonably accomplish, and there was too much going on to set aside the time needed to write seriously about what was then seventy years of life. But I knew that many autobiographies and memoirs start out as oral histories: the author *tells* his or her life story and a "collaborative memoirist" translates the audio recording onto the printed page. That seemed the way to go for me!

My first attempt to use audio tapes and a collaborating memoirist didn't work out, and that soured me on the idea. I set aside the project, and it languished for several years until 2013, when I was referred to Kathy Evans, a personal historian and memoirist. Kathy, a St. Louis resident, interviewed me every week or two—when I was in town—and began to convert the recordings into written material. The arrangement worked well and the process moved along.

As I talked about my childhood, and then my formal education and early professional career, I realized that the details of my life had the most meaning to me when considered against the background of what was happening in the wider world at the time. The demonization of Americans of

Japanese ancestry during the Second World War; the rise of suburbia, free-ways, and shopping centers in the 1950s; the questioning of long-standing mores and cultural assumptions during the 1960s; the tension and violence surrounding the anti-war movement: these events and changes seemed as much the stuff of my life as my interest in beetles and flowers, my grad-uation from high school, and my hiring at Stanford University. In telling my story with intention, I seemed to be making sense of my life in a holistic way that might not have been possible through any other means. It was very satisfying at the time; I hope that the result of the process is a life story that's more meaningful to others as well.

I also found that in narrating the stories of my life I gained some deeper insights into my own personality. This might seem surprising to those naturally possessed of good self-awareness, but to me it was remark-ably revelatory.

In talking about my life and then writing about it—and I did eventu-ally do a fair amount of writing—it was difficult to avoid the conclusion that much of what characterized my life and career could be seen as a re-sponse to inherent insecurity about my worth as a person. I suppose that most people, to the extent that they are being honest with themselves, are insecure; however, each person's insecurity is expressed in a very different way. As I moved through the story of my life, I became very much fasci-nated with the sources and expressions of my own.

The process of writing—I will now use the verb *write* after having ex-plained what lies behind it—consolidated my suspicions that I was born into a situation in which developing a solid sense of self-worth may have been difficult. Neither my mother nor my father had particularly happy childhoods, and my mother always considered herself to be in second place behind her brother. When I was born, in the middle of the Great Depression in Shanghai, my father had been out of work for a year, and his ongoing difficulty in keeping the family on a secure financial footing thereafter made my mother deeply anxious. I became more clearly aware in writing about my childhood that my mother responded to her worries about my father's lack of breadwinning abilities by pinning her hopes for the future on me. Although she was always loving and kind, she relent-lessly pushed me to succeed. Her expectations were always so high that I could never do well enough in school or have enough accomplishments. It certainly didn't help that after being cared for by a nanny—an *amah*—during my first year of life I was abruptly separated from her when my parents fled the advancing Japanese invasion. As an adult, I had a dream in which my amah passed me, as a baby, over the rail of a ship to my mother, and I slipped out of my mother's grasp, falling and sinking to the very depths of the sea. After the dream, I asked my mother about how I acted after being taken away from my amah, and she replied that I cried for most of the time we were on the ship and then ate very poorly for the next several years. That experience might have been enough to start me down the path to being inherently insecure.

As I grew up, I found three ways of establishing a sense of security and balance. These themes will be apparent as you read the book. One was making collections of things that captured my interest. A fascination with insects that began when I was six years old led me to collect butterflies and then beetles. Once I had a few boxes of neatly pinned insects, the impulse to collect was indelibly imprinted on my psyche when my parents praised my efforts effusively and showed my boxes to their suitably impressed friends. If it was a psychological need to seek validation that made me a collector, so be it! Collecting was what got me started in biology.

The second way I developed a more secure sense of self was to work on becoming the kind of person people wanted to be around. Those who have known me only as an adult may be surprised to learn that I was not always outgoing and usually jolly; as a child I was shy and reserved. As I began to adjust to life outside of my home, I turned naturally to jokes and storytelling as ways of making positive impressions in social situations. This trait has stayed with me all my life. It seems to fit my Irish heritage.

Third, I became highly motivated, particularly in the sense of establishing many goals and pursuing them all at the same time. This ambition may have arisen in response to my mother's high expectations: the more goals I had, the greater the chances of her viewing me favorably. In any case, an inner drive to accomplish a great deal established itself early on. I remember internalizing sayings like "We weren't put on this earth to rest." St. Augustine's familiar statement "Thou hast made us for thyself, O Lord, and our heart is restless until it finds its rest in thee" became a part of my plan.

As I wrote this book, I became painfully aware of how my ambition had many negative consequences, for both me and those closest to me. At the simplest level, the drive to succeed and to accomplish led me to cut corners. As I was growing up, this sometimes showed up as a neglect of personal hygiene and other aspects of self-care—not terribly consequential in the scheme of things but sometimes leading to embarrassment. As an adult, with the demands of building a career and raising one's status taking over, my ambition led me to focus my energies almost exclusively on work—often to the detriment of my relationships with my intimate partners and my children. Throughout much of my life I was certainly aware of shortcomings in my family life, but because what I was doing was so normal given the gendered division of labor at the time, I didn't think very much about the causes and consequences of my strained relationships with my family. My awareness of these problems developed later in life, but it was writing this book that brought it all home, sometimes in disturbing ways.

Please don't get the impression that in the book you are about to read I go on baring the deepest recesses of my heart. That's not the case at all. In fact, this preface will be the only portion of this book in which you will find me discussing at length the personal attributes of which I am not particularly proud. I have had the good fortune of leading a life directed primarily by my great affection for other members of my species—despite my ostensible focus on organisms containing chlorophyll—and I am happy to have made this a central theme of the book. Throughout my life, I have taken great joy in learning about and encouraging other people in any way that I could, to give attention and warmth to others, especially at the times when it matters most. It is my hope that this book inspires others to do likewise. It seems to me that encouraging others should be a central purpose of our lives.

It was in part my concern for the well-being of those who will inherit this earth and grapple with its challenges long after I am gone that led me to end this book with an epilogue that may seem atypical of an autobiography. During my life, I have witnessed an intensification of human impacts on our planet's ecosystems that is truly terrifying in its scope and potential consequences. Given what I have learned about the incredible diversity of life and the interconnectedness that ties humanity's fate to that of Earth's non-human species, I can't stand idly by while it is all threatened by our sad degree of ignorance and selfishness. Thus, I considered it necessary to do what I can, both in my life and in this book, to encourage the awareness, caring for one another, and love that it will take to preserve our planetary home.

— I —

Family Roots

MY ANCESTORS BUILT ROADS, operated banks and bakeries, owned inns and restaurants, farmed, prospected for gold, and rode for the Pony Express, but none of them were scientists. If my forebears didn't provide role models for my life as a botanist, however, what they did offer was just as valuable, perhaps even more so. On both sides of my family, they journeyed to faraway lands, leaving the familiar behind and taking with them not much more than their own resourcefulness and courage. They loved, lived, scraped by, prospered, and died far from their original homes. Thanks to their curiosity and pioneering spirit, conveyed to me from my earliest childhood in family stories and nurtured in the hearts of my parents, I grew up with a sense of being a citizen of the world. I gained an appreciation for the globe's endless wonders and a feeling that it was my destiny to travel and explore this amazing planet.

My own origins were tied directly to my great-uncle Frank Raven's journey across the Pacific at the very beginning of the twentieth century. A civil engineering graduate of the University of California, Berkeley, Frank had spent three years on the island of Maui, employed on civil

I

engineering contracts. He had done
well, but nursed dreams that Hawaii did
not seem to have the capacity to fulfill.
As a single, twenty-eight-year-old man
who must have felt he had nothing to
lose and much to gain, Frank boarded
a ship bound for Shanghai in January
1904.

Because he set out from Hawaii,
Frank's steamship voyage to China was
one third shorter than it would have
been for anyone traveling from the
American mainland. Unlike most Americans, he had already seen Shanghai,
having stopped there in 1898 on his
way to the Philippines as a soldier

My father's uncle, Frank Raven, built a
fortune in Shanghai from 1904 to 1935
and then lost it when the Great
Depression belatedly reached China.

during the Spanish-American War. His odyssey back there took place
amid an imperialist, expansionist age. It had been scarcely a decade since
the U.S. Census Bureau had officially declared the closing of the western
frontier in 1890. Although they had long before subjugated the native
people of North America and annexed large parts of Mexico, the imperialist sensibility was running at full tide among Americans, often mani-
fested as personal ambition. My great-uncle was one of many venturers
who set out for faraway places to see what they could find, use, create,
build, or appropriate. His choice of China as a place to make his mark
was not entirely surprising because the U.S. had been establishing a presence there for decades. While my great-uncle was getting his bearings in
Shanghai, millions of Americans were getting their own glimpse of the
Orient at the 1904 St. Louis World's Fair (also known as the Louisiana
Purchase Exposition), where China presented an elaborate, government-
funded pavilion.

Uncle Frank discovered a China in the process of transformation. The
country had increasingly come under the domination of Great Britain
and other foreign powers and was defeated in several wars during the
course of the nineteenth century. The Boxer Rebellion, an ardent but fu-
tile attempt to evict the hordes of Western missionaries and traders believed to be impoverishing and corrupting the country, had been brutally
put down just a few years before Frank arrived in Shanghai, and foreign

domination continued. The presiding regent, the Empress Dowager Cixi, found herself caught in the crossfire of old versus new as foreign powers pushed China for further modernization. Cixi was anxious to find ways to modernize China, but foreign powers left her little room to maneuver. Outside investment in China continued strong in a tumultuous environment that had enriched many foreigners and a few Chinese since the Opium Wars of the nineteenth century had built a generation of millions of addicts in China, while tightening foreign control of almost everything in the country. In this atmosphere, many foreigners became wealthy, and Frank Raven was one of them.

In 1904, Shanghai was a metropolis of more than one million inhabitants. She was the "Pearl of the Orient" and the "Paris of the East," a huge and diverse community where many refugees from other lands, particularly Jews, found refuge. Signs of modernization and international influence were everywhere. There was a call to translate the *Encyclopedia Britannica* into Chinese. Female foot binding had begun to lose favor, and many men cut off the queue, the long pigtail they had been forced to wear as a symbol of their submission to the Manchus, who, invading from the north, had founded the Qing Dynasty in 1644. Shanghai's so-called International Settlement, dominated by British and French interests, was increasingly entered by Americans and in effect governed from abroad. With its influence, a rapidly growing Shanghai was becoming increasingly Western in character. Passenger and merchant vessels crowded shoulder to shoulder for a mile along the city's waterfront area, called the Bund, which, dominated by tall buildings, developed as the economy expanded.

Soon after his arrival in Shanghai, Frank was hired as an engineer by the Shanghai Municipal Council, working on the network of roads that was under construction at that time. Shortly, however, he was, like many others, lured away from engineering by the opportunities associated with Shanghai's huge real estate boom. By 1914, Frank had made enough money to form the Raven Trust Company, the rock on which his subsequent banking and real estate ventures were built; collectively, "the Raven group" which grew to a total of $50 million by the 1930s, nearly $1 billion in today's dollars. Frank's assets grew rapidly into one of the largest foreign-controlled fortunes in China. In 1916, he opened a bank, the American Oriental Banking Corporation, legally a branch of an inactive corporation he founded in Connecticut, which ran with great financial success from 1916 until its demise in 1935.[1]

When a California businessman named Cornelius Vander Starr arrived in Shanghai in 1919, he found in Frank both a good friend and a willing partner in the development of his insurance empire. Raven and Starr remained partners and close friends into the 1930s. Starr's companies would eventually grow into what is now AIG, one of the Fortune 500 top ten companies before the Great Recession in 2008.

Sometime after his arrival in Shanghai, Frank met his future wife, Elsie Maud Sites, the Chinese-born daughter of Methodist missionaries. The Sites family had been long established in Fuzhou, a large coastal city, by the time Frank arrived in China. Elsie's parents had sailed to Xiamen (then known as Amoy) in Fujian Province aboard the *Cathay* in the summer of 1861. Elsie and Frank married in Yokohama, Japan, in 1909.

While Frank built his fortune in China, his brother Charles Hamilton Raven, or C. H.—my grandfather—made a life in Hawaii. A graduate of the University of California, Berkeley, like his brother, C. H. had moved to Maui just before Frank embarked for Shanghai. He had made his start in the islands by teaching school for a few months on the remote island of Molokai, after which he moved to Honolulu to take up work as an inspector at the Customs House. There he met my future grandmother, Lillian Stack; they were married in 1905. My father, Walter Francis Raven, was born in 1906, with his sister Helen following two years later. In 1910, C. H. left his job at the Customs House and moved with Lillian back to the San Francisco Bay Area, where their third child and last son, Charles, was born in 1912. The family shifted its place of residence several times during those years, sometimes living in Honolulu, sometimes in California. At times, C. H. traveled by himself on business, at other times with his family.

My father often likened the Honolulu in which he grew up to Thornton Wilder's *Our Town*. He remembered it as a pleasant Midwestern-style village, but set in the middle of the Pacific and with mango trees in the backyard. He recalled that in those days, before the Ala Wai Canal was dug, he and his family would sometimes travel out to Waikiki on a boardwalk. There they could buy ice-cream cones from the vendors who pulled their carts up and down the promenade. It was a pleasant life, though it wasn't to last for long.

In 1918, C. H. sailed alone to Shanghai and spent eight months there, visiting Frank and exploring the possibilities. The opportunities C. H. found in Shanghai clearly appealed to him; he returned to settle there in early 1921. While he got his bearings in China, Lillian and the children stayed in San Francisco waiting for the time when they could follow him there. My father never talked much about this part of his childhood. When he did, I got the impression that he was often unhappy during those years.

In Shanghai, C. H. bought Sullivan's Chocolate Shop, an ice cream parlor across from Frank Raven's American Oriental Banking Corporation on Nanjing Road. It was a popular gathering place where foreigners and Chinese alike celebrated special occasions. In 1922, Lillian sailed with the three children to reunite with C. H. in China, after my father had finished a year at Galileo High School in San Francisco. In Shanghai, the family settled into an apartment at the corner of Bubbling Wells Road and Nanjing Road in the French Concession, where Frank had made his home years earlier. Frank, by this time, was already wealthy and growing richer, and my grandfather's family was quickly introduced into Frank and Elsie's social and professional circles.

Shanghai was a flourishing metropolis of some three million people. Here on bustling lower Nanjing Road, the Chocolate Shop was located across the street from the American Oriental Banking Corporation. The Ravens owned both of these establishments. Today, Shanghai is the largest city in the world, with some twenty-five million residents.

In addition to running the Chocolate Shop, C. H. managed the first Western-style bakery in Shanghai, the Bakerite Company. Bakerite supplied American Navy ships with baked goods and over time earned a reputation among the military as offering "the closest thing to Stateside food" available in Shanghai. It was also said to be the first Western firm in Shanghai to hire a Chinese publicist. The motto that resulted—"Eat Bakerite bread and have only boy babies"—sounds strange to a Western ear, but to a Chinese person it simply meant "Eat Bakerite bread and have good luck."

My father Walter was enrolled in the Shanghai American School, a first-rate institution. After he graduated in 1925, Walter returned to California to attend Santa Clara University. There he met Jack Breen. It was a lucky introduction, as Jack's sister Isabelle would become my mother.

☙

While the members of my father's family made the journeys that defined Raven family history from a home base in California in the early twentieth century, my mother's ancestors undertook their most momentous treks two generations earlier on their way to California. Consciously or not, children look to their family history for clues to their own identity. As a boy, my mother's family pedigree deepened my boyhood sense of myself as a true Californian.

My maternal grandmother, born as Laura Stella Lenhart, was the seventh daughter and the thirteenth child of James Madison Lenhart, who had left Pottsville, Pennsylvania, for California in 1852. Once he had reached California—by sailing on the long, slow, and perilous voyage around the Horn—Lenhart occupied a number of different jobs, many of them connected with horses and transportation, in and around San Francisco. In due course, he met his future wife, Nanno Maria Barry. At the age of nine, Nanno had emigrated with her mother from Cobh in Ireland's County Cork, crossing the Isthmus of Panama by mule on her way to San Francisco and arriving in 1853. Her mother died within a year of their arrival, and after that date relatives presumably cared for her. She eloped with James Lenhart in 1860, when she was only sixteen years old.

James Lenhart signed up to ride for the Pony Express at about this time, possibly responding to a recruiting advertisement like this one: "Wanted— young, skinny, wiry fellows, not over 20. Must be expert riders, and are willing to risk their lives for the job. Orphans preferred. Wages twenty-five dollars a week." It is not known how many of these qualities James Lenhart possessed, but he got the position.[2] His part of the route took him some 120 miles from Pacific House to Strawberry Valley in Yuba County, on the western slope of the Sierra Nevada. Like his fellow Pony Express riders, James can be presumed to have done his job well; only one sack of mail was ever lost in the eighteen-month run of the Pony Express, and even this was found, with its mail delivered two years later.

James subsequently became an owner of the Lenhart & Brady Carriages Company, later called the United Carriage Company. He and Nanno lived at various places in San Francisco, with Nanno giving birth to thirteen children in rapid succession during that period. The last of these children, my maternal grandmother Laura, was born April 9, 1883.

Although James Lenhart was a respected figure in the community, he was accused in 1891 of "obtaining money under false pretenses." What transpired thereafter was reported that year in *The Morning Call*, a San Francisco newspaper. According to the paper's account, so distraught was he regarding the charges that the night before the hearing, he took morphine to "hasten his sleep." At the hearing the next morning, his name was read out three times to no avail. "Your Honor," said Prosecuting Attorney Stevens, "the accused is not in the land of the living. He died last night at his home, surrounded by his family." "Well," the paper reported the Judge as saying, "that ends all. In his lifetime Lenhart was well known to me as an honorable citizen. Mr. Clerk, strike the case from the calendar."[3]

My grandmother Laura was eight years old when her fifty-six-year-old father died. Nanno, left widowed and with children still in her care, never remarried. Fortunately, she received money from several of the associations of which her husband had been a member, allowing her to build a house for herself and her children. Within a few months, in June 1892, she bought a lot in Marin County, north across the Bay from San Francisco. My grandmother Laura grew up in the house Nanno built on that property in 1894, at a time when the area she chose, Mill Valley, was still sparsely populated and bucolic. Laura, friendly and outgoing, spent much of her time in the outdoors, pursuing activities such as catching mudpuppies in the creek behind the house. She liked to claim that as a seventh daughter in a family of thirteen children, she possessed certain mystical powers—but I never learned what they were!

In 1903, at the age of twenty, Laura married San Francisco attorney Peter A. Breen, nine years her senior. They lived in Mill Valley for a few years, Peter serving as the city attorney. Later, he commuted to his law practice in San Francisco. My mother Isabelle was born in 1908, a year after her brother Jack. In 1912, Laura and Peter and their two children moved to San Francisco. At the time, the city was energetically rebuilding from the devastation of the earthquake and fire six years earlier. In 1915, they moved to a new house built on 12th Avenue, one of the rapidly developing streets of the Inner Richmond District. There it was that my mother Isabelle reached adulthood. Many years hence, I would become intimately familiar with that very same house.

∽

I am grateful to know the details of my grandmother Laura's history, but it was an episode in her husband's family that had the most impact on me. My great-grandfather Patrick Breen Jr., the father of my grandfather Peter Breen, was a nine-year-old child in the Donner Party. That wagon train entered both history and legend when heavy snows prevented it from crossing the Sierra Nevada in the winter of 1846–47, just ahead of the California Gold Rush, which started soon after the discovery of gold along the American River in the foothills of the Sierra during the spring of 1848.

My great-grandfather's parents, Patrick Breen and Margaret Bulger Breen, were already seasoned travelers by the time they began their journey to California. They had left their home at Borris, in County Carlow, Ireland, to sail for Canada in 1828. They married in Ontario in 1831, then moved to the United States in 1834. After briefly trying out Springfield, Illinois, Patrick and "Peggy" joined the very first farmers in Iowa. Iowa would not officially open for homesteading and settlement for yet another five years, but a few families had settled around Keokuk and the Breens joined them in 1836, farming there for a decade. It was while the family was living in Keokuk that my great-grandfather Patrick Breen Jr. was born.

Many pioneers who emigrated west were enticed by the thought of free land, but the Breens also were seeking a place where they could openly be Catholic and have dependable access to the Sacraments. The Mexican province of California seemed to be just such a place, and so the Breens (among many others) set their sights across the Great Divide. For several years, they talked with Patrick Dolan, a bachelor farmer and neighbor in Keokuk, about traveling west. By the spring of 1846, they were ready to head for California.

Dolan and the Breens started out in fine spring weather on April 5, 1846. Of their train of five horse-drawn wagons, four carried the Breens with their seven children and all the possessions they brought along. Three months later, at the crossing of the Little Sandy River in what is now south-central Wyoming, they met and joined a caravan of wagons that included the Donner family, whom they knew from the time they had lived in Springfield.

Given the length of the journey, the group was understandably interested to learn of an alternative to the California Trail that would save them about 300 miles of travel. Now commonly known as the Hastings Cutoff, the shortcut was described by Lansford Warren Hastings's 1845 book *The Emigrant's Guide to Oregon and California*: "The most direct route, for the California emigrants, would be to leave the Oregon route about two hundred miles east from Fort Hall, thence bearing West Southwest, to the Salt Lake; and thence continuing down to the bay of St. Francisco, by the route just described." What the travelers were not told was that the highly touted route was actually little known or tested and involved crossing hundreds of miles of desert and ultimately crossing the high and largely uncharted Sierra Nevada at the eastern edge of California.

The members of the united wagon train left the Little Sandy on July 20, 1846, and sorted themselves eleven days later, when they reached Ft. Bridger, Wyoming. Some wagons departed onto the longer, better-known route to California, northward via Oregon. Patrick Dolan and the Breens, among others, joined with the Donner family in choosing the Cutoff instead. It was a fateful decision based on faulty information put forward by Hastings in his book; he even promised to travel with them as a guide, but left, instead, with another party of emigrants.

Hacking their way through the birches and alders in the Wasatch Mountains above Salt Lake City, the members of what would come to be called the Donner Party cleared the path that Brigham Young and his Mormon settlers would follow a year later. Getting through those mountains was arduous, taking the wagons about three weeks. They didn't reach the flatlands southwest of the Great Salt Lake until September 1, after five months of traveling and with the dangers of winter looming ever closer.

After losing so much time in the Wasatch Mountains and on the Cut-off across the Salt Desert, the Donner Party found itself nearing the Sierra Nevada too late to get across the range before the snows became too deep to traverse. The whole group, approximately eighty-seven people strong at that point, were caught with little choice but to stop well short of the crest, where they were soon buried by heavy early-winter snows. With their route west temporarily closed and necessities for survival in short supply, the Breens settled in for the long winter siege at the eastern end of what is now called Donner Lake.

They were fortunate, in that freezing cold, ominous November, to find a cabin that had been built two years earlier by pioneer Moses Schallenberger and then abandoned. They survived cooped up in that small cabin for four months of deep snows and increasing hardship. While Peggy Breen and Patrick Dolan had wisely had their cattle slaughtered and the meat dried early on, most families had not acted promptly enough to avoid losing their animals to the blizzards. Accounts of that winter differ in some of the details, but all agree on the bleakness of the experience: the cramped monotony, the biting cold, the bitter wind, the massive drifts, and the seemingly endless hunger, always under leaden skies that brought more and more snow. Hard as the members of the party fought for survival, they began to succumb, one by one, to hunger and cold.

My great-great-grandfather Patrick Breen made the only written record of the events of that terrible winter to have survived in its entirety. Written densely on folded paper, his diary spans the events from crippling snows of November 1846 through Patrick Breen's rescue in March 1847. Apparently for much of the time he was writing, Patrick Breen was laid up with serious gallstones, making his growing hunger all the worse. The University of California's Bancroft Library, which holds the diary today, notes that "The overall style displayed in Breen's diary is terse, prosaic, and, with respect to the actual conditions it records, understatedly matter-of-fact. The mood gradually develops from a mild anxiety concerning the weather in the early entries to a more exclamatory desperation as the crisis worsens."[4] In February of 1847, for example, he wrote, "We hope with the assistance of Almighty God to be able to live to see the bare surface of the earth once more. O God of mercy grant if it be thy holy will Amen."[5]

On December 16, Patrick Dolan left the Breens' cabin with a snow-shoe party of seventeen people attempting to cross the mountains to seek

A page from my great-great-grandfather Patrick Breen, who left the only written record from the disastrous Donner Party of 1846–47.

help. Sadly, he died on the western slope. By January, reports of the stranded pioneers had reached Sacramento. Rescue efforts were organized, a number of them by Captain John Sutter, who personally paid the people who were willing to help the starving victims escape their snow-bound prison. By that time, Sutter, a German-born Swiss who came to California in 1839, had amassed large landholdings and provided an important anchor to settlers in the lower Sacramento Valley. It took three months and four relief efforts to rescue all the survivors. Of the eighty-seven members of the Donner Party who had been stranded east of the crest, forty-eight survived to reach California, some of them having eaten the dead as a matter of survival.

Among the survivors were all nine Breens. Two of the Breen children, Edward and Simon, were rescued by the first relief effort, in February 1847, and led over the mountains and down to safety. Thanks to the providence of Peggy Breen, the family had more hides to boil and eat than others, and so only the two eldest and strongest children were taken out with that party. Subsequently, the rest of the family got part way out but became stranded in deep snows on the western slope, at a place that would come to be called Starved Camp, until the third relief party arrived. This group of rescuers left three of its members at Starved Camp to help those surviving there to try to get out to the Central Valley; the remainder of the party continued up and over the pass to try to do what they could for the others who were stranded at the original camps. The three men left at Starved Camp included two paid rescuers, together with John Stark, a mountain of a man who from the beginning had refused any compensation for joining the team.

The Breens and their five remaining children were in desperate condition and extremely weak from starvation. The rescuers discussed whether or not to try to get them out right then or to leave them for a future rescue party. The two paid rescuers decided to return to the settlements, but John Stark, who knew that such action amounted to a death sentence for the Breens, refused to go along with them. As recounted later by Patrick's son James, who was four at the time, John Stark said, "No, gentlemen, I will not abandon these people. I am here on a mission of mercy, and I will not half-do the work. You can all go if you want to, but I shall stay by these people while they and I live."[6] John Stark stood by his word, guiding the Breens to safety. According to John Breen, aged fourteen at the time, John Stark carried, "on his broad shoulders," the "provisions, most of the blankets, and most of the time some of the weaker children." He joked that had there been room on his back, he could have carried all the children "because they were so light from starvation."[7] My great-grandfather Patrick Jr. was among the children John Stark rescued and brought out of the rugged mountains to civilization. He was only nine years old at the time.

The Reed and Breen families alone would reach California with their entire nuclear families intact. Others who lived through the ordeal reported taking deep inspiration from Patrick Breen and his daily rituals of prayer. Some survivors' accounts of Peggy Breen were unfavorable, perhaps because, due to her husband's ill health, she "assumed a more prominent role in the family than some have thought seemly."[8] But her sons defended her vigorously. Certainly, she had been the only one in her family left with enough strength to keep making the fires, melting the snow, and taking care of them all, including her infant daughter Margaret Isabella, who was still nursing at the time. Hence, she can fairly be credited with the survival of the rest of her family and thus all their descendants. Peggy was a tiger who knew how to take care of her cubs, and would do so no matter what the obstacles might be.

Once down from the pass and after some months of recovery, the Breens arrived at the small settlement that had formed around Mission San Juan Bautista, in what is now San Benito County. "Once they arrived at San Juan Bautista and looked out over that valley," author Marjorie Pierce notes, "Patrick Breen knew they had reached their promised land."[9] They arrived in February 1848, the same month in which the treaty that ended the Mexican-American War was signed. General José Antonio Castro, who had served as Commandante General of the Mexican army in Alta California (an entity which then encompassed what we know as the state of California as well as considerable territory to the east), allowed the Breens to stay in his adobe house. It must have given the devout Breens much satisfaction that the house was diagonally across the street from the Mission church, where family members would be baptized and married in the decades to come.

No sooner had the Breens settled in than their first-born son, John Breen, a vigorous twenty-year-old, took off for the gold fields, won over by the tale of two prospectors who passed through town. He became one of the tens of thousands of seekers in the Gold Rush that followed the discovery of gold on Captain Sutter's land in January 1848. Luckier than most, John returned with a fortune then valued at $12,000, part of which he used to purchase the adobe and some 400 acres of farmland from General Castro. His father, Patrick Breen, enjoyed twenty-one years of ranching prosperity in California, finally passing away at the age of seventy-three in 1868; his wife Peggy, who had carried the day in Donner Pass, lived on for another six years.

My great-grandfather, Patrick Breen Jr., grew up in San Juan Bautista and established a successful ranch on the family property, raising cattle, sheep, wheat, and fruit trees. In 1864, he married Amelia Anderson, whose family story has its own share of courage and journeying. In 1833–34, her father and paternal grandmother had been transported from Armagh, in Ireland, to Sydney, Australia, as punishment for minor crimes. Amelia's father married her mother, Mary Nash, a free settler in Australia who had also emigrated from Ireland, near the end of his seven-year term of indenture. After Mary died in childbirth, Amelia's father remarried. Teenaged

General José Antonio Castro, the Mexican governor of Alta and Baja California, built this house in 1840. He allowed the Breen refugees from the Donner Party to stay in it when they arrived in San Juan Bautista in early 1848 and sold the house to them a few months later. The Castro-Breen Adobe is now part of the state park that includes the mission.

Amelia clashed with her new stepmother and left her family to sail across the Pacific by herself at the age of fifteen to join her late mother's sister near what is now Dublin, California, just east of San Francisco.

Amelia and Patrick Breen Jr. were married at the beautiful, newly constructed St. Mary's Cathedral in San Francisco. They settled down on the family farm at San Juan Bautista and later moved into the nearby new town of Hollister. Patrick Jr. and Amelia had seven children, including my grandfather Peter, their fifth, in 1874.

Peter Breen went on to attend Hastings College of the Law, part of the University of California located near the State Supreme Court in San Francisco. He met my grandmother Laura Lenhart while practicing law in the city. As noted earlier, he married Laura in 1903, becoming the father of my uncle Jack in 1907 and my mother Isabelle a year later.

*

My father Walter must have been smitten by Isabelle. A student at the University of California, Berkeley, she was vivacious, popular, highly intelligent, and avant-garde. These were heady times for independent young women, and Isabelle was taking full advantage of what was open to her. Proportionately few women attended university in the 1920s, and most who did pursued majors related to occupations traditionally filled by women, like teaching and nursing. In contrast, Isabelle was a psychology major. She thrived in the progressive environment of Berkeley. She and Walter dated while he finished his last two years at Santa Clara University.

After he graduated from Santa Clara in 1929, my father enrolled for a short time in the Harvard Business School. He then returned to Shanghai, where he accepted a position in his uncle Frank's bank, the American Oriental Banking Corporation, which was booming. (He was also offered a job by C. V. Starr, and later wondered what might have happened if he had accepted that position instead of the one in the family business.) The choice he made must have seemed almost inevitable, one that would have given him some feeling of stability as he was launching his life in China. He was a loving but insecure man; whether that was due to inborn character, the upheavals in his family life, or other factors is not clear. He certainly could not have seen what was ahead for the bank.

Walter and Isabelle corresponded across the seas, China to California, for more than two years until they became engaged in 1931. My mother later told me that Dad informed her by letter that there were two conditions she had to meet to be his bride: she must weigh less than 120 pounds and be able to play golf. It may have been written playfully; in any case, he can't have seemed too arrogant, I suppose, since my future mother accepted both his proposal and his provisos (and later bested him in golf with such regularity that he gave up sports entirely).

After graduating from Berkeley in 1932, Isabelle embarked for Shanghai with her mother, Laura Lenhart Breen, in June. One can only imagine how she must have felt at the age of twenty-four, taking her first trip to a foreign country, mysterious China, which was tumultuous and particularly threatened by Japan at that time. When Walter, in a white suit, and Isabelle, wearing a cape and elegant white turban, were married in historical St. Joseph's Catholic Church that summer, the society wedding made the *Shanghai Daily News*.

Making a new life in China was not easy for my mother. At home, she had been surrounded by friends who appreciated her gregarious nature and a family that supported her fully. And of course, she had lived in a modern culture, one that permitted her considerable independence. In Shanghai, she was immersed in a cliquish, colonial society. Suddenly, she had servants to manage, in-laws to please, and a new country and customs to try to make sense of. My father was busy with his job at his uncle's bank, and he and his family had well-established social connections and routines; she was a newcomer who found it hard to gain acceptance. There wasn't much personal freedom, and really no particular outlet for her skills and education (today she might have gone to work in one of the family businesses), and her hopes for a baby wouldn't be realized quickly. It was good that she was strong because she had to be.

My parents, Walter Francis Raven and Isabelle Marion Breen, were married at St. Joseph's Church in Shanghai, on August 28, 1932. It was a prosperous time for the city.

My Aunt Vivienne, who would become an important influence in my own life, was her main ally. The daughter of a Swiss cotton broker who had moved from Texas to Shanghai, Vivienne had married Charles Raven Jr., my father's younger brother, in 1933. She and Isabelle, fellow newlyweds, became fast friends and would remain so for the rest of my mother's life. Like my mother, Vivienne felt like an outsider, even within the expat American community; also like my mother, Vivienne was highly intelligent. Childbearing difficulties formed another bond between them. Vivienne and Charles's first child was stillborn, and my mother was at Vivienne's side during the birth. With my mother's own desire to start a family still unfulfilled, I'm sure that their close friendship consoled both women.

While they hoped for a child, my parents traveled extensively. In 1934, they departed Shanghai for a round-the-world cruise. On their return to China, they explored what they could of that country. The stories of their trips, shared when I was a child, were my first introduction to China's diversity. One place they particularly enjoyed was beautiful West Lake in Hangzhou, not far from Shanghai.

My parents also visited the old Chinese imperial capital of Peking (Beijing) in the early 1930s, bringing a servant along to cook food on the train. While freezing in under-heated hotel rooms by night at their destination, they were able to delight in the splendor of the Forbidden City by day. Although the sights were marvelous, travel could be worrisome in the China of the 1930s. Traveling through regions dominated by various strongmen, my parents had good reason to fear being attacked and robbed on the train—but the sense of peril only made their stories of these exotic places more exciting when I was told about them as a child.

The early 1930s were boom years for my great-uncle Frank Raven. In 1932, the year my parents were married, he had bought a mansion at the edge of the French Concession for $150,000, the equivalent of roughly $6 or $7 million today. The house, now the official Shanghai residence of the French Consul General, has a Spanish red-tiled roof accented by sunflower tiles under the soffit line, elaborate carved paneling inside, and lush, overgrown rosebushes in the garden. Frank's wife Elsie was the daughter of missionaries, so no liquor was ever served in their home. Still, Elsie did have her indulgences, according to family sources who knew her at the time, including having her servants peel grapes before serving them!

Heralded by *Fortune* magazine as having a "Midas touch," Frank Raven had assets of approximately $50 million by 1935 despite having arrived in Shanghai essentially penniless in 1904.[10] That year, however, his fortunes turned. When the impact of the Great Depression belatedly hit China in 1935, it was found that Frank Raven's bank had not been closely observing the same operating rules as other foreign banks in China.[11] Having purchased more securities than its cash reserves should have allowed and borrowed a sum greater than its assets, it failed in May 1935. Some sources blame recklessness and embezzlement for its untenable financial position, but there were many other destabilizing factors operating at the time, including the Depression, political shifts, and the Silver Purchase Act of 1934. Whatever the truth, my great-uncle, his family, and many of his customers were suddenly plummeted from great wealth to poverty.

Charged with a variety of misdeeds, Frank was jailed in Shanghai while he awaited trial. Under the rules governing American nationals doing business in treaty ports like Shanghai, an American judge in a Shanghai courtroom convicted him of "technical embezzlement" and sentenced him to a five-year prison term at McNeil Island Penitentiary in Washington State. When Frank was transferred to jail in the U.S., Elsie followed him back to America, where they would be reunited on his release—granted early—from imprisonment. I was never told anything about Elsie's response to her husband's fall, but both the dramatic change in her circumstances and the accompanying public shame must have proved very difficult given her deeply devout and upright character.

My grandfather, C. H. Raven, had put up the Bakerite Company and all three Chocolate Shops as collateral for his brother's bank. All those holdings had to be sold to help cover the bank's debts. The collapse left my grandfather, my father, and his brother Charles suddenly unemployed and with little money. Dad remained jobless for the remaining two years he lived in Shanghai. Soon exhausting his savings, he survived by borrowing money from Aunt Vivienne's father, Ted Schmid. The fall from wealth and high social standing must have been very painful for all concerned. But they continued to live in comfort in House 20, a unit in a small cluster of houses at 799 Avenue Haig in the French Concession, managing to get by somehow.

Babies being no respecters of timing, my parents finally conceived the child for which they had been praying—me—in the autumn of 1935, right in the midst of the economic crisis in the family and the financial chaos in Shanghai generally. My mother had a difficult pregnancy and birth, but after a few weeks in the hospital, all was well. The doctor who delivered her, Dr. Sun Ko-Chi, became a lifelong family friend. Supported by the Boxer Indemnity Scholarship Program, he had graduated from Johns Hopkins University in America in 1918 and then its medical school in 1922, after which he was appointed an assistant resident in obstetrics. Falling out with the Hopkins administration after a few years, he moved back to China, raising funds there and through friends in America to build his maternity hospital in Shanghai. It was in this institution, The

Woman's Hospital, where I was born on the sultry evening of June 13, 1936. That night, as my father walked back to the apartment where they lived, huge masses of insects, likely mayflies, swarmed under every streetlight. It seemed to my father that all the tiny flying creatures in the world had come out that night; he'd never seen anything like it. Dad always took this explosion of life as an omen of things to come for me.

My amah took good care of me from the time I was born until we left for San Francisco just before my first birthday.

A week later, my father reported to the American Consulate to file the record of my birth. On receiving the news by telegram, my maternal grandmother, Laura Lenhart Breen, promptly sailed from California to help my mother with the new baby. My mother also hired an amah—a nanny, or nursemaid—as was the Chinese custom, though she wasn't entirely happy to share her new baby with anyone. As was the tradition with Chinese babies, my amah tenderly swaddled me in a cocoon of warmth and security, carried me with her everywhere, and spoke to me in Shanghai dialect. When she smiled and murmured what a good baby I was, I'm sure I understood perfectly. After a few weeks, I was baptized at St. Joseph's Catholic Church, where my parents had been married, and set up for youth as an observant Catholic.

Meanwhile, the rest of the world was in tumult. Hitler was on the rise in Europe, America was suffering from the Dust Bowl and the Great Depression, and China's situation was perilous as well. As America's economic woes spread worldwide and the Japanese intensified their aggression toward China, prospects there grew dim economically and worse politically. Financial hardship was evident everywhere in Shanghai. My parents were forced to move out of their apartment and in with my grandparents C. H. and Lillian. With a new baby and an unemployed husband, this was an especially difficult time for my mother. The opportunities that had brought Frank and C. H., along with many others, to China were gone, and the Raven family's time in China was coming to a close.

In 1937, my parents borrowed money from Aunt Vivienne's parents, Ted and Mary Schmid, and booked passage for the U.S. We departed by steamer in early June, so my first birthday was celebrated aboard ship. The timing was fortuitous: the so-called "incidents" by which Japan had been taking control of China since 1931 finally resulted in full invasion and a de facto state of war by August 1937, only two months after we left. My grandparents left Shanghai in 1938, but my grandmother Lillian, who had been diagnosed with stomach cancer, would not live to see home soil; she died and was buried in Colombo, Sri Lanka. My grandfather C. H. headed for a widowed aunt's home on Stuart Street in Berkeley upon his return to California. My uncle Charles Raven Jr. and his wife Vivienne had the most perilous departure of all, suffering a dramatic separation during the mandatory evacuation of women and children from embattled Shanghai, but eventually they too arrived safely in the States.

By the time my mother, father, and I returned to California in the late 1930s, the exhilarated national mood of the twenties had turned dour. Americans were struggling desperately to survive the Depression years, and the frivolities of the flappers seemed another world. It wasn't an auspicious time to return. Nevertheless, my mother celebrated a homecoming to a place she loved with a loyal husband, a cherished young son, and hopes for a fresh new life.

A Strange Enchanted Boy

IN THE SHORT FIVE YEARS since my mother had sailed to Shanghai to join my father, San Francisco had become distinctly more modern. More people and buildings were crowded into its water-edged confines, and it sported two new bridges, one spanning the Bay and the other the Golden Gate. My mother's brother, Jack Breen, had taken part in building the Golden Gate Bridge, serving as one of the many engineers employed on the project. The structure was completed in 1937, just weeks before our return to the Western Hemisphere.

Upon our arrival in this newly updated shining city by the bay, our family lived for a short time with my mother's parents, Laura and Peter, on 12th Avenue, in the house that they had built and moved to with their children in 1915. After a year or so in the 12th Avenue house, we found our own flat at 26th and Cabrillo, nearby in the Richmond District. Nestled between Golden Gate Park and the Presidio, the Richmond District was then dominated by diverse populations of European origin. Later, it would accommodate many Chinese and Japanese homeowners as they moved out of their traditional enclaves in the more central parts of the city.

A neighborhood of small, neat row houses (most with common walls), its predominant design theme was Mediterranean. Most of the east-west streets were named in alphabetical order for Spanish explorers—Anza, Balboa, Cabrillo, and so forth—who had come centuries earlier, and other notable individuals. Those names added to my sense of a wide world beyond my own backyard. I never thought of myself as only American, even before I first traveled abroad as an adult. I felt like a citizen of the world, not of a single country. That sense was underscored by my early life in San Francisco, a large and vibrant city with immigrants from around the world, including a thriving Chinatown.

When we got settled into our own home, my parents unpacked the treasures they had carried from China. Among the objects they brought back were ornately carved headboards, ancestor paintings, pieces of lacquerware, a Japanese bronze vase with a shrimp on it given as a wedding present by an American judge, a rosewood chest, and a silver engraving of West Lake at Hangzhou. A small statue of Guanyin, a favorite and peaceful form of the Buddha, who held a special significance for my

My mother sailed out through an open Golden Gate in 1932, but over most of the next five years, the iconic Golden Gate Bridge was under construction, and completed just a few weeks before we passed under it on our way back to California. This image was taken from the spot where, as a teenager, I later discovered an exciting new kind of manzanita.

mother, was included. Growing up in a home filled with Chinese furniture and artifacts and listening to tales of our family's life in Shanghai, I would retain a sense that China was part of who I was. The stories I heard gave only a sketch of my family's life in China, but over time I was able to fill in many of the details for myself. The scraps of information my mother gave me joined the odd patchwork of lore I picked up from friends, school, films, and popular culture. Few Americans at that time had visited China, so it seemed incredibly exotic. I enthusiastically embraced an identity defined by birth in that faraway place. I remember my first-grade teacher asking each of us students our nationality. Without hesitation I replied, "Chinese, but with American relatives." Both the teacher and the other kids were confused, but it made perfect sense to me—a spiritual truth, even if it wasn't quite what the teacher had in mind.

In addition to her gift for storytelling, my mother had an eye for decorating and a love of order. She vacuumed, polished, and dusted on a strict schedule. The house was immaculate, right down to bureau drawers smelling of cedar. Some of the objects on display were tempting to a young boy, including a herd of black carved elephants, with ivory tusks, that stood on a low shelf. My mother would say, jokingly, "I'll give you a clap over the earhole if you touch those!" But of course I did touch them—luckily, without that "clap." I also remember her discussing with her friends the need to keep one's home clean lest it become necessary to summon a doctor in the night: a reflection of the times when doctors still made house calls, and of my mother's appreciation for order and propriety.

My father was a neat, tidy, gentle man, often garbed in a bow tie and cardigan sweater. He was a good man to his family and in the world beyond—a caring, liberal Democrat who was deeply aware of the sufferings of others. Dad read me stories, like *Treasure Island* and the tales from Kipling's *The Jungle Books*, as I was falling asleep. They filled my mind with visions of India: the fearsome Shere Khan and the mongoose Rikki-Tikki-Tavi, the little fur seal swimming home to the Pribilof Islands, the way the elephant got its trunk and the rhinoceros got its skin. In Kipling's *Just So Stories*, another favorite, the elephant wanted to know what the crocodile has for dinner, and the Kolokolo Bird answers him: "Go to the banks of the great grey-green, greasy Limpopo River, all set about with fever-trees, and find out!" This fueled my already strong curiosity, making me want, above all, to go find things out. It was my earliest education on the world's biodiversity, something I can only imagine that Kipling would have appreciated.

When I was still very young, my father began taking me for walks in Golden Gate Park, just a block from home, which to my youthful eyes was a gorgeous blur of green trees, shrubs, and grass, with animals including garter snakes, birds, rabbits, and squirrels, and blue water beyond. There were also the exciting exhibits at the California Academy of Sciences, which had an aquarium and such delights as a giant and imposing stuffed grizzly bear in the foyer of the North American Hall. In spite of the ongoing modernization, natural beauty still flourished in and around the city. The San Francisco of that time had plenty of wild places. I came to know the city in more detail as I grew older, enjoying its winter rains, often heavy summer fog, and the sonorous, repetitive booming of the foghorn, which was somehow comforting on cold, foggy nights.

My grandfather C. H. Raven—(as mentioned, now widowed and home from China)—moved in with us during the years we lived on Cabrillo Street. He was given what had been my bedroom, while I slept on a fold-out bed in the front room overlooking the street. I didn't mind; I adored having him in the house and always felt close to him. He was a kind and positive man, who spent summer afternoons relaxing and listening to Major League baseball games on the radio.

We were still living on Cabrillo Street when I started kindergarten, in 1941, at the Cabrillo School. I could easily walk around the corner to school, across just two streets, joining the parade of children all heading in the same direction. Along with most of the other children, on rainy days I wore a yellow rain slicker with a beaked hat. We looked like little mariners braving fierce New England waters. Watching me disappear into the mass of yellow on one of those rainy days, my mother felt a sudden and great sense of loss, as she would later tell me. But I enjoyed the sense of being one of the gang, from my rain boots to the dripping brim of my sou'wester, splashing through puddles with the best of them.

When Japan attacked Pearl Harbor, the worldwide conflict that had been ongoing for several years suddenly pulled America fully in; I was five years old and we were at war. The bombing brought the war much closer in practical terms, too. Worries that the Japanese might invade the West Coast by air or by sea were widespread. While those of us in San Francisco certainly didn't walk around all day expecting to be bombed, after

Pearl Harbor such attacks became a real possibility, and remained a concern during the first months of the war.

Since my parents had spent time in Japan (and had the knick-knacks to prove it) I'd always had a sense of that nation as one that was connected to my own family and life. But after Pearl Harbor, I experienced some confusion as I gradually learned that my parents had left China in the face of Japan's sweeping invasion in 1937. Further, everyone had begun talking about Japan as the enemy. Both San Francisco and California at large had sizable numbers of Japanese citizens, many descended from immigrants who had come to the U.S. as early as the nineteenth century. Starting a few months after Pearl Harbor, more than 110,000 of them, virtually all fiercely loyal Americans, were forced into internment camps. Unjustly, many of them lost their businesses, homes, and farms. Like many other citizens, my parents deplored the Japanese internment, but there was little they could do in the face of this mindless, racist response to the aggression of a country far across the Pacific.

Back in the Shanghai that we had left six years earlier, the Japanese took over the former Foreign Concession with the declaration of war with the U.S. and our allies. The unlucky foreigners who were still there were mostly transported to concentration camps. Some of my parents' friends died in the camps, while others suffered terribly but survived, joining us in San Francisco after the war.

At home, during the war years, we saw soldiers and sailors in uniform everywhere, getting ready to ship out or on their way home. They crowded the buses and streetcars, the train station, and the docks. Properly, in my view, everybody was very respectful to them, and helped them out where possible with small gestures like offering rides or picking up restaurant tabs.

Up and down the West Coast, residents were ordered by radio announcements, reinforced by government-issued posters, to enforce blackout during the evening hours when there seemed to be the possibility of an attack. During blackouts, we taped the edges of our curtains shut and used candlelight, which wouldn't shine out into the street and be visible to enemy forces. Signs hung everywhere reminding those on the home front to buy war bonds, conserve gasoline, and plant Victory Gardens. On the walls of grocery stores and post offices hung scary posters of Mussolini, Hirohito, and Hitler that said, "Enemy ears are listening." In addition to this demonization of the enemy, there was also a palpable sense of community, of everyone doing their part and of everyone's part being valued, no matter how small. Even a child could play a role. I received war bonds, issued to help finance the war, as gifts and brought a dime to school each week to buy stamps, exchanging my stamp book for a war bond once it filled up.

As it did for many others, World War II had a positive impact on my father's career. Considered too old for military service at 36, he worked as a personnel manager at Inland Steel, across the Bay in Richmond. Reflecting on the severe shortage of manpower during those years, he would say of those applying for work, "All you had to do is hold a mirror up under their noses to see if they were breathing. If they were, they got the job."

Of course, women were of central importance in the war effort. Rosie the Riveter was everywhere, in a thousand incarnations (including the lesser known "Wendy the Welder") as women entered the workforce in droves, keeping the country running while another generation of young men went to war. Growing up at a time when women all over the country were stepping into roles once dominated by men helped to shape my sense of women's equality and importance. Growing up with a strong, smart mother made it all seem perfectly natural.

✧

After my grandfather Peter Breen's death in March of 1943, my grandmother Laura, now alone, invited us to move back to 12th Avenue. My grandfather C. H. came with us and remained an important part of my life. I loved to join him, sitting on his bed up under the eaves on the third floor, to play cards or listen to baseball games on the radio. My second-floor room in that house had a view of the nearby homes and backyard gardens.

Arriving midway through second grade, I began attending Star of the Sea School, a parochial school located a few blocks east on Geary Street. My new school emphasized my Catholic heritage in a number of ways. I received First Communion and attended church every Sunday with my parents, absorbing Catholic doctrine as a major theme of my life. In those days, Catholic children were taught that they should not associate with non-Catholics lest they be led into the occasion of sin, but I was never quite sure what that danger might consist of.

A few boys in the neighborhood were friends and, with them, I explored our small world. Together, we even made an early foray into the world of entrepreneurship. Using a piece of white bread wrapped around a hook, we fished for carp in the then-pristine Mountain Lake, a small natural pool near the southern edge of the Presidio and the northern end of 12th Avenue. Then we sold our catch to a Chinese butcher on Clement Avenue. It was a modestly successful venture, though it inspired in me no abiding interest in commerce.

That winter, as World War II heated up and prospects began to look brighter for the Allies, I came down with the measles. At the time, the disease was endemic, with approximately 900,000 cases in the U.S. each year during the 1940s. In an attempt to slow its spread, measles sufferers were required to stay out of school for two weeks. Without modern distractions like television, it might have been a long and boring two weeks in bed. But my clever mother walked into my room and handed me a book with an orange cloth cover. It was a signed copy of a 1939 book on insects called *Six Feet*, by Oregon author Ruth Cooper Whitney.

The book title consisted of drawings of insects, each one shaped like a letter. I still recall propping myself up in bed, opening the book, reading one page, and then devouring another and another. The text of each chapter consisted of three or four pages about a particular kind of insect plus suggestions of related activities a child could do in his or her own backyard. The book gave nature a new face for me. I suddenly realized how many magical creatures were living right outside my own back door. *Six Feet* ignited my curiosity about all things natural, and in doing so transformed my life.

This was only the beginning. My mother soon gave me another book, about birds. Looking out the window at them, I saw them with new interest and attention. Suddenly I grasped how much living creatures varied, and how fascinating their differences were. It was a big surprise for me to realize, for example, that male and female English sparrows were the same species despite their striking differences. It was my initiation into the power of patient

As a second grader, I was so excited by the captivating tales of insects in this book, *Six Feet*, by Ruth Cooper Whitney, that I never looked back.

and careful observation. I couldn't wait to get over the measles, go out-side, and find all these beetles, butterflies, crickets, and birds for myself!

Many different kinds of insects and other animals lived right in our backyard, which was built on a remnant of the dunes that had originally covered the western half of San Francisco. Even better, one of the most beautiful and extensive parks in the whole country was right at my back door. Golden Gate Park was within walking distance, and back then adults weren't afraid to send their young children outside alone. Already famil-iar with the park from the weekend walks with my father, I was eager to dive deeper into its fascinating living world. In its planted pine, cypress, and eucalyptus forests; the scattered corners of undeveloped land; in and around its lakes and on its hills; and on the remnants of the dunes that survived in vacant lots both north and south of the park, there lived a diverse universe of plants, insects, and other animals. All of them de-lighted me as I searched for and learned about each species.

Upon my recovery from measles in the spring of 1944, at the age of eight, I started collecting butterflies. I was fascinated with the process of observing them, netting some of them, killing them with carbon tet-rachloride, mounting them on pins, and putting them in insect boxes to admire later. In my own backyard were painted ladies, white-winged cab-bage butterflies, yellow-and-black tiger swallowtails, orange monarchs, frisky little orange skippers of various kinds, and, at the right season, mourning cloaks, with their purple wings fringed with golden borders and blue dots. Yet more kinds could be found in different seasons across the habitats around the Bay Area.

A butterfly-collecting classmate named George Watson and I were thrilled to discover a colony of elegant pipevine swallowtails (*Battus phile-nor*) near the summit of the hill in the middle of Stow Lake in Golden Gate Park, along with a patch of their larval host plant, the native pipevine (*Aristolochia californica*). The adult butterflies were gloriously beautiful—dark, iridescent blue with seven large orange spots around the edge of the underside of their hind wings. In Marin County were orange tip butter-flies, a treat to find in spring, and brightly colored checkerspots, about which I was eventually to learn a great deal more. As I ranged further afield, I found more and more species of butterflies, but no matter how many I discovered my thirst wasn't quenched.

Aside from butterflies, other fascinating animals seemed to be every-where. In the sandy soils of our garden lived fearsome Jerusalem crickets, earwigs, wolf spiders, roly-poly pill bugs, darkling ground beetles, lady-bird beetles, snails, slimy slugs, and chirping crickets. In a nursery down the street from our house on 12th Avenue, I found two special kinds of beetles, both large and black, both introduced from Europe with nursery stock—a predaceous ground beetle (*Carabus nemoralis*) and a large rove beetle (*Tasgius ater*). I discovered many more kinds of beetles in special habitats in Golden Gate Park, on the beach, and all around as I learned about them and their characteristics.

Somewhat shy as a young boy, I found that collecting insects and showing the mounted specimens to others allowed me to forget myself and have something to talk about with people. I also felt proud to be doing something considered worthwhile, something my parents were proud of. As I searched for butterflies, beetles, and later, plants, I experienced the interconnections of nature firsthand. These experiences brought the joy that comes from doing something because you want to do it rather than because it is a requirement or obligation.

℃

Not all of my first explorations were in the city. During those early years, my mother and I, sometimes joined by my father, began to visit my maternal relatives in Hollister, a couple of hours drive south of San Francisco. There I was introduced to the rhythms of farm life. With my cousins John, Tom, and Peter, I dug fortifications under the row of pepper trees leading to their house. Nearby, in their orchards, I had my first taste of warm, juicy, tree-ripened plums, nectarines, and apricots. And of course, I looked everywhere for insects and, in later years, plants.

A significant, life-changing experience came about through another relative. Knowing my interests, my mother's cousin by marriage, Dick Hecht, cut out a newspaper article advertising a meeting of the Student Section of the California Academy of Sciences and brought it to my parents. The Academy (or CAS) was already familiar to me from visits with my father, but the fact that it had a special program for students came as a delightful surprise. The Student Section was designed to help develop children's interest in the natural sciences—and I'm living proof that it accomplished that job very well indeed.

Showing the clipping to me, my mother asked if she thought I might like to try becoming a student member of the Academy. Always eager to learn more about the natural world, I immediately answered yes. On my very first visit, I discovered that there were other kids who, just like me, were fascinated by beetles, plants, and other aspects of natural history. While I couldn't have put a name to it, in some way I knew I had found my true people, my "tribe." It was a turning point every bit as important as the discovery of *Six Feet*. I quickly became a regular participant in the Student Section's Saturday programs. I was much too engaged in what I was doing to realize that, at the age of eight, I was one of the youngest members who had ever enrolled there. (Shortly thereafter, the Academy raised the entry age to twelve—whether in retaliation or not I shall never know.)

The Academy was about four blocks from our home, close enough for me to walk to meetings there. As a new and eager member, I emphasized my newfound interest in beetles. By the time winter came in 1944, I had shifted most of my attention to collecting and learning about beetles rather than butterflies, though the latter remained an interest. Beetles were much more numerous and diverse than butterflies: they seemed to be everywhere, in every possible habitat that I could find and in astonishing variety.

The beetle enthusiasts that I came to know began giving me tips about where to find special ones. *Omus californicus*, a large, black, flightless tiger beetle, burrowed through the crumbling sandy cliffs above Ocean Beach. The buff, green-speckled *Omophron* lived in the moist sand at the edge of Mountain Lake in the Presidio. Under the bark of trees lived tiny and often indistinguishable members of the same family of predaceous ground beetles, Carabidae. (Several of my early discoveries are shown in the first group of color images.) Each of them thrilled me, and I eagerly sought for more different kinds.

The Student Section was under the leadership of Lois Fink and Joan Taylor, known at the Academy as "Tigger" and "Chip." Both were dedicated and passionate nature-lovers. They broadened our exposure to science, while also taking the care to discern what fascinated each of us individually and providing activities that matched those interests. A fellow Student Section member, Welton Lee, has written that "as if by magic," he would find himself talking to a career biologist who shared his absorption in a subject, deepened his understanding of it, and encouraged him to discover more.

Besides all they taught me directly, Joan Taylor and Lois Fink introduced me to the individual research departments of the Academy and the fascinating people who worked there every day. An especially vivid figure from my early time there was Ed Ross, who collected insects, took beautiful photographs of them, and educated the public about them in exhibitions and popular articles. It's no surprise that Dr. Ross and his colleagues in the Academy's Department of Entomology had such a dedicated following, a group of young enthusiasts who loved to visit the department and enhance their knowledge there.

With instructor Joan Taylor in the Student Section of the California Academy of Sciences, I posed for a publicity shot, together looking at a mounted wood duck. Four years after my enrollment in the Student Section, this image was taken just before my twelfth birthday (1948).

As the months and years passed, my interest in insects became part of my parents' show-and-tell. While she served coffee to dinner guests, my mother inevitably brought up my collection. "Peter, why don't you get a box of pinned insects to show our guests?" Happy that my parents were proud of me, I overcame my inherent shyness and brought the boxes out.

My parents responded to my hobby in characteristically different ways. Mother embraced my fascination with nature, but her approval was not without other expectations. She sought excellence on all fronts and I conscientiously tried to provide it. Her questions were endless. If I got an A-minus, why not an A? If an A, fine, but what about my friends—was I making the right kind of friends? It was the same with beetles and, later, plants. What was I learning? What had I found that day? Was I doing it well? Mom kept me at arm's length in a way, and she very much wanted me to succeed as well as be happy. Although she was proud of me, she had very high expectations and didn't do well at recognizing the limits of pushing people. Much later, I realized that by prodding me about my friends, she was trying to help me value friendships along with personal accomplishments, and not to rely too heavily on the latter in measuring my self-satisfaction.

Early on, I came to understand that if I cultivated a variety of skills, I would build up a kind of insurance for maternal approval. If I got a bad grade, at least I was good at telling jokes. If the jokes fell flat, well, just look at this collection of beetles and butterflies! My mother's demands created in me a restlessness of sorts, a drive that impelled me to keep moving in order to feel legitimate, to stoke the fires of every engine lest one of them slow down or derail. Like so many things one learns from one's parents, this has been both a strength and a weakness, but certainly honed my ability to multitask successfully.

For my father, it was enough for me to be happy. He was always warm and accepting. My love for beetles didn't have to seem impressive to others for him to embrace it: it was enough that I was enthusiastic. Softening my mother's drive for excellence and making it more humane, he gladly drove me far and wide so that I could see insects and plants different from those that occurred in our neighborhood. On these drives, Dad told me some stories about his own life as a child. He hadn't been particularly happy, but did look back with joy on his younger years in Honolulu. His recollections helped me get to know him better and gave me another window to faraway times and places. I loved being with him. He was invariably gentle and understanding to me, giving me a kind of reassurance and security that has been important to me all my life.

If my times with my father and my friends at the Student Section were happy ones, my days at Star of the Sea School were less so. Thanks to a variety of mentors, among them my grandfather C. H., by the time I finished second grade I already knew how to read well, how to write stories, and even how to type (I used a small, portable Remington typewriter). It seemed to my teachers that third grade had little to offer me, and so the adults involved summarily decided—with what later seemed to me little consideration of the social or developmental aspects—that I should skip that grade.

When I started fourth grade, my schoolmates were mostly a year older than me. They were physically bigger than me, knew one another already, didn't laugh at my jokes, and didn't share my interests. Since I was an only child, I already was used to spending time alone, amusing myself. But feeling excluded in a group was painful, and the loneliness and self-consciousness it caused didn't begin to leave me for years. The nuns who taught at Star of the Sea may have been fascinating human beings, but as educators, most were typical of the period and thus quite strict. They believed that teaching children required rapping knuckles and making those who misbehaved stand in the corner. I certainly didn't escape those punishments. The nuns tried mightily to teach me the Palmer method of cursive writing, but I never learned it, instead developing my own crabbed script and unique way of holding a pen, both of which they hated. Rebuked by my teachers and largely ignored by my peers, my grades suffered as school became just a dreary seven hours to be endured each day.

Love of the outdoors was my consolation. I felt a personal relationship with nature, a sort of kinship; though I tried to be precise in my collecting, my connection with the outdoors was deep and emotional rather than simply intellectual. The "book of nature," with all its mystery and surprise, continued to open to me, and that was a joyous experience.

Music was my other chief source of happiness. During those long enjoyable drives with my dad I would fiddle with the radio dial, finding my favorite tunes. Among these was a song by eden ahbez (he insisted the name be spelled without upper-case initial letters) called "Nature Boy" that captured so much of what I felt—isolation, curiosity, a sense of the world as a wonderfully wide and mysterious place, and a desire to be seen and appreciated. Its lyrics seemed to symbolize well many aspects of my childhood—a kind of inherent sadness, but always feeling that I had a purpose, even if I didn't always seem to understand it well.

At the same time that I was busy collecting insects and exploring the outdoors, I had become an avid reader. For example, I scoured the used copies of *National Geographic* that I was able to buy for knowledge of nature farther afield. I didn't really think about the people behind the magazine or its stories, I just loved its extraordinary color photographs and its detailed discussions about people, places, and creatures all around the world. In addition to expanding my knowledge of nature, *National Geographic* also kept China on my mind through stories like "Along the Yangtze: Main Street of China" and "Pirate-Fighters of the South China Sea."

A frequent visitor to some of San Francisco's branch libraries, I would look for books on natural history. There weren't many that were relatively simple or suitable for children. I took many handwritten notes on what I found in those neighborhood libraries, since there was as yet no simple or inexpensive way of making photocopies. San Francisco's Main Library was another valued resource. I would bring request slips to the reference desk there and, after a wait that seemed hours long, the treasured volumes would be brought out to me. In these ways, I gradually built up my knowledge of natural history, but what I could learn for myself outdoors always seemed more important.

ℰ/ᴐ

By the time I had attended Saturday classes at the Student Section for a year, my parents decided I was ready for a bigger adventure. In the summer of 1945, just before the end of World War II, I spent a month with the family of Eleanor Everall Goldshire, my mother's Chi Omega sorority sister. The Everall Goldshire clan—Eleanor, her husband, and their two daughters, who were about my age—had a summer home among the redwoods in the Santa Cruz Mountains about fifty miles south of San Francisco. It offered a whole new world plants and animals for a nine-year-old boy to explore and became a place of life-changing experiences for me. Bright blue Steller's jays, large orange banana slugs, steelhead trout, and other magical creatures added to this enchanted forest.

Their rustic cabin was located above Zayante Creek, in the San Lorenzo River watershed. Since the cabin was in a dense grove of redwood trees across the creek from the road, we transported groceries and other supplies across the creek using a line with a pulley. I slept outside on a cot, waiting to be devoured by grizzlies or, more likely, mountain lions. Luckily, they never came. The shaded stream, gliding smoothly around rocks and bends, with beds of coltsfoot leaves in the shallows, was home

The majestic, fog-moistened redwood forests of Central California proved a true wonderland for an inquisitive nine-year-old boy encountering them for the first time.

to exciting new beetles around the rocks along its edges, and many other creatures in the water. Muted orange mudpuppies, the same salamanders that my grandmother Laura Breen had played with as a child in Mill Valley, were abundant and fascinating. It was a whole new world for me.

In this biological wonderland, there were many kinds of beetles that I hadn't seen before. I discovered them in and under rotting logs and stumps, under rocks, along and in the creek, and in sunny places outside of the redwood groves, where there were different species. Many insects were attracted at night by the cabin lights, among them the spectacular root-boring *Prionus californicus*, a bright brown beetle that could be more than two inches long. When I picked one up to show young Peggy Goldshire, I quickly realized she did not share my fascination! The Oregon stag beetle, *Platycerus oregonensis*, was a special treat: the larger males were all black, with striking large mandibles, and the females had bright blue wing covers.

I was excited to encounter these beautiful insects and to add specimens to my collection. What I collected had no monetary worth, but all of it had personal value to me and brought me great satisfaction. Slowly, I also came to understand that some of my specimens might have scientific value as well. Realizing that I would be able to tell others about those exciting specimens helped motivate me always to go on another hike, climb another hill, flip over another log.

With the Japanese surrender on September 2, 1945, the war ended and there was general jubilation. An America at war was basically the only America I had ever known. The conflict that had formed the backdrop to my entire early childhood was suddenly gone. With the war over, the civilian sector in the United States began rapidly to regain its life and vitality. Unfortunately, this also meant the end of my father's defense position. After a stint with Pabco Paint, Dad went to work for his brother Charles, the co-owner of Raven-Wager Motors on Mission Street. The dealership had sold used cars during the war, putting them in line for a license to sell new cars when Chrysler Motors resumed manufacturing civilian vehicles in late 1945. With the eagerly awaited shiny new Dodges and Plymouths arriving in numbers, the dealership prospered. It provided a good job for my father, giving the family a decent income, at least for a time.

But generally, over the course of my childhood, my father never seemed to hit his stride in the business world. Despite his intelligence and generosity of spirit, he never seemed to find out who he was or what he really wanted to do. (Later in life, however, he did find satisfaction when he went to work for the city of San Francisco.) Though he never complained, I'm sure the uncertain times along the way were hard for him. It was certainly a disappointment for Mom that he wasn't doing particularly well financially—it drove her crazy that Dad was often out of work. Given my mother's worries, it is not surprising that I grew up with another unspoken message: the sense, true or not, that it was my job to make up for my father's lack of success. Mom poured her hopes, dreams, and expectations onto me, her only child.

His career ups and downs notwithstanding, my father remained a crucial influence on me. Dad continued to wax enthusiastic about specimens I had assembled and take me on trips to find new ones, and I continued to appreciate his support for my interests. This kind of hands-on involvement was not something that fathers of that era often did, but I mostly took it for granted. His genuine interest in others, keen sense of humor, and unfailing kindness made him a real role model. He encouraged me to be good at what I undertook more gently than Mom did, and to care about the world beyond my own concerns.

In the postwar years and right through my time in high school, the CAS Student Section continued to absorb, entertain, and educate me. Instead of simply delivering lectures to us, the Student Section instructors expected us students to write short articles for the *Bulletin* and give talks, reasoning that this would strengthen our ability to express ourselves and build our self-confidence. They were right, of course, and I would later recognize their expectations as a wonderful gift. Besides the regular Saturday classes, there were opportunities to visit the Academy after school and to go on weekend field trips to natural places like the slopes of Mount Tamalpais. The occasional overnight trips were even more exciting: terrific ways to observe more of the varied habitats of Central California, a living jigsaw puzzle of interlocking communities and species. I appreciated how lucky I was to grow up on the edge of one of the great natural harbors of the world, surrounded by lands rich with redwood forests, chaparral, tawny dunes, coastal scrub, grasslands that turned golden-brown in summer, oak forest, reeds and rushes, and serpentine soils home to an incredible diversity of plants and animals. It was the perfect incubator for a future botanist.

The Student Section was the core of my social life. Returning on buses from our trips, we would sing together from songbooks that our leaders brought along—traditional songs like "Red River Valley," "On Ilkla Moor Baht 'at," and "The Eddystone Light." Outside of formal Student Section activities, we kids got together in smaller groups, visited Playland at the Beach, went trick-or-treating, and otherwise just amused ourselves. Our bond extended beyond the love of science; we built strong friendships with one another and enjoyed them greatly. In the years when I felt like an outsider at school, the Student Section—where age mattered less than common interests—was a crucial source of companionship and support. Here, I was an insider.

During those years, I began to visit the Academy's entomology department with increasing frequency. I particularly appreciated Ed Ross, the department's chair, who supervised the activities of the students who came in and pestered the curators. But even as he managed us, he was friendly too, and anxious to introduce us to the joys of field exploration that were so important to him. One of the other "stars" there was Edwin Cooper Van Dyke, one of the world's leading experts on beetles. Dr. Van Dyke was already eighty-five years old when I first joined the Student Section, but his long career still had a ways to go. Old-school, formal, polite, and immensely knowledgeable, he was a formidable figure for a young boy to approach, to say the least. But he always had a moment for me, a kindness I appreciate even more today than I did then. Hugh Leech, an eminent specialist in water beetles, arrived at the Academy in 1947, bringing his collections with him. All of us who came to know him loved to collect these sleek, streamlined, swimming creatures and then bring them to Mr. Leech for his identifications and comments. He inspired many of us with his friendly nature and his patience with our endless questions. Even later, after I began to concentrate on plants, I continued collecting water beetles for Mr. Leech because of his warm encouragement.

Even though I had a particular focus on beetles, my childhood interests started me out in life as a generalist in biology: a naturalist in the traditional mold, someone with an appreciation for the astonishing, awe-inspiring number and variety of living organisms on our planet. My experiences as a child were a meaningful introduction to the concept of biological diversity that was to become so important to me and my work over the years to come.

– 3 –

Botanical Prodigy

IN THE SUMMER OF 1946, when I was ten years old, I returned for another stay with the Goldshires. This time, I looked at other kinds of organisms beyond the beetles and butterflies that were my early passion. On a visit to the nearby wharf in Santa Cruz, I obtained and pickled a small octopus to add to the growing store of treasures in my bedroom. In addition, I took my first steps in the direction of botany by pressing specimens of some of the plants that grew in the vicinity of the Goldshires' cabin, some in the shade and some along the sunny abandoned railroad tracks up the hill from us, where the trees had been cleared. In the shallow waters along the creek's banks grew extensive patches of a plant with large, maple-shaped leaves a hand-span across that I later determined to be coltsfoot (*Petasites frigidus* var. *palmatus*), a member of the sunflower family that sent up white flowering stalks before its leaves, in early spring.

Handsome clumps of sword ferns, with their long, dark green fronds, were scattered on the open floor of the redwood forest, and delicate goldback ferns, with much smaller leaves forming an overall triangular shape, grew on exposed dry rocky slopes. On moist slopes in the shade, masses of maidenhair ferns, with exquisitely delicate, shining, dark brown stems and deeply divided leaves, flourished. Growing in clumps in the shade were trilliums, with three leaflets on each stem surrounding a single three-petaled white or maroon flower, depending on the species. In dappled shade grew a delicate wild rose with soft pink flowers. I collected specimens of all these and more.

On the dry banks of the railroad bed, on the slope above the creek, I found *Clintonia*, a beautiful plant with bright blue berries clustered at the top of the stems and large, thick, oval leaves. I probably didn't have those words for it at that age, but I felt such awe, wonder, and gratitude in seeing this plant. Here it was, growing in its own special place beside the decaying railroad ties. It had grown there before I came, for who knows how many decades or centuries. Given the right conditions, it would keep coming up every spring whether I saw it or not, and it would be beautiful whether I saw it or not. I suddenly felt the smallness of myself in relation to the grandeur of the world and of the Earth's natural treasures.

Soon after I returned from Zayante that summer, my parents moved again. My grandmother Laura sold the house on 12th Avenue and moved into a tall apartment building on Broadway with her sister Carmelita Lenhart. My parents and I moved to a lovely rented apartment at 2108 California Street, at the corner of Laguna. Our home was at the head of a long slope that led eastward down to the Bay. It was within easy walking distance—about a mile—from the Greyhound Bus Station, which would become a central hub for me when I began making trips on my own to different parts of the Bay Area. A few blocks from our house, on Bush Street, was beautiful St. Dominic's Catholic Church, which we began to attend faithfully.

Naturally, not everything in my life had to do with science. I became a junior member of the Olympic Club, which I often visited to use its pool. I got my teeth straightened. From time to time, my mother took me shopping for clothes downtown at department stores like Roos Brothers, The White House, and City of Paris, always garbed in white gloves and a neat white hat. The time we lived on California Street turned out to be particularly happy for all of us. My parents and I would walk a block to the small grocery store on Octavia Street to get Washington Square Mint Patties for dessert and return home holding hands, my father joking about the round shape of the "square" patties. The cable car line ran past our apartment, running all the way west to Divisadero, and I became so accustomed to its continual humming sound that I would awaken abruptly in the middle of each night when the cable was turned off for a few hours.

My bedroom at this time summed up much of what my life was like. It was relatively neat and contained the typical props of a young boy growing up: comic books, a pocket knife, a baseball bat and glove. Next to my bed was a Bakelite radio, small enough that I could tuck it under the covers to listen to *Radio Mystery Theater* while I was supposed to be sleeping. Other possessions were less typical. A tall spear made of dark wood, brought back from Tanganyika Territory by Aunt Vivienne's mother, leaned in the corner, inspiring thoughts and questions about Africa. My desktop, shelf space, and bureau drawers were populated with specimens from the natural world: agates, feathers, shells, animal bones, shark's teeth, geodes, the shed exoskeletons of arthropods, interesting driftwood, all of them fascinating objects that I had found.

By then, music had become an important part of my life. In our new home, a phonograph and an eclectic stack of LPs sat among the furniture and Asian artifacts. My parents had recordings of the popular music of the day, such as soundtracks for Broadway shows like *Porgy and Bess*, *Oklahoma!*, and *South Pacific*. Our family favorite, however, was Cole Porter's delightful *Kiss Me Kate*. I loved the music of the classical composers too, especially the Romantics, including Chopin, Grieg, Sibelius, Bizet, Dvořák, and de Falla, listening to it avidly on the radio. Hours were lost in the wonders of de Falla's *Night in the Gardens of Spain*, Bedřich Smetana's *The Moldau*, and Nikolai Rimsky-Korsakov's *Scheherazade*. I felt fortunate to have music around me and shared my enjoyment of it with many of my young friends. It was a wonderful way to begin to experience the different cultures and voices of the world.

Because of our move, I transferred to the St. Brigid School, located at Broadway and Franklin about nine blocks from our house. This landmark institution was run by the Sisters of Charity of the Blessed Virgin Mary, who had operated it since its founding in 1888. To my surprise and relief, at St. Brigid's I found a group of teaching nuns who inspired me, expected me to do well, and treated me with respect. They saw me as a straight-A student waiting to be born. They truly believed in me, and that transformative confidence lifted both my grades and my spirits. Their kind example helped to teach me how important it is to affirm and encourage others.

When it came time to try to identify some of the plants that I had collected in the summer around Zayante, I naturally turned to the botany department at the California Academy of Sciences. My first visit to the department that autumn was quite intimidating. I became even more awed when I was ushered into the venerable presence of Alice Eastwood. One of California's most famous pioneer botanists, Miss Eastwood had served as Curator of the CAS Department of Botany since 1892, famously saving the most important specimens during the San

Although the famous botanist Alice Eastwood was seventy-seven years my senior when we met in 1946, she treated me very kindly, thus whetting my enthusiasm for plants.

Francisco earthquake and fire in 1906. She would hold the position until her retirement in 1950 at the age of ninety. Since I was only ten, I didn't fully appreciate how remarkable she was, but I certainly knew enough to understand how nice she was to me.

Despite her prestige, Miss Eastwood seemed pleased to make time for me. "You did a fine job in making that collection," she remarked as she looked at my specimen of a wood rose. "You collected a branch that had a fruit on it. As you can see, the sepals have fallen, and that tells us that it is *Rosa gymnocarpa*, which means 'naked fruit.' The more complete your specimens are, the easier it will always be to name them." With encouragement like that, I was bound and determined to continue finding interesting plant specimens and preparing them with care. I still had a few boxes of insects, but these months marked the germination of my deep interest in the green world of botany.

In spite of Alice Eastwood's praise and encouragement, my collecting techniques remained primitive. I dried the small snips of plants I collected between the pages of magazines, telephone books, or anything similar that I could find. When they were dry, I mounted them with Scotch tape on small sheets taken from a loose-leaf notebook. At the bottom of each page I would print the name of the plant, the place where I had

Willis Lynn Jepson's *A Manual of the Flowering Plants of California* (1925) was an authoritative reference for me as I learned about the plants of the region where I lived.

found it, and the date. I could identify many plants using the dichotomous keys in my blue-covered *A Manual of the Flowering Plants of California*, informally known to botanists simply as Jepson after its illustrious author. The manual's descriptions of species' habitats and localities also told me if a species had been previously documented in the place I had found it, and gave me clues about where to look for plants I hadn't yet seen.[1] It was all a great jigsaw puzzle, and working steadily to solve it proved immensely satisfying.

I gathered specimens from virtually every place I was fortunate enough to visit: Zayante, Hollister, the waste and weedy places around San Francisco, the destinations we visited on Student Section field trips. As each of the specimens was pressed and mounted, my technique improved. On one memorable family outing, my parents and I drove up the back road from Livermore to the summit of Mount Hamilton, with me collecting plants along the way. On another occasion, my father drove me to the Rocks Road section of the Breen Ranch near Hollister. There we left the car and wended our way on foot up the hills, past menacing cows and through the thick, spring-green grass, admiring and collecting some of the bright-colored wildflowers for which California is so justifiably famous. My collection gradually grew to encompass hundreds of plant species.

Knowing that my summer visits to the Goldshires had been wondrous for me, my parents decided to look for a new outdoor option for the summer of 1947. They found one at Camp Santa Teresita, a summer camp operated by the Catholic Youth Organization at Bass Lake, in the Madera County foothills of the Sierra Nevada. The drive to Bass Lake allowed me to appreciate for the first time the scope of the summer-dry foothills that bordered both sides of the largely agricultural valley—the Inner Coast Ranges on the west side and the lower foothills of the Sierra Nevada to the east. I had seen similar, tan hills in my earlier, more local travels, but never over such striking expanses.

As we made our way up the gentle lower slopes of the Sierra, the dry, grassy hills slowly gave way to scattered pines and oaks. Still higher, as around Bass Lake, at 3,400 feet elevation, we saw groves of ponderosa pines, incense cedars, and other trees. It was to be another two years before I had the opportunity to ascend the range still higher and experience first-hand the marvelous forests that grew there. But for now, Bass Lake was magnificent enough!

Camp Santa Teresita accommodated about two hundred boys in ten wooden-floored dorm tents grouped around a large mess hall. The dorms were segregated by age, with the younger campers on the lower slopes nearer the lake. While our accommodations were plain, there was nothing ordinary about the diversity of insects and plants that populated this beautiful area.

With green meadows filled with colorful wildflowers, clear streams, and the blue lake itself, it was a tremendously inviting place for an eleven-year-old biologist. I was still very much interested in beetles, and there were plenty of kinds there to admire, sometimes very close at hand. Small buprestids (metallic-colored wood-borers), medium-sized long-horned beetles, blister beetles, and scarabs could all be found in abundance; June bugs buzzed through the twilight dusk around our campfires. In the lake were many kinds of water beetles that I hadn't encountered around San Francisco. I eagerly caught those in large numbers, to bring back to Mr. Leech for what I hoped would be his admiration. The caring counselors mostly encouraged my interest in natural history. Barr Healy, a young seminary student who really understood my love of nature, gave me a box of beautifully varied Florida tree snail shells that he had collected, a wonderful gift.

By the time I returned to Camp Santa Teresita the following summer I had switched my focus almost completely to plants. I now had totally different priorities. At eleven, flowers were places where I could often find beetles; at twelve, I brushed beetles off to get to the plants on which they had gathered! Interesting plants were abundant all around the camp. On a hike to Angel Falls, above Bass Lake, I discovered two kinds of plants that stood out vividly: the lovely Sierra lupine, *Lupinus stiversii*, with its purple and gold flowers and, hanging over the falls, the stream orchid, *Epipactis gigantea*. They were completely fascinating to me and omens of many wonderful discoveries to come.

Charles Hamilton Raven (C. H.), my parental grandfather, taught me many things during the years he lived with us in San Francisco. He was an exceptionally kind and gentle man who had led an adventurous life in California, Hawaii, and China.

That thoroughly enjoyable summer was abruptly interrupted on July 10, when a counselor took me aside to tell me that my grandfather, C. H. Raven, had died the day before. Of course, no one at camp knew who my grandfather was or how much he meant to me. I understood that my parents were not sending for me to come home; they were just letting me know of the news. I was twelve years old, understanding the significance of a family death for the first time, and I really didn't know what to do. The counselors were kind, but C. H.'s death made me very sad. He had been a very special friend and mentor for as long as I could remember, and I missed him greatly. Camp lasted a little longer and I went on with it, but rather forlornly. Flashes of memories about him would come to me throughout the day and while I lay in my bunk at night, staring at the canvas ceiling of our tent: C. H. teaching me to type, encouraging me to write little stories, always inspiring me to learn and improve.

When camp ended that summer, my parents picked me up and the three of us drove the fourteen miles north to the entrance of Yosemite National Park. Perhaps they knew that I needed consolation and distraction from my grief; Mom wasn't especially outdoorsy, and Dad didn't care about bugs or plants, he just cared about me. I saw some remarkable things on the trip. As we wound around the side of a slope covered with chaparral on our way to the park, we were amazed to see the gorgeous, five-foot-tall flower stalks of golden eardrops (*Ehrendorferia chrysantha*), growing on a hillside that had been burned a year or two earlier.

In a meadow just in front of the store at Camp Curry in the Yosemite Valley I spotted a clump of milkweed plants in flower, with a number of brightly colored insects that included red milkweed bugs and cerambycid beetles, feeding on their leaves and stems. The plants also attracted bright orange monarch butterflies. It was a spectacular show, almost as exciting to me as the exhilarating beauty of the tall, gray, luminous, glaciated granite walls that surround Yosemite Valley. Leaving Yosemite, we followed the winding road down along the meandering course of the lovely Merced River, and I noted the fascinating plants along its banks, many of which I had not seen before.

Late that summer, I was delighted to receive a full-sized plant press, complete with blotters and crossed slats of wood, the frame a bit larger than a folded sheet of newspaper. It had belonged to my grand-aunt Josephine Breen, another of my grandmother Laura Breen's sisters, who had died a couple of years earlier. Excited to have this more professional way to preserve my specimens, I would keep that press and its successors very busy. Encouraged by the acquisition, I began to expand my sense of what I could accomplish in the realm of plant collecting. The Academy's Student Section also remained an important source of inspiration and learning, as well as a social context for many of my activities.

By that time, John Thomas Howell, about to succeed Alice Eastwood as the Chair of the CAS Department of Botany, had taken me under his wing. A prim, gentle, and profoundly kind person, Howell was utterly devoted to his field and to the students, colleagues, and volunteers who came to the department and stayed to join in his projects. I didn't understand enough about botany to appreciate his scientific excellence, but I was among the many who understood his gift for inspiring, encouraging, and motivating people. Just as important, he thoroughly enjoyed plants and possessed a deep ability to communicate about them to others.

At Mr. Howell's suggestion, I started to prepare and use separate paper labels for my dried plant specimens, rather than simply writing the information on the page to which the plant was attached. At first, it seemed logical to type the labels so that they would be clear, but Mr. Howell explained that,

My greatest encouragement in the pursuit of botany came from John Thomas Howell of the California Academy of Sciences. Tom was always friendly, helping me in many different ways.

to him, the labels on plant specimens were the equivalent of a signature on a painting and should always be handwritten. Only then did he point out gently that he was not able to read my handwriting! With this inspiration, Howell achieved what the nuns never could: I worked hard on improving my penmanship so I could make fairly legible labels.

Mr. Howell also suggested that I start numbering my specimens in the order in which I collected them so that it would be easy to refer to the individual samples. It was good advice given at the right time, since I was kicking up my plant collecting into high gear. I loaded my press with blotters and a stack of newspaper (folded to fit the press) and headed off with my parents for Hollister in early September 1948. As soon as we reached Uncle Jack and Aunt Aileen's farm just outside of town, I collected a specimen of bindweed (*Convolvulus arvensis*), a nasty European weed that grew along the edges of the fields. It was specimen number 85. I had assigned my first 84 collecting numbers to specimens that I had collected earlier that year and were worth carrying forward.

Now in possession of a real plant press, I began to collect larger and therefore more useful specimens than I was able to manage earlier. In drying them, I was soon taught first-hand lessons about plants' high water content. Specimens will last a long time if dried properly, but proper drying can be tricky. Some of my collections, not adequately dried, rotted; blackened and covered with mold, they were essentially worthless. I found that the more frequently I changed the blotters, the faster the drying and the better the preservation of color in the dried specimen. It became a habit to spread the blotters around the floor of my room at night to dry, change the blotting papers in the morning, and stack it all back under my bed for the day.

Sponsored by Mr. Howell, I joined the Sierra Club as a junior member at twelve, the minimum age for membership. My first trip with the Club, made with a local group, was a late-November outing to Bethel Island, in the delta of the San Joaquin River. At the time, I was just excited to gather more specimens, but my connection with the Club would turn out to be both enduring and rewarding.

In the spring of 1949, as plants again came into bloom, I began using my copy of Jepson's *Manual* in a more intensive, focused way. Poring over its pages, I learned that a unique kind of shooting star then called *Dodecatheon hendersonii* var. *bernalianum* grew on the slopes of Bernal Heights, a steep, partly undeveloped hill in the southeast quarter of San Francisco.

With my father, I searched for and located it there in March. I took particular pleasure in entering the name and locality for this specimen in my notebook. It was new to me and quite attractive.

At about the same time that I visited Bernal Heights, my parents started allowing me to go by myself to a wider variety of places in and around San Francisco. They let me venture out alone on Muni streetcars, which offered three fares for twenty-five cents and reached all parts of the city efficiently. On weekends I could take Greyhound buses to particular localities outside the city that promised special plant collections. Once a bus had dropped me off, I would walk all around that area, observing and collecting samples of the more interesting plants. Rambling San Francisco locations like Twin Peaks, Mount Davidson, the Presidio, the beaches, and the Lake Merced area, I sought records for a flora of San Francisco that Mr. Howell and I had started to prepare for publication. In April I participated in a Sierra Club trip to the Tiburon Peninsula in Marin County, a lovely series of rolling hills that extended eastward into the Bay just north of San Francisco. The peninsula was largely covered with greenish serpentine soils on which grew open grassland crowded with flowers, including a number of species that grew only on that type of soil. A week later, I returned to Tiburon on my own, again with the blessing of my parents, and collected sixty more specimens; two weeks later, I added another fifty. A variety of short trips followed with the Student Section, the Sierra Club, and my parents. At the end of June, my parents and I made a nostalgic visit to the Santa Cruz Mountains around Zayante, where I was able to make "grown-up" specimens of some of the plants I had seen earlier, as well as many others.

I was collecting plants because I wanted to know them, all of them. Making a specimen of a plant was the best way of impressing it into my permanent memory. Once each specimen was identified, mounted, and labeled, I turned it over to the Academy. Most of them were accepted and incorporated into the herbarium, which made me very happy. As I learned more and spent more time with the professional botanists at the Academy, I came to understand that collecting served an important scientific function because a record without a voucher specimen was just a rumor. In addition to learning botanical fundamentals such as this, I was gradually acquiring a knowledge of the flora of California and a feeling for the patterns that plants formed on the landscape while also—importantly for a thirteen-year-old boy—having a great deal of fun.

❧

In the midst of this informal apprenticeship, I was growing up. Right around my thirteenth birthday, Dr. Sun Ko-Chi, the doctor who had presided over my birth in Shanghai, came to visit us in San Francisco. He had made periodic visits in the years after the war, often as part of fundraising efforts for his hospital in Shanghai. This time he was here to pick up a block of radium, fortunately enclosed in a lead case, that my parents had stored for him in their basement. My parents apparently asked for him to help with a delicate matter while he was there. He took me aside and, after an awkward silence, opened one of my books on insects. Pointing at the male and female sex organs in the illustrations, he asked in his thick accent if I knew how insects mated. Of course I did, and I said as much. Dr. Sun then concluded our mercifully brief conversation by saying that it was the same way for people. I stared at him, or maybe at the floor, taking in what he had just said. As what he was trying to communicate dawned on me, I felt mortified. Did my parents breed like insects to have me? Why couldn't they have just explained it to me themselves? I felt a bit resentful that I had to get such information so awkwardly from Dr. Sun, whom I really didn't know very well. I was beginning to have the same stirrings as any adolescent boy, so the incident made me very self-conscious. I'm sure I must have blushed!

That was the last time we would see Dr. Sun. In the tumult surrounding the formation of the People's Republic of China later that year, all official communication with the U.S. was halted. Dr. Sun disappeared from our lives, but not before he had made one final request for my parents' help. Dr. Sun expected his two children, Yobonne and Yi, to receive their college educations in the United States. He gave my parents a sum of money—I think it was about $5,000—and asked that they hold it for him against those future expenses. Glad to be of use, they banked it in a separate account. But after China closed to the West, there was no way for the children to come to the U.S. to study, to get the money to them, or to contact Dr. Sun.

❧

By writing more than three articles for the Student Section newsletter during the previous year, I had earned the opportunity to take part in a

special outing in June of 1949. It was a trip to Norden, near Donner Pass—
my first venture into the High Sierra and first exposure to the fascinating
plants that grew there. Patches of snow remained on the ground when we
arrived on June 28. More snow lay on Boreal Ridge, which sloped up
some 2,000 feet above us to the north. Once the mountain sickness that
plagued me the first few days abated, I began to really enjoy the sunny,
plant-filled slopes. Despite the patchy snow, many kinds of plants were
flowering. Among them, I enjoyed finding many that were new to me,
and even rounded up a few that had eluded Mr. Howell during his studies there early in the same decade. As my fellow Student Section members
and I discovered the fascinating animals and plants that lived in these
magnificent mountains, our friendships grew closer. For me, both kinds
of familiarity—with the new mountain plants and with the students who
accompanied me—were sources of great happiness. Surrounded by those
who shared my interest in the natural world, I found kinship and the sense
of belonging that had eluded me in school. Here, I was in my element.

Another significant aspect of my first high Sierra experience was being
in close proximity to Donner Lake, the place where my great-grandfather
Patrick Breen Jr. and his family had nearly starved and frozen to death
just over a century earlier. Reflecting on the fact that they had survived
when others had not, I began to feel that I owed it to their memory to
make something of myself. I felt special gratitude for the compassionate
John Stark, who had rescued my great-grandfather with his family from
the snow pit at Starved Camp so long before.

I returned home with nearly 300 collections of plants—a bonanza
even in that time of avid collecting. Mr. Howell and I had lots to talk
about, and that delighted me. The more I worked with him, the more
I came to appreciate this extraordinary man. Each detail I learned about
his career deepened my admiration, including his work with Willis Linn
Jepson—author of the manual I was finding so useful—and his years of
collecting plants around California and other Western states with Miss
Eastwood. I also began to grasp the value of Mr. Howell's approach to
understanding the plants of California, an incredibly diverse state botanically. He believed that intensive collecting and detailed study of the
plants found in particular areas would become critical building blocks to
understanding the plants of the whole state. Accordingly, he worked tirelessly to write local floras and encouraged others to do so as well. He urged
amateur botanists who lived in places like the Ojai Valley, the plains of

eastern San Luis Obispo County, Kern County, Monterey County, and other areas to discover their local plants, determine their distributions, and catalog them in floras. His astute comments helped them to understand the plants they were collecting, and a good number of the species they found by scouring their own localities intensely proved to be new to science.

Tom Howell's particular interest in the plants of Marin County resonated with me, as this was where my mother was born and several generations of my ancestors had lived. Tom had started hiking there in the 1930s to gain the endurance he would need to reach several localities where he wanted to find certain plants for his studies. His 1949 publication, *Marin Flora: Manual of the Flowering Plants and Ferns of Marin County, California*, was the first scientific publication that mentioned one of my own collections, a tiny aster relative called *Pentachaeta bellidiflora* I had spotted on the rocky slopes among the buildings at Marin City while waiting for a bus transfer to Tiburon early one spring morning. Having it included in this estimable mentor's book was heady stuff for a thirteen-year-old nature nerd!

As memorable as his knowledge and kindness was Mr. Howell's personal touch. He never made me feel he was too busy to talk with me. Time after time, as if he had nothing better in the world to do than to inspire young botanists, he patiently gave me help, guidance, and encouragement. He would smile and then bend down to look at the specimen I was working on. "That's a beauty," he might say. "Where did you find it?" In addition to urging me to improve my penmanship, Mr. Howell helped me refine the techniques I used for drying, taught me about important botanical relationships, and gave me tips for distinguishing species in plant groups that were "difficult." He pointed out that professional botanists make mistakes too, and that the identifications of specimens filed in herbaria were often unreliable—information that saved me from much frustration. Thanks to him, I gained the benefits of outstanding scientific mentorship and meaningful collaboration at an unusually young age.

It was during this phase of my life that I began to realize for the first time that the natural areas around me were disappearing. After the war ended, people had begun to flood into California, building industry and often setting precedents for the rest of the country. The Stonestown Shopping Center, developed near Lake Merced beginning in 1949, was an example. My parents doubted that this new venture could succeed, as people were used to going downtown to shop rather than out to the suburbs. But the ubiquity of the automobile and the growth in suburban housing developments were changing that habit, and my parents were soon proved wrong. Eventually a new campus was built for San Francisco State University near the shopping mall, and gradually most of the once-wild dunes and slopes in the area were lost to developments for housing or commerce. Watching such changes, I very slowly became aware of the need for conservation, for a mindset that treasured unspoiled land and its diverse habitats. Full awareness of the need for conservation was still decades away for me and for most others as well; in any case, as a boy heading into his teens I had many other things on my mind.

⋐⋑

I graduated from St. Brigid's in June 1949. That same year, a few months after turning thirteen, I entered St. Ignatius College Preparatory School (SI), a venerable Jesuit institution related to the University of San Francisco, with which it could trace a common history dating back to 1855. SI was located in a large renovated commercial laundry building at Turk and Stanyan streets, between the east end of Golden Gate Park and the west side of the University of San Francisco campus. I won a scholarship that covered my $15-per-month tuition. SI was an ideal place for me, with a rather rigid scholastic atmosphere, and I did well. I took four years of Latin and two of Classical Greek; I knew enough about biology to understand that standard plant and animal names were given in Latin, and sometimes in Latinized forms of Greek or other languages, and so I knew just how useful it was to learn these "dead" languages. I also had excellent classes in chemistry and physics from experienced, committed teachers. Oddly, no biology was offered—but that was all right, as I was starting to learn about it on my own.

Socially, I felt very much on the outside, contending with the ever-present age gap that separated me from my classmates among the inevitable

At the time of my graduation from St. Brigid School in June 1949, I was a gangly 6 feet 2 inches tall and about 140 pounds.

woes, doubts, and insecurities of adolescence. My isolation eventually eased as I formed friendships at St. Ignatius, but it would take a while. In the meantime, Sally Barrett, a close friend from the Academy's Student Section, began to occupy an important place in my life. Sally loved music as much as I did, understood my passion for plants and collecting, and always gave me her warm support. We spent a lot of time together, and I began to call her my girlfriend.

After entering St. Ignatius, I began to volunteer in the botany department at the Academy after school, preparing specimens for mounting and seeing that their labels were in order. I didn't need extra motivation to work hard on anything botanical, but Mr. Howell often left a candy bar for me in the drawer of the desk where I was working. In later years, discussing it, he called it "the bait." My tasks involved working with others much of the time; being around those with similar interests in a community that fostered learning, dialogue, and collaboration was profoundly inspiring.

During those years, my aunt Vivienne Raven was the source of opportunities and insights that I wouldn't otherwise have experienced. My parents lived as solid upper middle-class people no matter what their financial circumstances were at a given time, but Vivienne and her husband Charles (my father's brother) were considerably wealthier than we were. Aunt Vivienne, who had already arranged for me to join the Olympic Club, also enrolled me in dancing lessons. She broadened my cultural horizons as well, taking me with her children Carla and Charlie Jr. to occasional performances of operas and other musical and stage events, such as a performance of *Peter and the Wolf* at the Jewish Community Center. All these experiences helped build my confidence and polish my social skills.

Distracted by school, I didn't do much plant collecting in the autumn of 1949, but my efforts picked up steam as winter gave way to spring early in 1950. In late February of that year, I decided to explore the San Bruno Mountains, the tall range of hills that lies immediately south of San Francisco. To get there, I rode a Greyhound bus to Brisbane, along San Francisco Bay. I left so early that I arrived there before sunrise and in a dense fog. Once the sky began to brighten, I found my way to the base of the mountain and began climbing, making it to the top of the ridge by midday and hiking west to the highest summit by early afternoon. There I was surprised to find a beautiful low-growing species of manzanita, *Arc-*

tostaphylos imbricata, in full bloom around the rocks that surrounded the TV tower at the mountain's peak—the same place where it had been found by Alice Eastwood a half century earlier, sans the TV tower of course.

In late March, I accompanied a Sierra Club outing to Corral Hollow in the hills east of Livermore. Called by the Spanish explorer Juan Bautista "canyon of the good winds," Corral Hollow was once a haven for outlaws and home to wild horses. It consists of a streambed in a relatively narrow canyon flanked on both sides by steep, grassy hillsides. Many steep side canyons enter into the main canyon all along its path down to the plains of the San Joaquin Valley. The hills around Corral Hollow generally received far less rainfall than areas closer to the coast, but in the winter and early spring of 1950, the winding valley had received unusually copious rainfall. Its slopes glowed with an incredible range of colors, far more vivid than anything I had ever seen, flowing down and filling the hills flanking the canyon-bottom road. It was as if the hills had been splashed at random with paint from many different pots. From a distance, the scene was one of broad strokes of color—an Impressionist landscape. But each time we stopped to investigate, the individual species and plants came into focus: deep-orange fiddlenecks and other borages, lavender gilias, blue phacelias, golden California poppies, pastel lupines, mauve owl's clover, daisies, tiny goldfields, and yellow tidy tips, ray flowers neatly edged in white. Adapted to these often dry hills and slopes, these annuals have seeds that can remain alive in the soil for many years, ready to reproduce when enough rain falls.

One of the most interesting plants that we collected on this outing was the extremely rare *Eschscholzia rhombipetala*, a small-flowed species of California poppy. With flowers the size of a dime, this poppy germinates and grows only in those rare springs when its moisture requirements are fully met. That day, its compact clumps formed a small colony on a clay slope at the top of the Tesla Hills. In a small, narrow canyon leading to the north at the upper end of Corral Hollow, we also found the rare and beautiful large-flowered fiddleneck, *Amsinckia grandiflora*, forming a small patch.[2] Though many of the plants we saw that day were not as rare as these two, each seemed more interesting than the one that came before it. This trip was a turning point for me, showing as it did that plants were even more varied and wonderful than I had ever imagined, and that there were endless delightful discoveries to be made among them.

Later that spring, I decided to climb Montara Mountain, another tall ridge that rose 1,901 feet from the Pacific above the town of Colma just south of San Francisco. I wanted to see how the plants there compared with those I had found a couple of months earlier on San Bruno Mountain to the east. It was a beautiful, exciting day, highlighted by my discovery of a colony of *Salvia spathacea*, hummingbird sage, a magnificent plant with large rose-red flowers that attracted hummingbirds, their primary pollinators. What I remember most about that day, though, was looking down on a huge colony of coyote bush, *Baccharis pilularis* subsp. *consanguinea*, and wondering seriously how many herbarium specimens would be needed to document the variability within that one colony adequately.[3]

That spring I also went on a driving trip with Tom Howell so that he could introduce me to the plants of Napa County. As we drove from Calistoga to Mount St. Helena beyond it, I became enthusiastic about the possibilities of studying the plants of Napa County in detail. Even with my fledgling botanist's eyes, I could see that the county supported a great diversity of plants. The vegetation was extremely varied, growing in habitats ranging from the salt marshes where the Napa River flows into the northern arm of San Francisco Bay to the top of 4,341-foot Mount St. Helena at the valley's northern end. The western edge is rimmed by the largely forested Mayacamas Range, which separates the Napa Valley from the Sonoma Valley to the west. The hills that rim the Napa Valley to the east are much drier, although Howell Mountain, part of that ridge, boasts a grove of redwoods much further inland than where these trees normally occur. What particularly intrigued me was the way that plants like ones I had so enjoyed in Corral Hollow, ones with affinities with the arid interior or even the desert, ranged up along the eastern side of the hills east of the valley, while coastal forests dominated the mountains to the west—a pattern characteristic along the length of the Coast Ranges.

As a geographically varied region within reach of San Francisco, rich in different plant communities and species of plants, Napa County came to seem a very suitable place for which to write a flora like the ones Mr. Howell was creating. It was an ambitious goal, but one that captured my imagination. For the time being, however, the logistical realities of being a fourteen-year-old meant that I couldn't yet explore and collect in earnest.

❧

The summer of 1950 brought the most exciting opportunity yet to learn about the plants of a part of California completely new to me. Mr. Howell had found that he would not be able to participate in the Sierra Club Base Camp outing that summer, the first time he would miss the trip since 1940. He had been using these annual trips to systematically increase the information available on the plants of high elevations along the Sierra Nevada, and wanted his effort to continue without interruption. Would I go in his stead, he asked, and help fill in details about the area being visited that year? Naturally, I said yes, making no effort to hide my excitement.

Base Camp outings had been a Sierra Club tradition for many years. During a two-week-or-longer timeframe, up to 300 participants would hike in to a remote location with the help of pack animals, set up a base camp, and from there make hikes and side trips of various lengths. The participants included individuals and families who just want to spend relaxed time together, dedicated outdoor enthusiasts and mountaineers, and expert scientists. That year, Base Camp would be located on the upper

My future wife Sally Barrett and I, with California Academy of Sciences director Robert Miller and Tom Groody, host on the Academy's then new TV program, *Science in Action*, on the beach above the tide pools at Moss Beach, just south of San Francisco in 1950.

Middle Fork of Bishop Creek on the east side of the Sierra. I was thrilled at the prospect of heading up into the far reaches of the Sierra, eager to see and collect the plants that grew there, and anxious to live up to the faith that Mr. Howell had placed in me.

It was arranged that I would get a ride to the base camp trailhead with a man named Stebbins. Making plans with Dr. Stebbins by telephone, I explained that I would like to bring a plant press, containing cut pieces of corrugated paper, blotters, and newspaper. Without much ado, he explained that he knew what I was talking about, and that he had plenty of room for my supplies in his car. When my parents dropped me off at his home in Berkeley, I still had no idea who he was. As we drove along on our way eastward toward the Sierra, I began to learn a little about him and his twenty-year-old daughter Edie, whom he was bringing along for a family bonding experience. During the course of the conversation, I finally connected the name of a plant species that I had collected, *Hordeum stebbinsii*, with the name of the driver of the car in which I was now riding. Ledyard Stebbins was modest about his accomplishments and I wasn't yet aware of the nuances of his career, but it dawned on me that he was an important man in his field. And yes, he certainly knew about plant presses.[4]

After an overnight stay at a Forest Service campground in Tuolumne County, we arrived in the early afternoon at the pack station at the lower end of Lake Sabrina, where we were to leave our personal camping gear to the mule convoy. Dr. Stebbins and I investigated the plants in the vicinity during the remainder of that day. Early the next morning, Stebbins, Edie, and I began our hike past Lake Sabrina, a man-made lake located on the Middle Fork of Bishop Creek that had been constructed just over forty years earlier to provide hydropower. As we made our way around the water's edge, Stebbins called my attention to a tiny gem of a saxifrage, scarcely two inches in height and with minute white flowers, that was growing among the huge boulder slides along the trail. It was *Saxifraga hyperborea*, a species that ranges all around the Northern Hemisphere at high latitudes and southward locally along the mountains. I raced over the boulders to see more of the diminutive saxifrages hiding under the large stones and to discover whatever else might be growing there with them. Leaving the reservoir, we wound our way, constantly climbing over switchbacks past some breathtakingly beautiful mountain lakes, making our way toward the campsite. It was located in the green, flowery meadows at the lower end of Dingleberry Lake, a gorgeous small lake at about 10,200 feet elevation.

A variety of guided hikes and climbs were available each day for the duration of Base Camp. Accompanying now this group and now that one, I hiked along mountain paths, across Powell Glacier, and beside glacial step-down lakes with playful names like Topsy Turvy, Blue Heaven, Hell Diver, Hungry Packer, Baboon, Burro, and Fishgut. We discussed the fish population in the different lakes and snow evaporation versus melting and runoff in the mountains, learned about the history and evolution of the pine trees that grew locally (both lodgepole and white-bark pines), and considered the geological history of Dingleberry Lake, a classical glacial tarn. Nature was our study, but also our living quarters, the place where we ate, slept, and bathed. Completely immersed in one of the most beautiful mountain ranges in the world, I was filled with the same kind of spirit of discovery and wonder that the great naturalist John Muir recorded in his classic *My First Summer in the Sierra*.

As the days passed, Dr. Stebbins began to teach me about the evolutionary relationships among the plants we were seeing, as well as the role of chromosome numbers in understanding those connections. He showed me a type of daisy relative with golden-yellow flowers that formed clumps among the rocks in the drier slopes of the area, identifying it as *Haplopappus eximius* subsp. *peirsonii*. Earlier, he explained, he had investigated what was then considered to be a second subspecies of the same species, subspecies *eximius*, on Mount Tallac at the south end of Lake Tahoe, some 200 miles north along the range. These separate populations looked different to him, and, having determined earlier that the northern ones had a chromosome number (the number of chromosomes in each of their cells) of $2n = 18$, wanted to see if the southern ones would have a different chromosome number. Dr. Stebbins had brought along a fixative to prepare the buds of these plants for subsequent examination of their chromosomes. Together, we fixed several bud collections of the populations that we found along Bishop Creek. Little as I understood the details of what was going on at the time, it was exciting to be involved in this quest.[5]

Dr. Stebbins and I both searched for plants with equal enthusiasm for as many hours as the sun would allow. During the course of the outing, I began to make a crucial conceptual leap. Previously, I had viewed plant species as individual "items," and collecting specimens had been about finding and learning about ones I had not seen earlier. Working alongside

Ledyard Stebbins, I began to understand that the same plant species often vary according to where you find them, and that underlying genetic differences were often expressed as subtle differences in form. In addition, I was beginning to learn about plants as members of an ecosystem: an assemblage of plants, animals, and other organisms that had its own logic. Ecosystems differed, and it was of great interest to measure the differences between them. For a budding botanist, these were extraordinarily valuable gifts of understanding.

One day on the trail, we happened upon a nice man with an enormous pack that reached from his shoulders to his knees, stuffed with frying pans, pickaxes, and all manner of other gear. He turned out to be the famous mountaineer Norman Clyde. With Clyde as our guide, Stebbins and I set out one day to explore Coyote Ridge, a high limestone ridge east of the main backbone of the largely granitic Sierra Nevada. Tom Howell, who had collected plants there earlier, had suggested that we visit the place because he had found so many interesting plants there. Once we were on the open slopes atop Coyote Ridge, I darted from one kind of plant to another, taking samples and dropping loads of them into my butterfly net (which I was also using, in a desultory manner, to collect a few of the relatively scarce pollinators). I delivered the loads of unpressed plants to Ledyard, whom I blithely assumed would press them as his contribution to our joint venture. Not surprisingly, after about two loads, Dr. Stebbins said, "Hey! You take some turns pressing. I want to look at the plants too!" But it was all entirely convivial, and as always he was very good-natured with me. When Clyde realized that we were finding very unusual plants on the limestone rocks on Coyote Ridge, he recalled another major limestone outcrop about thirty miles north in the drainage of Convict Creek. We couldn't make the journey on this outing, but later that same summer, Dr. Stebbins would make a series of interesting discoveries there.[6]

Throughout the trip, I took careful descriptive notes on the distribution, abundance, and habitat of the plants I collected, keeping in mind the assignment Mr. Howell had given me. Using these notes, I summed up all that we had learned and observed about the plants in the area in an article entitled "Base Camp Botany" that was published in the mimeographed Sierra Club book commemorating the trip, *1950 Base Camp*.

I was sorry to see Base Camp end. It had been a time of intense learning and adventure. And importantly, in taking my work and my interest in plants seriously, Dr. Stebbins had given me the idea that I might have a career in botany. At the end of the drive home, Dr. Stebbins dropped me off at our new house in the Sea Cliff district of San Francisco, a home I had yet to see because my parents had moved there during my absence. He stepped in to talk with my folks and help carry in the bundles of dried plant specimens we had collected. I had a deep sense of pride listening to this great man tell them about our trip, referring to me as if I were a colleague, and talking about me in the most inspiring and encouraging way.

ↄ∕ↄ

The summer of 1951 began auspiciously with a trip to San Luis Obispo on my fifteenth birthday. Family friend Madge McGregor, whom my mother had met several years earlier when Madge served as her nurse following a hysterectomy, was traveling to attend a nursing meeting and volunteered to take me along. Never having botanized that far south, I was eager to see what plants were growing there, so I jumped at the chance. At first light on the morning of my birthday I was out on the coastal sage-covered slopes near the motel where we were staying. It was my introduction to the plants of Southern California, or at least the northern outpost of that region: black sage, white sage, bladder sage, bushmallow, and many flowering annuals all growing together, appearing in their full glory as the sun rose higher in the sky. After she finished with her meeting, Madge drove me on a nice loop into the interior of San Luis Obispo County, where I was able to find many more plants of interest.

Soon thereafter, my high school years took a turn for the better when a classmate encouraged me to join the school track team. I tried out and was pegged as a distance runner, so cross-country and running became central activities for me. Being on a team was a great way to make friends, and I did so with increasing feelings of acceptance and opportunity. I also made friends within what was playfully called the "Irish Mafia" at St. Ignatius.

Then came the 1951 Sierra Club Base Camp trip. On the strength of my previous year's experience, I was invited to serve as Camp Naturalist. That summer, we camped at Ediza Lake in lovely alpine meadows facing the Minarets, west of the Sierra crest. As Camp Naturalist, I took over Tom Howell's custom of preparing an account of the plants collected and recorded from the area, recording 527 species in that year's edition of the

Base Camp yearbook. A highlight of Base Camp that year was a chance to visit with Japanese artist Chiura Obata, a master of *sumi-e*, Japanese ink and brush painting, who attended the second two-week session with his wife Haruko. He had depicted landscapes all over California, but one of his favorite places to paint was the Sierra Nevada. He had come to the United States at seventeen and become a professor of art at the University of California, Berkeley (though his tenure had been interrupted by his internment during the Second World War). I was deeply impressed by the beauty of his work and his vision. Twenty years later, their son, the famous architect Gyo Obata, was to become my friend in St. Louis, a turn of events that would have been unimaginable that summer.

In the spring of 1952, I visited Mount Tamalpais on a trip organized by Ledyard Stebbins. In addition to a group of students, he brought along Edgar Anderson. A few years older than his good friend Dr. Stebbins, Dr. Anderson was an eminent botanist with a long career at the Missouri Botanical Garden, having been hired as a geneticist there in 1923. Dr. Anderson was spending time in California studying local examples of plant hybridization, a field of study for which he was noted. It was great to hear him speak about the subject, though at the age of fifteen and with no formal instruction in botany I understood next to nothing of what he said.

When I turned sixteen, in June 1952, Madge McGregor taught me how to drive, probably with fewer conflicts than would have arisen had one of my own parents tried! Driving with a stick shift over the steep, hilly streets of San Francisco is not the easiest way to learn how to operate an automobile, but as far as I know, I didn't wreck any clutches. Of course, driving opened up many new opportunities for me. Whenever I could borrow a car, I ranged further into the countryside to collect plants than I had been able to manage earlier.

For the Base Camp trip in 1952 I once again served as Camp Naturalist. That year's Base Camp was located in the Evolution Basin, another magically beautiful area that lay directly west across the mountains from Bishop Creek. Now sixteen years old, I climbed 13,800-foot Mount Darwin, a feat that Ledyard Stebbins had accomplished from the other side two years previously. As I had the year before, I recorded our collections and observations of the plants. Around that time, I began to write short scientific notes on the plants I was collecting in addition to the general articles in the Base Camp volumes—good practice for later botanical and scientific writing.

Beginning in 1953, I started to go on car-camping trips with my friend and classmate at St. Ignatius, John O'Rourke, who had joined the California Academy of Sciences Student Section in 1948. Six months younger than me, John had an easy manner that made him a pleasure to be around. By the time I met him, he had already developed the interest in geology that would become a lifetime fascination and occupation for him. John had a 1936 Chevrolet coupe given to him by a family friend, allowing us to travel to many places that would otherwise have been inaccessible. Our trips were fun and relaxed. Rather than fussing with fancy camping gear, we just threw our sleeping bags out wherever we wanted to stop and cooked our dinners on a Coleman stove. Since uncommon plants often grow selectively on unusual geological substrates, we always had lots to see and to discuss. Our many trips greatly broadened the knowledge of plants and plant distribution I had already gained.

Meanwhile, I went on surveying the plants of the city, with the nearby Presidio an especially fertile field. My first important discovery, in the spring of 1951, was a single spreading clump of manzanita about fifteen feet across, rising no more than six inches from the ground, on a hilltop above Baker Beach (see color image, #17). Collected some forty-five years earlier but remaining as an unclassified specimen in the California Academy of Sciences herbarium, it had been effectively "lost." Ultimately it was described as an unrecognized subspecies of *Arctostaphylos montana*, an erect shrub that I had come to know during hikes on Mount Tamalpais, across the Bay to the north. Once its unique nature had been recognized, its discovery became part of the lore of the park.

I thoroughly enjoyed my work as naturalist on a Sierra Club Base Camp outing in the mid-1950s.

Flanked by my future wife, Sally Barrett, and my friend, John O'Rourke, fellow students from the Academy's Student Section, having a good time together.

During my last semester in high school, I made another notable Presidio discovery, finding an unfamiliar species of *Clarkia* (*Godetia*), in April 1953. The specimens I brought in to the Academy were mounted and placed with the unidentified specimens of the genus in the herbarium, but the story would not end there.

On Sunday, June 14, 1953, I graduated from St. Ignatius with high honors, a much happier teenager now that I had participated in track over my last two years. At graduation, I was awarded a scholarship to the University of San Francisco. With my college education secure, I looked forward to moving on.

Soon after graduation, a family tragedy overshadowed my feelings of accomplishment. The auto dealership where my father worked, Raven-Wager Motors, failed. Three days later, my Uncle Charlie, who had managed the company for his silent partner, jumped to his death off the Bay Bridge. The family was shocked and deeply saddened. Aunt Vivienne, who played such an important role in our lives, was suddenly left to raise her two children alone. Since she and Uncle Charles were prominent socially, the news of his suicide was featured in all the papers, adding to Vivienne's pain.

A week after Uncle Charlie died, still deeply shaken and grieving, I joined the 1953 Base Camp in the Mono Recesses, a lovely series of hanging valleys carved out by the tributaries of the main glacier moving down the valley of what is now called Mono Creek. Our camp was in the Second Recess, and I stayed there for the full three periods Base Camp was in session, for a total of six weeks. Being busy, close to nature, and without the distractions of radios and newspapers helped me collect myself emotionally and prepare to enter college in the fall. Back at home, my father suffered deeply in the wake of his brother's suicide. Their lives had run in tandem for years—childhood in Honolulu and California, time together in Shanghai, houses only a few blocks apart from one another in Sea Cliff—so his brother's absence must have caused wrenching pain. In addition, of course, my father was now unemployed . . . again.

Stepping into the breach, my mother found a job at the California Department of Public Welfare in downtown San Francisco. She became a full-time wage earner, providing support for the family. There she used her psychology background in working with families eligible for aid to their children and in finding job placements for those who needed them. She enjoyed the work, regaling my father and me with tales of her adventures. Focusing on his own future, my father decided to partner with a distant relative, Moore Devin, who owned a small furniture-refinishing business. The two got along well and Moore Devin was a likable sort. Business wasn't great, but it was something.

By the time I was preparing to enter college at the end of the summer I had fallen fully in love with my longtime sweetheart Sally Barrett. There was no question of marrying yet, but both of us knew we were heading in that direction. The world seemed spread before me, a promising future forming on the horizon.

— 4 —

Formal Training

Enrolling at the University of San Francisco (USF) in the early autumn of 1953, a couple of weeks after returning from Base Camp, I planned for a major in biology. My ambitions were modest: I expected to have a career teaching high school, with botany as a hobby on the side. Despite having experienced the company of first-rank scientists, collecting thousands of specimens, and hearing Dr. Stebbins talk about graduate studies, I still couldn't quite imagine science as a full-time career. Had anyone in my family worked in research or academia, I might have gained a different perspective. But as I've said, while my family was close and supportive, it had no scientist role models for me to emulate.

Ordinary Americans were feeling optimistic about the future, with its promise of technology and growth. Higher education certainly seemed a wise investment, no matter the limits I was placing on what it might make possible. The shortages and constraints of war were becoming distant memories, the economy was strong, and Americans seemed intent on enjoying themselves. A sign of the times, *TV Guide* had gone on sale in the spring of that year, affirming that television was here to stay; my parents purchased their first set soon afterward.

I started USF as a lanky seventeen-year-old kid in khakis and sweaters, with the usual adolescent dreams and a lot on my mind. Since our family was still getting back on its feet after Uncle Charlie's suicide, I took on part-time jobs to help out a bit with finances. Among these, I especially enjoyed my job as a runner for a small securities firm in downtown San Francisco called General American Securities. In that era, the actual stock certificates, checks, and invoices needed to be picked up and delivered by close of business each day, and the only way to get them around in the dense traffic of the closely packed financial district was to carry them on foot. I typed many of the checks and invoices myself, and even learned a tiny bit about the vagaries of the market, watching prices rise and fall from day to day. I also worked briefly as an usher at the Alexandria Theatre on Geary Street in the Richmond District. That was a grubby job that I really didn't care for, but it did provide some cash.

Continuing to live at home with my parents, with whom I had moved to the Sea Cliff district at the north end of 26th Avenue in 1950, I rode the bus to school each day. I filled my days with many different activities: courting Sally, volunteering at the botany department at the Academy, working the part-time jobs, and, of course, doing schoolwork. At the same time, I continued to explore the Presidio, rich in native plants, for the flora of San Francisco I was creating with Mr. Howell.

Academically, I had a relatively difficult time with physics and statistics, but I was happy to start taking my first biology classes. It was an exciting time to be studying the life sciences. James Watson and Francis Crick had recently suggested a possible role for DNA as the code of life, setting in motion what would become a revolution in the field of biology. One of my instructors, Dr. William Hovanitz, was preparing some of the first electron micrographs of biological materials. We enjoyed hearing about his efforts to interpret the lines, hollows, and squiggles that showed up on the images. With these sorts of discoveries and innovations in scientific instrumentation, it was evident that new vistas were opening up.

A week or so after completing my first year of college, in the early summer of 1954, I was standing by the door to the Academy's botany department when a middle-aged couple walked up. Since I didn't recognize them, I helpfully informed them that the door led to a section of the museum that was not open to the public. Smiling, they set me straight about who they were—botanists from UCLA named Harlan and Margaret Lewis.

Harlan and Margaret, as they urged me to call them, asked if I was the young botanist named Peter Raven. I answered affirmatively, and they explained that I might be able to help them sort out a question they had about a specimen of the genus *Clarkia* in the Onagraceae that was in the Academy herbarium, one that I had collected the previous year in the Presidio. They were engaged in a major study of that group of plants and had recently published a monograph that included all its known species.[1] In the course of completing this work, they had come across a pressed *Clarkia* specimen that Katharine Brandegee, based at the Academy, had collected four decades earlier. They hadn't been able to place her specimen among the thirty-three species of the genus that they had recognized as distinct. The problem was that living plants were crucial to their approach to classification and, since there were no recent sightings of plants like the one that Brandegee had collected, they didn't have any with which to work. The specimen of mine they had reviewed the day before, however, seemed to be identical to Brandegee's, so now they had a lead on how to get their hands on some seeds. Thus they immediately asked me exactly where I had found those plants. If they could acquire seeds, they could grow plants in their lath house in Los Angeles, hybridize them with similar species, and thus work out the plant's relationship with other members of the genus.

This was all very exciting, but where exactly *had* I collected the specimen? Well, it turned out that I didn't really know. Hiking through the 1,500 acres of the Presidio looking for unfamiliar plants was one thing, but driving excitedly along the hilly roads with the Lewises and trying to remember where I had found that specific plant turned out to be quite another. The fact that the plants had finished flowering for the year by the time we started looking made our task even more difficult, and we failed to find what we were seeking. I resolved, nevertheless, to relocate the colony from which I had made my collection when the plants came back into bloom the following year.

That summer, 1954, brought with it the usual adventures in the outdoors to which I had become accustomed. They began with a CAS Student Section trip to the east side of the White Mountains, beyond the Sierra Nevada near the border separating California and Nevada. There, we camped for several days in a high desert valley, Deep Springs, where

salty marshes occupied much of the valley floor. We were led by Fred
Schuierer, head of the Student Section at that time, who invited us to
help him with an investigation of the habits of an endangered amphib-
ian. The handsome species on which we focused, the Deep Springs black
toad (*Bufo exsul*), occurs only in that one location.

My eighteenth birthday arrived while we were camping at Deep
Springs. I had to register for the draft within a couple of days, not an
especially convenient task coming from the wilderness. We drove for an
hour or so over the White Mountains, through the fascinating array of
life zones that occur on that desert range. In Bishop, the principal city
of the Owens Valley, we found the draft board, where I registered for the
selective service. (Happily, and partly because the armistice ending the
Korean conflict had been signed during the preceding summer, I was
never drafted.)

Returning home from Deep Springs, I prepared to head up to Base
Camp once again. By then I had personally observed the plants of the
High Sierra from Donner Summit to south of Mount Whitney, a swath
of mountainous terrain approximately 260 miles long. My experiences
had given me a broad view of the plants of the region, and the collections
I had made had already begun to prove useful to other botanists. Many
of these plants were outstandingly beautiful.

I was starting to get just slightly bored with the repetition of plants
that I could find on Base Camp trips within that portion of the Sierra
Nevada. I had learned that with only a few exceptions, the plants that
grew at high elevations near to or above timberline in the Sierra Nevada
were largely the same all along the crest of the range. Given that similarity, I would typically prepare a single specimen from each area we visited
to document the existence of a given species in that particular region.
Only in species that were difficult to distinguish in the field, like the sedges
(*Carex*), bluegrasses (*Poa*), and rock mustards (*Draba*), did I collect a
number of specimens, to be sure of which species actually occurred in that
particular area.

In that year's summer, however, my interest was reignited as the number and kind of Sierra Club outings multiplied and I played an increasing
role in their organization and management. That year, 1954, there were
three two-week Base Camp sessions at Bear Creek, on the west side of the
Sierra north of the Evolution Basin, as well as a Wilderness Base Camp in
the far southern Sierra in the region of Crabtree Creek near Mount Whit-

Dazzled by the beauty around me, I gaze out on the white peaks and glaciers of the Sierra Nevada.

ney. During the main Base Camp, I found quite a few interesting plants, especially around Mono Springs where our pack station was located. In fact, I found a few plants new to the region in the dusty flats near the station in the last minutes as I waited for my father to pick me up for the drive home.

Leaving the middle two-week session at Bear Creek, I went to join the Wilderness Base Camp near Mount Whitney. On this trip, there was more moving from place to place and more climbing than on the main Base Camp trips. So far to the south, in this more arid part of the range, I enjoyed the distinctive landscapes. I encountered several plant species that were new to me, some of them growing in the relatively dry sandy meadows of the area, a special kind of habitat that didn't occur farther north.

A highlight of the Crabtree Creek outing was climbing Mount Whitney, the highest peak in the continental United States. I made the ascent from one of our camps near the base, recording the several species of plants that I found growing above the 14,000-foot level along the way. I contributed a note about them to Mr. Howell's journal, *Leaflets of Western Botany*, whose pages I had continued to populate with snippets about the plants I was observing on my various outings.

Returning to USF for my sophomore year, I elected a major in biology. My father was still a little concerned about my pursuing a career in science. He couldn't imagine what exactly that might look like and it didn't seem like a sure thing that I would be able to earn a living wage. He generously proposed that on graduation I might consider joining him in the furniture-refinishing business. I was sure by then that, in some way, I could find a career teaching science and continue with the study of insects or plants. Nonetheless, I really appreciated his offer.

That year in school, I began to realize that USF didn't offer classes in what I was most interested in learning: detailed information about plants and insects. The faculty wasn't large or diverse enough to offer majors in either entomology or botany, both of which I considered possible choices for my further academic training and eventual profession. Attending the

University of California, Berkeley, became my dream. My mother, her father, and my paternal grandfather C. H. had all proudly attended Cal Berkeley; Ledyard Stebbins had taught at Berkeley before departing for UC Davis, and he had introduced me to some of the university's other botanists. With its connection to family tradition as well as its reputation as a famous scientific center, UC Berkeley seemed to me a school in which my more specialized interests could thrive. However, any thought of a transfer seemed doomed by the lack of finances. Though Berkeley did not charge in-state students tuition, I would have to live away from home, and our family's thin budget was too tight for that.

My desire to attend Berkeley and pursue science in a more serious way received a boost in the spring of 1955 when, during determined searching at the Presidio, I finally found the population of *Clarkia* from which I had collected my specimen two years previously. I waited until they set seeds and gathered a packet to send to the Lewises, brimming with satisfaction and an awakening sense that it *was* possible to make a career in botany.

Then, following my sophomore year at USF, I got the break I needed when a family friend gifted me the money I would need to attend Cal. The offer was precipitated by a tragedy: the furniture business my father ran with Moore Devin failed. The firm had dealt primarily with insurance claims on damaged furniture, and too many people simply pocketed the checks and left their furniture with the shop unclaimed. For me, the silver lining was that Moore Devin's sister-in-law, Claire Fitzgerald, who remained single all her life, stepped in at that difficult time. A strong believer in higher education, she offered me $2,000 per year to help pay for my two remaining years of college. It was more than enough to make attending Berkeley possible. I was elated! The kind of professional career that Dr. Stebbins had described to me had suddenly become a realistic possibility. Thanks to him, I had learned that most graduate students in the sciences were supported by fellowships; now knowing that I could afford to attend Berkeley for my final two years as an undergraduate, the way forward seemed open for the first time. It was a truly exciting prospect!

Once I had qualified to transfer to Berkeley in the autumn of 1955, I began thinking about the choice of an academic major. Berkeley had more specialized departments in the sciences than USF, making both entomology and botany possible majors. I was torn. Although I was spending virtually all my free time with plants, my interest in the study of insects hadn't entirely left me. Before making my decision, I consulted Professor J. W. MacSwain, an advisor in the Entomology Department at Berkeley. Genial and helpful, "Mac" helped me to understand the set of courses that I would take if I majored in entomology. He whetted my appetite by telling me about a summer entomology camp the students attended at the northern end of the Sierra Nevada and showing me fascinating boxes of insects that the students had collected the preceding year. It all gave me much to think about over the summer before my junior year. Meanwhile, my father started working for the city of San Francisco, beginning a long period of steady and satisfying work that would carry him into retirement.

That summer, I headed back to the mountains for my sixth summer of Base Camp outings. I skipped the main Base Camp, which was held near the Minarets, as it had been five years earlier. Instead, I attended the two-week Wilderness Camp on the upper Kern River, near where we had camped the summer before and with many of the same plants. The trip began with a two-day hike that started on the east side of the Sierra, above Owens Valley. During the trip, I led the ascent of 14,025-foot Mount Tyndall and then another of 13,990-foot Mount Barnard. One of the participants on the latter hike wrote of the ascent as follows: "Peter and Robin amazed us by talking continually all the way up the mountain, just as if they were strolling down the avenue. Most of us could hardly gasp out a single word, or at most a short sentence, while we were climbing." My high school participation in track plus endless nature hikes around hilly San Francisco had given me good wind!

One day on this same trip, I hiked forty miles down the Kern River Canyon seeking plant novelties like the Tulare County rockcress (*Boechera pygmaea*), a compact perennial plant with flat fruits on short erect stems, and mountain pincushion (*Orochaenactis thysanocarpha*), a yellow-flowered annual daisy that thrived in the sandy soils of the region. It was the longest hike I had ever done. Feeling driven, it began to seem to me as if my purpose in life was to push on as far and as fast as I could, but the ultimate goal of this activity was anything but clear to me.

Later that summer, I took part in my first out-of-state Base Camp outing, at Yakima Park on the eastern slopes of Mount Rainier in the state of Washington. The excursion included a particularly marvelous experience that I shall never forget. One afternoon I hiked alone around the mountain, following a trail that led to a fire lookout on its northwest side.

As I ate the food I had brought for dinner, I drank in the majestic view of Puget Sound and the Olympic Peninsula that lay before me. In the distance, lights twinkled on as darkness gathered, revealing the serene beauty of the clear night sky. After spreading out my mummy bag on the porch of the lookout, I slept soundly and was treated to another version of that same glorious view in the spreading sunlight of the next morning's dawn. The almost ecstatic awe, the sense of moving beyond the scientific realm into the poetic and even the spiritual, was something I experienced repeatedly during my summer trips. In a time before "awesome" was used to describe a delicious dessert or a new pair of shoes, each day in the mountains seemed to bring me face to face with *truly* awesome sights, what John Muir called "beauty beyond thought." It was important for me to feel that profound emotional connection to the natural world as I went deeper into the science of plants and the environment.

Although I greatly enjoyed the scenery of the Cascades and the people with whom I was camping, that trip had its "down" moments. Some of them related to Sally; away in the mountains, I began to miss her more and more. I also missed my beloved California plants and the focus of the projects through which I arranged my activities. The flora of Napa County, the flora of San Francisco, knowing the plants of California more generally: these goals seemed logical to me. At first, I couldn't find the same purpose in this introduction to the plants of the Cascades—and without that, my thinking went, why deal with them? I gradually got over my funk, however, and began to appreciate the plants of this very different mountain environment, much moister than the ones I knew.

ↄ

Returning to the Bay Area and preparing to enter Berkeley, I found a boarding house on Durant Avenue where students lived near the campus; it provided a room and two meals every day except Sunday for a very reasonable sum. The books I needed would rarely cost more than $30 total a semester, and there were few other expenses—no tuition was charged at that time—so Mrs. Fitzgerald's funds covered things nicely without the necessity of taking part-time jobs. Moving in, I embraced my new life with gratitude. I had realized during the summer that plants were my true love. Also, since there were more entomologists than botanists, I thought that my contributions might be more useful in botany. So, despite Professor MacSwain's alluring description of the entomology program, I enrolled at Berkeley as a junior botany major.

My early concentration on plants and field science probably made me an unusual student, but I hugely enjoyed the ordinary pleasures of campus life, which were all the more fun now that I was living away from home for the first time. Those were the years of panty raids, water bag fights, and other pranks, and the pleasures of football games on lazy fall afternoons, followed by a return to the house to drink beer on the balcony and watch the crowds moving by. Not coincidentally, I stopped attending Mass on Sundays; I still felt embedded in Catholic culture and appreciated the way it had shaped my values and outlook, but the ritual of Mass had ceased to be important enough to pull me away from all the excitement surrounding me.

Diving into my academic studies with enthusiasm, I quickly forged friendships with many botany professors, graduate students, and researchers at Berkeley: Janet Stein, a student of tiny algae; Marion Cave, a student of plant cells and their chromosomes; Don Stone, a teaching assistant for my course in advanced plant taxonomy; Annetta Carter, Bob Ornduff, and Helen Sharsmith—to name just a few. At the University's Jepson Herbarium, which had been left behind by W. L. Jepson and maintained separate from the main herbarium, I continued my friendships with Rimo Bacigalupi and Tom Robbins, whom I had gotten to know earlier thanks to Ledyard Stebbins.

The very highly regarded Berkeley Botany Department had many outstanding professors. First on the list, for me at least, was Lincoln Constance, who supported and encouraged me the whole time I was at Berkeley. A plant taxonomist specializing in the parsley family, he had earned his Ph.D. degree under Jepson a couple of decades earlier. Always well dressed and proper, he had tremendous life force underneath his propriety and formality. Though not a scientific groundbreaker in the way that Ledyard Stebbins was, he was nonetheless a gifted scientist. He also excelled at academic politics, leading to his appointment as Dean of the College of Arts and Sciences at the same time I entered Berkeley. Much to my advantage and that of others, he kept up with many of his botanical and mentoring activities despite his new administrative role. Lincoln clearly accepted me as a professional botanist in the making, letting me work part-time on checking a manuscript he had jointly completed on Japanese parsleys and generally helping me to feel that I had a future in the

field. As I found my place in the bigger scientific "pond" that was Berkeley, his belief in me made me proud and offered welcome reassurance.

Another Berkeley professor from whom I learned a great deal was Herbert L. Mason. Unlike Dr. Constance, Dr. Mason was stocky, rumpled,
and rather eccentric. Well known for his unpredictable and lengthy trains
of thought, he was a typical "absent-minded professor" in many ways. But
he was also a person who thought with exceptional depth about plants,
ecology, and the environment. He cared very much about his students
and how best to train them. Dr. Mason was a born philosopher who
helped us all to learn to think—the essence of what a university is about,
but something too often lost in the scramble to memorize necessary facts
and definitions. He was an expert in taxonomy, ecology, and paleobot-
any, the study of fossil plants and their histories. His classes, including
Advanced Taxonomy and Phylogenetic Systematics, were crucial learning
experiences for me. Mason would have us read important botanical papers, repeat earlier experiments, examine various kinds of evidence, and
reconcile differing results. He didn't just care about discoveries or facts.
He wanted us to understand the process through which people had arrived at their published conclusions, not simply to memorize the results
they had reported. He trained us to be able to replicate what others had
done, and to thoughtfully appraise the results. Prompting us to ask "why"
as well as "what" and "how," Herbert Mason constantly reminded us to
look at the big picture, the larger meaning of things.

A highlight of Dr. Mason's plant taxonomy course was its series of
field trips to the Mendocino Coast, the Sierra foothills, the serpentine
areas surrounding the New Idria Mine in San Benito County, and the
Central Valley. Exciting plants were found and studied on every trip, and
lasting friendships cemented as well. Gathering in the evenings before
turning in, we would open a jug of cheap red wine and discuss the fine
points of evolution, plant systematics, and life itself far into the night. It
was energizing and inspiring to work every day with real botanical colleagues, and I reached out to them all. No matter what someone's level
of study or experience might be, I wanted to make friends, and to learn
about what my new friends were doing. For me, friendship fueled collaboration, making it possible to work toward common goals with ease and
mutual trust—and appreciation for a good joke or two. I felt completely
at home in this community of fellow plant lovers.

Under the guidance of Constance, Mason, and other professors, I broadened my appreciation of plants as living organisms. I stopped looking at them as simply objects—exotic, delicate, beautiful, fascinating, but nonetheless objects—and began to pursue an understanding of the bigger picture: the conduction of water, the production of fragrances, the adaptations to habitat, the means by which species had achieved their current geographical distributions. I started to ask questions, and those questions guided my continuing exploration of the plant world. What does a particular number or configuration of chromosomes *mean*? Why do so many leaves have three layers of tissue rather than some other number? How do plants change over time? How does where a plant grows affect its appearance? How do new species evolve?

After my junior year at Berkeley ended, I spent six weeks of that summer at UCLA at the invitation of Harlan Lewis. I lived in an almost-deserted dorm during the weeks I was on campus, driving north on a couple of the weekends to visit Sally. Once settled, I began working with the plants of the Presidio *Clarkia* that were being grown in Harlan's lath house from the seeds I had sent the year before, comparing them with the related species grown alongside them. I made hybrids between these species so that we could study their relationships, use the information gained to work out whether the Presidio plants were distinct genetically, and discern which species were their closest relatives. Working beside Harlan Lewis in the lath house, I absorbed a great deal about botany, as if by osmosis.

Just as educational were the field trips I took with Harlan and some of his students, trips that helped me get to know the plants of Southern California. Dick Snow, an extremely nice man who had just completed his dissertation under Harlan's guidance, was among those who took me around the ranges and other natural areas near Los Angeles and beyond. On one of these trips, Dick and I were driving along a road in the chaparral in San Diego County late one afternoon when we came on a population of a member of the four-o'clock family, *Mirabilis multiflora* var. *pubescens*. Its spectacular flowers, just opening then, were larger than silver dollars and bright rose-purple, with the anthers and stigma protruding far out of the flower. In the gathering dusk, they were being visited and pollinated by hummingbird-like hawkmoths, *Hyles lineata*. To me, these plants seemed magical. On returning home, I began to read about them and their relatives, wondering if that group might be the one on which I could base the research that I was starting to plan as a basis for my dissertation when I entered graduate school the following year.

After six weeks in Los Angeles, I left to attend what would turn out to be my last two Base Camp outings. The first was the 1956 Wilderness Base Camp, held on the headwaters of the Kings River. We entered the area from the west, crossing the high-elevation Sawmill Pass on the way in to our camp. Among the rocks on the plateau just north of the pass, I chanced upon a species of locoweed (*Astragalus*) that seemed different from any other I had seen. I collected a few of the plants, which were straggling out among the small stones that covered the pass, at about 11,300 feet elevation.

A few weeks later I headed north for another Base Camp in the Cascade Range, this one in the vicinity of Glacier Peak. Following the long boat ride up the beautifully glacier-carved Lake Chelan, we hiked into our pastoral campsite deep in the rainy mountains. The northern plants were becoming more familiar now, and contrary to my lack of interest the prior year, I was starting to enjoy them. Slowly, my botanical sights were broadening.

Back at home, I brought my Sierran *Astragalus* specimens to the Academy for Tom Howell to see. He recognized them as unusual and sent them off to Rupert Barneby, the expert on the genus, at The New York Botanical Garden. Barneby promptly described my find as a new species and honored me by naming it *Astragalus ravenii*, Raven's milk vetch. Having my collection valued in that way by an expert increased my already strong plant hunting fever.

I returned to Berkeley for my senior year feeling optimistic. I continued to enjoy campus life, throw myself into my classes, take trips or walks to collect plants when and where I could, and spend as much time with Sally, now a home economics major at Berkeley, as possible. Music, both classical and popular, continued to be a source of inspiration and enjoyment despite my busy academic schedule. *My Fair Lady* had opened on Broadway earlier in the year. Its ebullient music and story—focused in part on themes like education, language, and adaptation, all dear to my heart—made it an immediate favorite for me and my parents when we heard the original cast recording. For a time, I walked around campus humming or whistling the tunes.

Bolstered by what I had learned during my stay at UCLA that summer and the encouragement of my professors at Berkeley, a research and teaching career in botany seemed clearly to lie within my grasp. That made graduate school the obvious next step. But I was faced with a difficult choice. Lincoln Constance, my ceaseless advocate during my time at Berkeley, was eager for me to carry out my doctoral work with him there. Berkeley was the logical choice, any way I looked at it: I was already there, I valued Dr. Constance, and the Berkeley botany department was an excellent one. Yet I felt powerfully drawn to The University of California at Los Angeles. A onetime "streetcar" college, it had fewer departments, professors, and students and a less impressive academic history than Berkeley. But after my enjoyable and productive summer working there with Dr. Lewis, it seemed natural for me to pursue graduate work with him at UCLA. I admired Harlan Lewis greatly as both person and professor, and I sensed I would grow under his tutelage.

After careful thought, and with the gracious blessing of Dr. Constance, I decided to say goodbye to Berkeley. Happily, money was no longer an issue. Over the course of my senior year, I won two fellowships offering support for my first year of graduate work. Since the one offered by the Turtox Biological Supply House was the more lucrative of the two, I accepted it instead of a National Science Foundation Fellowship.

My parents attended my graduation, joining the huge crowd of happy families and friends of the graduated attending the ceremony in Memorial Stadium. I departed full of gratitude for all I had learned at Berkeley, the extraordinary people I had learned it all from, and the financial support that made it possible, both from Mrs. Fitzgerald and the University of California system.

❧

Many botanists first become interested in the field of botany as a particular branch of science and then come to appreciate plants as the beautiful and interesting organisms that they are. For me it was the opposite. From the beginning, I had found plants to be marvels of loveliness, pattern, texture, and color, down to their minutest details. When I entered UCLA as a doctoral student in the fall of 1957, I still related to plants more as a lover than as a scientist. Only as my graduate-level training progressed did

I fully begin to understand them from a scientific perspective—although mentors and teachers beginning with Ledyard Stebbins had consistently nudged me in that direction.

Having been at UCLA before, it wasn't difficult to settle in. I shared an apartment with Lee Wedberg, a graduate student I had met the summer before, and his older brother. Lee's brother was a stockbroker who rolled out of bed at 3:30 a.m. to get to his office in time for the opening bell of the New York Stock Exchange three time zones away. Since I rose about 11 a.m. and then worked until about 1:00 to 2:00 a.m. on a typical day, Lee's brother and I almost never actually saw one another. Lee and I sometimes went into the field together, on our own or with different groups of the other graduate students, and we always had fun.

My true home was the UCLA botany department. When I first arrived, graduate botany students shared small offices on the top floor of Royce Hall, which at that time was the physics building. A saying by the English philosopher Joseph Priestley, the discoverer of the gas now known as oxygen, was engraved over one of the doors: "Nothing is too wonderful to be true." I was inspired by it every day—including the day in November when we climbed up to the roof of the Physics Building to see Sputnik, the path-breaking Soviet satellite, sailing across a cold starry sky, signaling to us that a new world had arrived.

The UCLA botany department was filled with remarkable—and colorful—professors, visiting fellows, and graduate students. For me, of course, the key figure was Harlan Lewis, my major professor. Though Harlan had a scholarly demeanor, he could dish out some sarcasm as well as jokes. We students had to work hard to keep up with him, but dominating as he was, he enjoyed being challenged. Harlan guided me toward topics that would form the basis of my doctoral work and, as it turned out, decades of study thereafter. In particular, he got me thinking more deeply about the dynamics of plant populations and the ways in which they evolved. By the time I arrived at UCLA, Harlan was already starting to move in the direction of academic administration, so I'm grateful that, during my time there, he was still fully hands-on in the botany department.

Professor Mildred Mathias was another one of my most important teachers and friends at UCLA. Born in rural Missouri, Mildred had earned her doctoral degree from the Henry Shaw School of Botany at Washington University in St. Louis at the age of only twenty-three; the school had been

Mildred Mathias, professor of botany at UCLA, was a generous mentor and friend to me from the time I first met her in 1956 until her death four decades later.

formed in 1885 in collaboration with the Missouri Botanical Garden. She did important studies on the carrot family (Apiaceae) and eventually collaborated with Lincoln Constance on these fascinating plants. Mildred had come to UCLA as the director of the herbarium, later joining the botany faculty and becoming a full professor in 1955.

With an easy manner and a genuine interest in what others had to say, Mildred was naturally likeable, but I found her especially endearing because she seemed a kindred spirit: like me, her mind was always going off in all directions and she was prone to following many tracks simultaneously. I served as a teaching assistant in her course on plant families and learned a great deal from that experience amongst the rich assemblage of cultivated plants that thrive in the gentle climate of Southern California.

It was unusual for women to chair science departments in the 1950s, but the chair of our department was Professor Flora Murray Scott, whose diminutive stature was combined with a scholarly, rigorous mind. Scotty, as she was known to friends, was appreciated for her kindness, her precise use of the English language, and her tartan skirts and hair ribbons. Another vivid personality in the department was Professor Carl Epling, a specialist in biological systematics and evolution. A veteran of World War I, Epling had joined the staff at UCLA in 1924, immediately upon his graduation from the Henry Shaw School of Botany, the same program in which Mildred Mathias enrolled a couple of years later. Epling was the world's leading expert on the mint family, Lamiaceae (Labiatae), producing monographs on many of the groups of mints that occurred in North and South America. Epling could sometimes be seen in a Hawaiian shirt and sandals and occasionally even in a bathrobe, always with a cigarette between his fingers. He was a brilliant person who made a deep impression on me, particularly with his keen appreciation of the process of evolution. Like Ledyard Stebbins, he had met and been influenced by the great evolutionist Theodosius Dobzhansky, the Russian émigré who contributed so much to shaping the modern evolutionary synthesis.

At UCLA, I also worked closely with Professor Henry J. Thompson, always called Harry. Harry had grown up in Los Angeles and earned his Ph.D. degree at Stanford. He had joined the faculty at UCLA after serving as a Navy pilot in the Pacific during World War II. At UCLA he helped launch innovative courses in integrated biology, courses that spanned the fields of botany and zoology together. On field trips to the desert, often with Harlan, Harry became a student of the genus *Mentzelia*, the blazing stars, finding that they presented a pattern of intricate relationships that would occupy him for the rest of his career.

Among my first tasks at UCLA was to choose a topic for my dissertation. Like many beginning graduate students, I had been thinking about possible topics even before I arrived to begin my studies. Top of mind for me was a study of the genus *Mirabilis*, the group to which belonged the beautiful plant that Dick Snow and I had encountered the summer before. That idea came to an abrupt end, however, when Dick mentioned the possibility to Harlan Lewis. Because of his extensive and ongoing studies of *Clarkia*, Harlan had been awarded a National Science grant to conduct research on the plant family Onagraceae. He knew that he would be able to support my work more easily—for example, reimbursing travel and lab costs—if I were to pursue one of the members of that family as my dissertation topic. The practical advantages were compelling, so I kissed *Mirabilis* goodbye.

The group I took up at Harlan's suggestion was *Chylismia*, included at the time within the evening primrose genus, *Oenothera*. Harlan's interest in *Chylismia* had been sparked when he observed that the widespread species *Chylismia claviformis* had both white- and yellow-flowered geographical races. The white-flowered ones, called "brown eyes" by locals, formed occasional natural hybrids with a related yellow-flowered species, *Chylismia brevipes* ("suncups"), and the hybrids also had yellow flowers. Observing this relationship, Harlan hypothesized that the yellow-flowered populations of *Chylismia claviformis* might have originated following hybridization between the two distinct species, followed by natural selection and eventual stabilization of the intermediate populations. The flowers of both subspecies of brown eyes that Harlan observed opened in late afternoon, while those of the suncups open near sunrise.

I began studying *Chylismia* during my first semester at UCLA. That autumn, I took a field trip to eastern California and Nevada to observe populations of a species, *Chylismia heterochroma*, which bloomed at that time of year and had unique lavender-purple flowers. I had good luck finding those plants and began to prepare for field studies of their spring-blooming relatives. Naturally, the amount and distribution of the rain that might fall on the deserts that winter was a major concern: if there was little or none, there would be no plants to study! As it turned out, the spring of 1958 was a very wet one on the desert, with fields of flowers coming up everywhere. Since the plants I wanted to study were doing so well, I raced as rapidly as I could all over their range, driving as far as Arizona, Oregon, and Sonora in northwestern Mexico. In another, drier year I would not have been able to accomplish anything. Between January 23 and April 1, 1958, I made 712 collections, most of them *Chylismia*, gaining a good deal of insight into the group in the process. I sampled hundreds of populations of *Chylismia* and carefully observed their variation and their relationship with one another; later I determined their chromosome numbers. In order to observe the fertility, chromosome pairing, and other aspects of chromosome behavior in hybrids between the species, it was necessary to find hybrids that had formed spontaneously in nature, since the plants were virtually impossible to cultivate successfully, most seedlings falling early victims to a damping-off fungus. So I studied the natural hybrids I found here and there in the deserts with special eagerness.

Aside from morphological comparisons, the tools available for the study of plant evolution in the late 1950s included chromosome numbers and chromosome pairing and fertility in hybrids, and determining whether distinct species or races remained largely distinct (that is, didn't hybridize much or at all) where their distribution areas came together. Since then, we have gained a far deeper understanding of relationships by learning to compare the sequences of bases in the hereditary molecules DNA and RNA, but the simpler tools we had available sixty years ago were certainly helpful before we had the ones that we have developed since.

℘

The regional collecting trips that I had begun during my first stay on campus in 1956 were resumed with enthusiasm once I had enrolled as a full-time graduate student. First and foremost, of course, were the trips I took to investigate *Chylismia*; in the first half of 1958 they took up almost all the time I had free from classes and other obligations. As the spring flowering season for most species of *Chylismia* came to an end, I began to collect different kinds of plants and, as I was doing for *Chylismia*, often fixed their buds in special preservatives for chromosome studies as well. Such studies were useful, since chromosome number and morphology taken together proved excellent ways to gain insight into plant evolution. In contrast to earlier years, I made general collections of plants only under special circumstances, which continued to open up at times or on trips made especially for that purpose.

In the spring of 1958, the ample rains of the preceding winter led to the abundant flowering of native plants throughout the West. My colleagues and I saw exceptional opportunities for investigation in particular areas, one of them the northern part of the Baja California peninsula in Mexico, where many interesting plants—quite a few of which are found nowhere else—grow less than a day's drive south of Los Angeles. In April 1958, Harlan Lewis, Harry Thompson, and I set out for this botanical wonderland, filled with amazing species—all of them new to me. We drove through extensive flowery fields and collected as many kinds of plants as we could find. We made 193 collections over the course of three days before our southward progress was halted by the high waters of the Santo Domingo River. Unable to cross the flooded river in our university station wagon, we were forced to return home earlier than we had planned.

I very much wanted to see what botanical treasures might be flowering to the south, on the other side of that flooded arroyo. Harlan and Harry were otherwise occupied and were thus unable to make a second trip, so I persuaded Mildred Mathias and Jane Turner, Mildred's herbarium assistant, to accompany me on a second trip down the peninsula a few weeks later. Both proved great companions for the further adventure. Remarkably, in view of her extensive later travels, this was Mildred's first foray beyond the boundaries of the United States. She was fifty-one years old and I was twenty-one, but the differences in our ages and experience disappeared in the midst of our mutual passion for plants.

From the campus motor pool, we signed out a stake-bed truck with high ground clearance to make certain we would be capable of making the river crossing. By the time we got there, however, the Santo Domingo River was nearly dry, so the size of our vehicle turned out to be overkill. The fields of flowers continued for as far as we drove on to the south and southeast, with the region beyond El Rosario populated with even more interesting plants than those we had encountered earlier. The collecting was so good and the road so bad that one day we drove a grand total of six miles!

Among the many choice plants, we found a kind of sage that had been named (by Carl Epling) *Salvia chionopeplica*. It is found only in a small area of Baja California, though it is a member of the same group that fills the coastal slopes of Southern California with fragrance. As a result of our collections, we were able to determine its chromosome number and study its characteristics in the field. Another special find was *Sanicula deserticola*, the southernmost representative of a genus that had fascinated me since my earliest days in botany, and a member of the family (Apiaceae, the parsley family) in which Mildred specialized. Botanically, the trip was marvelous, a joyous and invigorating time during which we made 350 collections to add to the earlier bounty from the peninsula.

During the summer of 1958, I continued to collect plants in Southern California. Although I'm not one to particularly enjoy going off by myself, during those years my passion for plants consistently overcame my lack of enthusiasm for being alone. Earlier in the year, when I was concentrating on *Chylismia*, I made several two-week trips alone to collect samples. There was really no alternative as it was rare for anyone else to want to follow the same routes that I had chosen at the time I needed to go. On these treks, I would usually become intensely lonely by the second or third day. But then, on around the fourth day, I would adjust, reach equilibrium, and grow content with desert, self, and work.

❧

Though doctoral work was my main priority, I enjoyed collaboration and was something of a glutton for punishment in devising and launching additional projects. By the time I entered UCLA as a graduate student, I had already collaborated with Tom Howell and Peter Rubtzoff, a Russian emigre who had arrived in California with a deep interest in botany, on a flora of San Francisco. It was published by the University of San Francisco in 1958 with the three of us listed as coauthors—my first book publication. At UCLA, new collaborative opportunities seemed to be everywhere. One of the most significant for me involved Harry Thompson. As a native of West Los Angeles, Harry understood the plants of the region well. As he and I became closer, we decided to collaborate on a flora of the

Santa Monica Mountains. This impressive range, which forms a natural boundary between urban Los Angeles and the San Fernando Valley, extends from the Hollywood Hills in Los Angeles west to Point Mugu in Ventura County, with elevations varying from sea level in the beaches and low areas along the Pacific coast to Sandstone Peak at about 3,000 feet elevation. One of its most significant features is the Malibu Lagoon, located at the mouth of Malibu Creek, a critical stop for birds migrating along the Pacific flyway. Fortunately for the environment and for generations of Southern California residents to come, most of the range is unsuitable for development, allowing large areas to be set aside as wilderness.

As I began working with Harry on the flora of the range, I was aware of the lesson Tom Howell had taught me as a teenager: the best way to get to know the flora of California generally was to concentrate study on the plants of particular local areas. Harry and I made the first collections for our flora in the autumn of 1958, after I had done all the collecting of *Chylismia* that was possible that year. Harry would drive through canyon after canyon while I wrote down the names of the plants that we passed— places to stop in the canyons were few and far between—so that we could work out the overall distribution of each plant species in the range. Using all of the herbarium specimens we could find at UCLA and elsewhere, together with our own observations and collections, Harry and I began putting together our book. It was too big a job to complete quickly; our work and explorations would extend into the next decade.

At UCLA, I loved being among graduate students just as passionate about plants as I was. I was learning that, in some important sense, graduate students educate one another, offering dialogue, help, and challenge. Scientists are often regarded as reserved or solitary types, and some indeed are just that; but many of us thrive on collegial relationships, friendships, and the ability to communicate the results of scientific studies to others. I was firmly planted in this more extroverted category. The interchanges with other graduate students allowed me to explore subjects outside my specializations and enjoy the pleasure of helping and encouraging others. Many of our conversations and musings morphed into actual collaborative projects. In fact, I participated in so many of these collaborations that Harlan Lewis occasionally kidded me about being the only graduate student who had graduate students. Among the many graduate students with whom I worked were Stan Davis, Lee Wedberg, Ted Mosquin, David Bates, Kunjamma Mathew, and Donald Kyhos. Our collaborations felt immensely rewarding to me and cemented what would become lifelong friendships.

Two collaborative projects stood out as particularly notable. Kunjamma and I collaborated on a paper separating some closely related species of *Cryptantha*, small white-flowered annuals that belong to the borage family. Observing plants in the desert, we noticed sharp differences within what was at that time considered a single species. Those differences were confirmed when we determined their chromosome numbers, and on this basis we divided the complex we were studying into three distinct species. On returning home to Kerala, India, after finishing her Ph.D., Kunjamma was pressured into an arranged marriage, which meant the end of her botanical career. It saddened me to see this very bright and capable botanist unable to fulfill her scientific potential.

With Don Kyhos, I worked on a substantial project to count the chromosome numbers of as many species of the sunflower family, Asteraceae, as we had the time and resources to collect and analyze. I did most of the collecting while Don, an excellent cytologist and rabid perfectionist, did most of the chromosome counting. Full of enthusiasm, I filled hundreds of the specimen bottles in which we preserved buds in a mixture of ethyl alcohol and acetic acid. This stopped the replicating chromosomes in their tracks and made it possible to observe and count them under a microscope once they had been properly stained. (This was the same process that I used on *Chylismia* for my dissertation, and for many other plants as well.) Over time, we determined the chromosome numbers of hundreds of species of Asteraceae for which the chromosome numbers had previously been unknown. We would publish the first of fourteen papers on the subject in 1960.

❧

During my second year of graduate school in 1958, I decided to tie up one loose end in my work: the identification of the relatively few specimens of insects that I had caught visiting the flowers of *Chylismia* and a few other Onagraceae during the course of my trips through the West. I had observed so few insects at the flowers that I could draw no particular conclusions about how the pollination systems of *Chylismia* operated and simply assumed that there was nothing special about them. How wrong I soon turned out to be!

Having chosen botany rather than entomology as my major at Berkeley, I had temporarily lost touch with J. W. MacSwain. But I remembered his expertise and cordiality well, so I decided that contacting him would be a good place to start on my insect identification quest. Welcoming me warmly, he watched me separate the layers of cotton on which I placed the insects I had collected. When I got to one layer from southern Idaho, a large black bee grabbed his attention. He picked it up, caught his breath, and asked, "What's this doing in here?"

It turned out to be a member of *Onagrandrena*, a subgenus of the very large bee genus *Andrena* and one of the very group that he and fellow Berkeley professor E. G. Linsley were studying at the time. My bee was lustrous black, slightly longer and more slender than a honeybee; I had caught it in Idaho one early morning the previous June on a white-flowered evening primrose (*Oenothera*), on flowers that had opened the night before. Taking me over to one of the cabinets in his office, Mac pulled out a drawer of pinned insects to show me row after row of bees. They resembled my specimen, but all were from the Colorado Desert. Since these bees are large and lustrous black, they are as obvious as they possibly could be as they cruise among the flowers. I felt certain that I had researched that region carefully while studying populations of *Chylismia* earlier that year, but I hadn't seen a single one of these beautiful bees anywhere. What had been wrong with my observations of pollinators?

At its simplest, the answer was that one of the bee species represented in Mac's collection finishes its activities before 8:00 in the morning, while another doesn't appear until late afternoon, around 4:30 or so. Abundant as the bees are under the right circumstances, my collecting habits hadn't been timed to see them. Once I figured this out, the possibility of finding them had me hustling all over the West looking for more specimens. In each new locality, I had to wait patiently, never

A female *Andrena* (*Onagrandrena*) *rozeni,* collecting pollen from the flowers of the afternoon-blooming *Chylismia claviformis* in California's Colorado Desert. When I realized that these handsome bees were frequent visitors to the plants I was studying, I began a systematic hunt for more of them throughout the western deserts.

knowing whether the bees would appear or not. It was an exciting, unpredictable quest, one that I enjoyed as much as anything I had done in the field. As I brought my samples back to Berkeley, we were able to extend the range of the group to several new regions, including Nevada. During my travels in the Silver State, I found a bluish-black new species of *Onagrandrena* that my colleagues named *Andrena raveni* in my honor.

Ɛ⌀Ɔ

During my first full year at UCLA, Sally and I had kept in touch through telephone calls, letters, and as many visits as my busy schedule and the 380-mile distance separating us would allow. Neither of us wanted this untenable situation to continue, so we decided to get married as soon as she finished her degree work at Berkeley. We tied the knot on August 16, 1958, at Holy Name of Jesus Church, in the Sunset District of San Francisco. She was eight days short of her twenty-first birthday, I was twenty-two, and we had already been good friends for ten years. John O'Rourke, a mutual dear friend from Academy Student Section days, served as my best man.

After a week-long honeymoon, we returned to Los Angeles to set up house in an apartment on Wilshire Boulevard near the UCLA campus. The following month, Sally traveled with me on a collecting trip to southern Nevada. We enjoyed Las Vegas and had a great time, though as always I was focused on botany. In the Spring Mountains near Las Vegas, a unique fall-blooming species of Onagraceae, the Nevada willow herb (*Epilobium nevadense*), grew on rockslides among the pines at middle elevations. We were successful in our quest to find the colony and collect material for what proved to be a very interesting subsequent analysis.

With my parents Walter and Isabelle, after my marriage to Sally Barrett in 1958.

Soft, straightforward, and sweet, Sally was in many ways a refuge from the intensity of the work I was doing. She was the opposite of my ambitious mother with her occasional hard edges. With Sally so supportive of my career, it seemed that marriage would only make life better, and I assumed without thinking that it would offer little distraction from my graduate and professional work. That assumption was supported by the popular culture of the time. From films to television, magazines to popular fiction, 1950s American culture celebrated the man as the provider, the person who gave the household money and status and for whom women and children waited at home. Unthinkingly—and, contrary to the example my mother had set—I leaned on the stereotype, allowing my home life to take a distant second place to my career.

Sally became pregnant early in 1959. We were excited and pleased at the prospect of becoming parents, but largely ignorant of all it would involve. It surprised me to find out that Harlan Lewis was not particularly enamored with the idea of our having a baby. He and my other professors clearly wanted the best for their protégés, but they also wanted us to be dedicated to botany rather than distracted by family responsibilities. Harlan was only half joking when he said one day, "You know, we gave you permission to get married while you're still in graduate school, but not to have children!"

In the summer of 1959, Mildred Mathias secured for me an invitation to join a six-week trip to Colombia. Since the trip would afford my first exposure to South America and to the tropics, I jumped at the chance. Driven as always by my urgent desire to see more plants, learn more, and move forward in botany, I took for granted that Sally wouldn't mind being by herself in the latter stages of pregnancy. I was completely unrealistic about the situation, which seemed so normal and reasonable to me then that I scarcely gave the effect of my absence on Sally a thought.

The trip, part of an exchange program between UCLA and the country of Colombia, was organized with Professor Henry Bruman of the UCLA Department of Geography and a group of the department's undergraduate students. The flight east was my first experience on a jet and my first glimpse of our country east of the Rocky Mountains. After several days of briefing by the State Department in Washington, D.C., we departed for Colombia on their national airline, Avianca. When our plane stopped for repairs in Miami, we were able to spend a couple of days in that city, enjoying its impressively international character.

The eight-hour flight to Colombia—made in a propeller plane, dodging storms over the Caribbean—ended in Bogotá, the nation's capital. As I stepped off the plane, I expected to find myself surrounded by steaming tropical jungles or something of the sort. Instead, I saw forests of *Euca-*

lyptus trees (which had played in important role in drying out the marshes where modern Bogotá is situated) and grassy fields flanked by rows of the same Monterey pines I had known growing up. I hadn't registered the fact that Bogotá is located high in the Andes, on a dry lake bed at an elevation of 8,660 feet, and so has an essentially temperate climate. But the tropical forests were nearby, below and off the plateau, and we were to see a lot of them before leaving the country. There were plenty of unusual plants to see at higher elevations too. The mountains above Bogotá, such as the Nevado del Ruiz, are home to cloud forest communities. Above them grow strange plant formations called páramos, dominated by large clumped grasses and giant single-trunked members of the sunflower family in the genus *Espeletia*, which stand out in the low páramo vegetation like gray-clad monks (*frailejones* in Spanish, their vernacular name).

In Bogotá, we attended classes for two weeks at the Universidad Nacional. In addition to the chance to improve our Spanish, those sessions familiarized us with the history, geography, people, music, literature, and art of Colombia. While they were necessarily overviews, the classes gave us a good sense of the diversity of the country, and we enjoyed them thoroughly. Our introduction completed, we set out to travel around the country, visiting universities in Medellín, Cali, Popayán, Bucaramanga, Norte de Santander, Santa Marta, and other centers. Traveling by bus, we got a good look at the varied countryside along the way. We were invited to parties with Colombians of our age, and at one point I struggled through a radio interview, trying to explain in my limited Spanish why I thought the plants of Colombia were so wonderful.

From Colombia, I flew straight to Montreal, Canada, to participate in the Ninth International Botanical Congress (IBC), my first experience attending an international meeting. I was scheduled to present a paper on the evolution of Onagraceae with Harlan Lewis. That huge honor made me proud, but I would have been thrilled to attend even if a presentation opportunity hadn't been included. Convened every few years, the Congress is an occasion for botanists from around the world to come together, report their research results, share ideas, and make decisions regarding such things as revisions to the rules for naming plants. It was exhilarating to meet with and listen to distinguished botanists from all over the world. I enjoyed the lighter side of the gathering—chat, gossip, teasing and jokes—almost as much as the science. Botanists are as competitive as any other professionals; as international experts re-encountered each other, there was a certain amount of sparring, most of it good-natured. A day or two after our presentation, I was walking along with Harlan when we ran into an eminent botanist, David Keck. "Harlan, I see you're tagging along with Peter trying to meet some important people," Dave quipped. The witticism notwithstanding, it was I who had the honor of meeting "important people," and I left the Congress feeling grateful and energized.

All in all, I was gone for nearly two months. Naturally, I was thrilled to see Sally after so long a time away. Our first child was born soon after I returned, on October 10, 1959. The days of dads being present in the delivery room were still in the future; while Sally endured pain I could not imagine, I sat for hours in a small, smoke-filled waiting room at the UCLA Hospital with other nervous expectant fathers. I was on tenterhooks for the moment when the nurse would step into the room, call my name, inform me of the baby's gender, and, I hoped, finish with the comforting words, "Mother and baby are doing just fine." After about six hours, I was given the good news that our firstborn was a daughter, and yes, mother and baby *were* doing fine. We named our child Alice Catherine Raven, for Alice Eastwood and Katharine Brandegee (we preferred an alternate spelling of Brandegee's first name). Alice was a beautiful baby with a playful attitude, and we were thrilled at the prospect of watching her grow up.

During my studies of the Onagraceae, I had come to know Dr. Philip A. Munz, director of The Rancho Santa Ana Botanical Garden in Claremont, California. He was the world expert on Onagraceae, and so it was natural that I had gotten to know both him and his graduate students Bill Klein and David Gregory, who were helping him finish up his life's work on that plant family. In the middle of my last year of graduate school at UCLA, Dr. Munz offered me a job as director of the herbarium at Rancho Santa Ana. Anxious to achieve a degree of stability in my career, I accepted, with the understanding that I would spend a postdoctoral year somewhere else before reporting for duty. Rancho Santa Ana seemed a perfect place for me to pursue my further studies of Onagraceae and the plants of California, the latter being the stated research objective of the institution.

In the spring of 1960, I completed my doctoral dissertation on the systematics and evolution of *Chylismia*. (It would be published as one of the

University of California Publications in Botany in 1962 under the title "The Systematics of *Oenothera* subgenus *Chylismia*.") In the dissertation, I recognized fourteen species, all confined to the deserts of the western United States and adjacent northwestern Mexico.

With my dissertation completed, only the doctoral defense remained. It was conducted by a committee of five members. Harlan and Mildred were of course among that number. Also on the panel was paleobotanist Daniel I. Axelrod, a gruff and meticulous professor from whom I had taken a particularly fine course in introductory geology during my first year in graduate school. Since I had had a relatively easy time two years earlier passing my qualifying examination, I wasn't particularly afraid of what might come up in the course of my doctoral defense. My delusions were shattered when Axelrod asked, "How long ago would the separation have come about between populations of plants and animals in the Sierra Madre Oriental in southern Mexico from their relatives in the Appalachians?"

"That would have been following the early Miocene," I answered, and then added without thinking, "Wouldn't it?"

Axelrod glared, pounded the desk with his fist, and said sternly, "You're not asking me the questions, I'm asking you the questions!" Axelrod, twenty-six years my senior, would later become both a close friend and a research collaborator, but you couldn't have convinced me of that at the time. Fortunately, I was able to catch my breath and soldier on, grateful when my defense came to a successful end.

My dissertation would turn out to be the beginning of a research focus that would last for decades. My study of Onagraceae was based at its start on the work of Harlan Lewis and others, including a number of doctoral students and other researchers; step by step, we would be able, in the following years, to build the Onagraceae into a model of plant evolution, a model that would be helpful in establishing principles of interest for the study of evolution in general.

In June 1960, I was awarded the degree of Doctor of Plant Sciences on UCLA's Dickson Plaza, a grassy area surrounded by shade-giving coast live oaks. The commencement address was given by the outstanding physicist Vern Knudsen, who had served as UCLA chancellor that year after long and distinguished service as Dean of the Graduate Division. As for my graduation from Berkeley, my parents were in the audience, but this time Sally and Alice, still a tiny baby, were there with them. My past and my future seemed to come together; while I felt some sadness about reaching the end of my wonderful and fulfilling time at UCLA, I felt deeply optimistic about what was to come.

– 5 –

England

HAVING EARNED MY DOCTORAL DEGREE, I had a career to build and a young family to provide for. Knowing that there was a secure position waiting for me at Rancho Santa Ana Botanic Garden, I turned my attention toward the question of where to spend a year of postdoctoral work. Wanting to broaden my experience, I flirted with the idea of using the year to learn more about the rich living world of the tropics. Toward that end, I corresponded with the famous botanist E. J. Corner, who remained at his post at the Singapore Botanic Gardens after a remarkable career that included guiding the institution smoothly through the period of Japanese occupation. I also thought of spending the year with Professor Jack Heslop-Harrison, a noted scholar of plant genetics and evolution at the Queen's University in Belfast; he had written a short book on plant systematics and evolution that I very much admired. Neither option seemed quite right, and so I searched on. Eventually, with Harlan Lewis's advice, I decided to enlarge my understanding of Onagraceae as a whole by taking up the study of Old World species of the family. For that endeavor, there was no better place than London. Its two large herbaria, at the Natural History Museum and the Royal Botanic Gardens, Kew, held large numbers of specimens, outstanding resources on which to base my proposed studies.

Once I had made the decision, I solicited invitations from the authorities at each institution in order to apply for a Postdoctoral Fellowship from the National Science Foundation. When the fellowship, generously supporting an entire year of work, was granted, the deal was sealed. Sally and I were excited about what we expected to be a great adventure: neither of us had been to Europe, and we loved the idea of seeing new places and meeting new people. There was another advantage to spending the year in London: my aunt Vivienne was now living there for half of each year. In 1956, she had married Ralph Blechynden Moller, affectionately known as "Budgie" to his friends. He was the scion of a family that had amassed considerable wealth in steamships (the Moller Line plied the waters between Hong Kong and Europe). He and Vivienne moved seasonally between their apartment on London's Berkeley Square, a home in The Peak in Hong Kong, and an estate near Cambridge where Budgie, a horse-racing aficionado, maintained the stables that his Danish father had built there. As Sally and I planned our move, it was reassuring to know that the always helpful and knowledgeable Vivienne could help us find our bearings.

My studies in London would be a natural outgrowth of those I had carried out at UCLA. I was in effect deciding to make the study of Onagraceae the leading focus of my research from this point forward. It was a choice I would never regret. Over the years to come, in studies involving the anatomy, pollination, production of secondary plant compounds, geography, and evolution of the members of this family, I would have opportunities to address scientific puzzles about the relationships of plants, collaborate with some of the most cooperative and fascinating colleagues imaginable, mentor dozens of talented graduate students, visit interesting and remote parts of the world, and set the stage for later developments that I couldn't possibly have foreseen in 1960.

With our plans confirmed, Sally and I got busy wrapping things up in Los Angeles. I had to abandon some of the smaller projects I had underway in California in order to clear the decks for what was to come. I did, however, continue collecting in the Santa Monica Mountains through late July, as I intended to finish up that flora when we returned to the U.S. As I returned the borrowed specimens that were still in my care to their permanent homes in various herbaria, I couldn't yet know that this was the last time in my life that I would ever be truly caught up.

In August, Sally and I, with our ten-month-old baby Alice, found ourselves winging our way toward London, filled with excitement about what awaited us. I had written to Vivienne to let her know of our arrival time at Heathrow Airport. She had replied, "Chauffeur Tyler will be awaiting you, carrying a furled umbrella, at the barricade." To welcome us to England, Vivienne took us out for lunch, where we became acquainted with potted shrimp and other traditional English foods.

With Vivienne's help, Sally and I found an apartment in Twickenham, near the famous rugby grounds there. (After a few months we would relocate to a somewhat larger home in the same area, the bottom floor of a house in Pope's Grove, right on the banks of the Thames.) England had neither fully overcome the hardships of World War II nor modernized to the extent that California had. Wages and prices were low, supermarkets were nonexistent, no stores did business on Sundays, the first few miles of freeway in the country would not open until the following year, credit cards were unknown, and petrol stations were few and far between. The war, which had brought so much privation and destruction to British home ground, still scarred the national psyche. One afternoon, Sally was standing in line buying groceries when a British woman (who had presumably heard Sally's American accent) complained loudly, "You didn't have to go through the war like we did!" Not a comfortable moment, but the difference could not be disputed. That incident and others gave us a glimpse of the world through English eyes.

More usually, our American accents and language got lighthearted reactions. The way I pronounced the word "mountain" was always good for a laugh, and when I told my first English audience that I had studied "herbs" with a silent h, the entire group burst out laughing—and I had no idea why. And then there was the time I wrote "Thames River" on a plant collection label and had it pointed out to me that the word "river" was redundant and, if ever used in such a name, should come first. Sally and I found surprises around every corner—and I'm sure we constantly puzzled and amused the English people we encountered in turn.

I had never journeyed beyond the Americas except through the pages of *National Geographic*, movies like *King Solomon's Mines*, and musicals like *South Pacific*. England was not just a new country to me but also a door to a new world. We loved it, and took as many chances as we could to savor its history, customs, landscape, and culture. We especially appreciated the antiquity that was brought to life by its historic buildings and ruins. Everywhere we went we found evidence of the distant past. Near York Cathedral, we came across a statue to a man whose illustrious grandson, George Washington, "found fame in a land as yet unknown to him." We were astonished by the ruins of Hadrian's Wall, Bath's surviving Roman construction, and Stonehenge, standing in lonely splendor on Salisbury Plains. On my way back from collecting plants in Kent, I was surprised and delighted to come across a roadside historical marker that recounted the landing of William of Normandy in 1066; in an encounter with still earlier times, a colleague and I walked the ramparts of Maiden Castle, a Neolithic hill fort in Dorset some 2,600 years old that occupied nearly fifty acres. It was thrilling and also humbling, a reminder of how long humans had inhabited these lands.

℃

As originally envisioned, I pursued my studies at two different institutions, the Natural History Museum in London and the Royal Botanic

Gardens, Kew. Being able to divide my time between them allowed me to study in different herbaria and libraries and work with two different sets of people. The specimens of Onagraceae held by these two institutions were so numerous that it was a challenge to examine as many of them as I wanted during the time available. Knowing that my time in England was limited only made the temptation to work long hours stronger; unfortunately, my focus on work made Sally's adjustment to our new home that much harder.

At the Museum, a genial and interesting man named John F. M. Cannon was in charge of the section of the herbarium that included Onagraceae, and so I fell naturally under his care. He was

John F. M. Cannon, who later became Keeper (head) of Botany at the Natural History Museum. John was a friend and important collaborator during my postdoctoral work at the museum and for many years to come.

busy coordinating work on a large public exhibition gallery dedicated to plants that was slated to open the year after my departure. John and I got along famously and quickly became friends.

The Natural History Museum's beautiful building in South Kensington was designed by the noted architect Alfred Waterhouse and dedicated in 1881. At that time, it had provided the first truly suitable housing for the British Museum's natural history specimens. But the collections had expanded steadily over the ensuing eighty years and were now overcrowding the spaces where they were housed. By the time I arrived there, a large gallery initially used for exhibition space had been commandeered for plant research collections. Offices, desks, and other working spaces had been squeezed in among the polished wooden specimen cases and newer metal ones; the botanical library was housed in a two-story room at the far end of the gallery.

John Cannon's office was located in an alcove not far away from the desk I was assigned, and two eminent botanists, Frank Ludlow and William Stearn, worked at desks beside mine. The department's relatively informal working atmosphere allowed for plenty of interchange. Seventy-five-year-old Frank Ludlow had been a lieutenant in the Great War, a teacher and educator in 1920s India, and an explorer and collector of plants and animals in the Himalaya from that time onward. In addition to being a fine botanist who contributed thousands of plant specimens to the Museum's herbarium, Ludlow was a noted ornithologist who also gave it an impressive 7,000 specimens of birds. I soon discovered why he was beloved by so many, including the royal families of both Britain and Bhutan: he was a friendly, congenial man and a goldmine of information.

The Natural History Museum, London, where I spent many happy and productive months studying plants in 1960–61.

Bill Stearn had a brilliant gift for understanding and explaining the rules of the nomenclature (naming) system for plants—he was sometimes called "the modern Linnaeus" —coupled with an unstinting willingness to share what he knew with others. A polymath whose genius was first recognized while he was working in a bookshop in Cambridge at the age of twenty-one, he took his first botanical job in the library of the Royal Horticultural Society in London. Posted to Southeast Asia during World War II, he took advantage of the opportunity to observe and collect many of the local plants. Tall, with a somewhat flamboyant white moustache and an outgoing manner, he was always a pleasure to converse with and a great source of knowledge.

J. E. Dandy, the Museum's Keeper of Botany, also contributed much to my time at the museum. Dandy was a robust personality who maintained high standards and was broadly knowledgeable about plants. He was always willing to interrupt his own work and help those who wanted botanical matters to

J. E. Dandy (1903–1976) was Keeper of Botany at the Museum during my postdoctoral year there. Expert on magnolias and pondweeds, he was unfailingly generous and kind to me.

be worked out in the best way possible. We held innumerable discussions and I always benefitted from his intelligence, meticulousness, and knowledge. One weekend day in the spring, Mr. Dandy invited me to his home to watch a cricket match on television, explaining the rules as the game progressed. We then went out collecting plants in the meadows near his home. An expert on water plants, he lowered a hooked weight on the end of a string into a pond and pulled it out dripping with aquatic plants, a low-tech but very effective means of collecting them.

As I looked for a project that would make use of the Museum's rich material, John Cannon suggested that I might study and identify the herbarium's accumulated Himalayan material of the genera *Epilobium* and *Chamaenerion*, the willow herbs and fireweeds. Those collections were fascinating, there was material to study in quantity, and Frank Ludlow had personally collected many of the specimens. As I began to study the material, I felt the soundness of my decision to study Onagraceae being affirmed once again. The more I came to know about this plant family, the more I wanted to know. Of course, having Mr. Ludlow working in such proximity made the plants I was examining all the more interesting.

The Royal Botanic Gardens at Kew was also an excellent place for increasing my knowledge, but the setting at Kew was considerably more formal than at the Museum. It took months before I could get an appointment with the director, Sir George Taylor, to introduce myself. I spent many days at Kew without conversing with anyone, and it was quite a while before I found out that coffee and tea breaks were held, apparently by invitation only, at a table accessible only by way of a gangplank that led out of one of the herbarium windows. Such breaks were much less ceremonial at the Museum, being held right off the main herbarium with all comers welcomed.

I worked in the Kew herbarium once or twice a week, building fast friendships and working relationships with several of the staff members there despite the general air of exclusivity. I already knew of C. E. Hubbard, the Keeper, before I arrived at Kew, having consulted his excellent short guide to British grasses while still in California. A short, soft-spoken man of about sixty, he helped me to find my workspace near the Onagraceae and responded generously to my queries. Another eminent figure at Kew, J. P. M. Brenan, "Pat" to his friends, was especially kind to me. An expert in African plants and the Principal Scientific Officer at Kew, Brenan did work that was particularly relevant to me: he had revised the African species of the genus *Ludwigia*, the water primroses, a primarily tropical group of the family Onagraceae. Brenan's excellent treatment formed the basis of my own studies, completed while in England, of all the *Ludwigia* species that occurred in the Eastern Hemisphere.

Keith Jones, a Welsh cytologist, worked at the Jodrell Laboratory, the site of laboratory research at Kew. Like me, he was a newcomer to the institution. We had met when he visited with Harlan Lewis briefly during my graduate years at UCLA. Now, he provided a generous welcome to his laboratory and made possible continuity in the chromosome studies that I was still pursuing. He was a cheerful, supportive

The Herbarium at the Royal Botanic Gardens, Kew, where I alternated days with those spent at the Natural History Museum during my postdoctoral year in London.

person, who encouraged me during the lonely times at Kew before I made many friends.

❧

By the time winter set in, Sally and I had begun to feel at home in London, if not entirely at ease. We had learned during the darkening days of autumn that Sally was pregnant with our second child, so we contacted the National Health Service in order to seek advice from a physician and lined up a midwife for the birth, expected in May. By now, I had made quite a few friends at both the British Museum and the Royal Botanic Gardens, and we sometimes socialized with them after work and on weekends. John Cannon and his wife Margaret were particularly warm and sympathetic. We also became close to visiting Americans Bob and Mae Thorne. Bob was a tall, outspoken, pleasant man on sabbatical from the University of Iowa; he was working at Kew on the relationships between the families of flowering plants and had just returned from time spent on that project in Australia and New Caledonia. He and his family were filled with interesting stories of their time in the Southern Hemisphere.

Sally and I were both appreciative of such friendships, and they were especially welcome for her. While I enjoyed social interactions through much of each workday, Sally was alone at home with a young child and experiencing the hormonal disruptions of pregnancy. She had left her entire support network of family and friends behind in the U.S. Unfamiliar with English culture, she had not built up an adequate substitute in London. For her, our little network of friends provided her only companionship outside of our family and was her principal defense against the despondency and loneliness with which she often struggled. Since I was always busy, often away, mostly oblivious to her needs, and less sensitive than I might have been to her situation, I wasn't much help.

A welcome break came at Christmas, when Sally and I had the opportunity to enjoy the festivities in grand English style with Aunt Vivienne and her husband Budgie at their elegant estate in Cambridgeshire. We toured the stables, enjoyed a Christmas feast at the majestic dining table in the presence of numerous impeccably dressed servants, and participated in the traditional gift-giving ritual of Boxing Day. It was unlike any kind of Christmas celebration I had ever experienced.

Come spring, I began making field trips alone and with Brian Styles and Terry Pennington, graduate students from Oxford who shared my interest in plants of the Onagraceae, in pursuit of the dozen or so *Epilobium* species that occur naturally in England. The species of *Epilobium* were generally regarded as difficult to tell apart, and so we took special pleasure in figuring out their distinctive characteristics and learning to recognize them as well as the natural hybrids that formed occasionally between them. The results of these studies were applied directly to my forthcoming treatment of Onagraceae for *Flora Europaea*, the consolidated flora of all of Europe that was being prepared at that time. I also undertook a special study of the genus *Circaea* (enchanter's nightshade) and ranged widely and with great pleasure throughout much of England and Wales in search of these plants, getting to know the countryside and its delightful towns and sweet shops in the process.

Sally and I had bought a blue English Ford sedan—one with left-hand steering so that we could bring it home with us—and it was in this vehicle that I made my solo collecting trips. After mastering Britain's left-side-of-the-road driving, I began learning the vagaries of English and Welsh roads. Going out for two or more days at a time, I would usually find bed-and-breakfast lodging each evening. In particular, I found myself unprepared for the great beauty of the Lake District, which is even more majestic than Wordsworth's poetry had suggested. My Kew friend Keith Jones had told me about his native Wales, but I was still surprised by its distinctive character. I enjoyed its musical if unpronounceable language, unique customs, striking scenery, and warm people.

In addition to traveling within Britain, I was able to take trips to the major European herbaria to examine their material of Onagraceae pertinent to my studies and meet their people. I visited Paris, Geneva, Stockholm, and Uppsala for a few days each. As I did for the field trips, I traveled alone; Sally, growing larger every day with the baby we were expecting, remained at home with Alice.

Out of doors in England, I had the opportunity to collect European representatives of groups of plants other than Onagraceae. At first, I had the same questions that had made it difficult for me to get excited about collecting in the Cascades of Washington State when my earlier framework had been solely California. How would the general collections that I might make relate usefully to my existing knowledge and projects? I started by collecting native British plants that were either the same as, or related to, the ones that I knew in California. There were also many kinds of exotic weeds, some ultimately of Mediterranean origin, that occurred in both places. Gradually I began to collect more widely, especially in groups like the sedges (*Carex*, beloved by Mr. Howell) and grasses, where the genera and species seemed to run together with the ones I had known earlier. As it had always been, it was satisfying to get to know new plants in new landscapes.

Walking into the Museum's herbarium one day, toting a plant press filled with specimens I had collected the preceding weekend under one arm, I happened to encounter Mr. Dandy. "Instead of carrying that press," he said, "why don't you just hang a sign around your neck that says, 'I'm nuts'?" Not every botanist felt the compulsion to collect as intensely as I did. But Mr. Dandy and I supported each other, eccentricities and all.

∽

On May 11, 1961, a clear spring day, Sally gave birth to our second daughter. The birth took place at home with our midwife in attendance. During the delivery, beautiful white swans were sailing gracefully up and down the flooding Thames not far outside the window—though Sally, of course, had other matters occupying her attention. We named the baby Elizabeth, after the Queen, but she was called Lizzy from the start, and later Liz. Both mother and child received good care, virtually free under the provisions of the National Health Service, and there were no complications.

After Lizzy's birth, I began to wrap up my work in the U.K. and prepare for the return home. Our year abroad—a time of great scientific productivity for me—was coming to a close. I prepared my account of the work I had completed on *Epilobium*, wrote a paper on the Old World species of *Ludwigia*, finished a paper on the British species of *Circaea*, and tied up a few other loose ends. Mr. Dandy informally offered me a job in the Museum's Botany Department if I would agree to take up the study of mosses, a group of plants of which they had the finest collection in the world. I appreciated his offer greatly, but a variety of factors—from the increasing strains on my marriage to the job I had waiting at Rancho Santa Ana—made it clear to both Sally and me that it was time to head home.

On the sunny August morning of our departure, we visited the meadows at Runnymede, where Magna Carta had been sealed in 1215, and I collected a few plants as we bid goodbye to John and Margaret Cannon. John laughingly challenged me to collect a few more plants in Boston on arrival so that I'd have made collections on both sides of the Atlantic on the same day. Tempting as the challenge was, we were too tired when we touched down to take him up on it.

After arriving in Boston, Sally, Alice, and Lizzy flew on to San Francisco, where Sally, eager to reunite with family, would stay with her parents for a few weeks. I rendezvoused with my father and the blue Ford, which had been shipped by boat for pickup in Boston. Dad and I had decided that driving across the country together would be a great way to reconnect; it would also be an excellent opportunity to visit some major botanical institutions, meet their scientists, and look at some key specimens of Onagraceae in their herbaria. I could also collect plants along the way.

It was a great pleasure to see my father again after so long a time away. We started our trip at Harvard University, where many botanists I knew, including Ledyard Stebbins and Edgar Anderson, had received their graduate training. Driving south, we visited my father's divorced sister Helen Bradford, who was living alone in Manhattan. The first visit I'd had with her since my early childhood, it was all too short. After visiting Aunt Helen, we drove to see The New York Botanical Garden. On the way, we took a detour to visit Frank Lloyd Wright's newly opened Guggenheim Museum, its design inspired by the shell of a chambered nautilus and the geometry of the golden ratio. I saw it in musical terms as well, almost as a symphony of architecture. I hadn't really known that architecture could be so original, and visiting the museum was an absolute delight.

Moving on to the Midwest, we stopped in at the Field Museum of Natural History in Chicago. When Louis Williams, curator of the botany department there, asked where we had parked, I nonchalantly replied that we had left our car on the street. "You must move it immediately!" he responded. "It will be broken into or stolen!" Noted for his studies of the plants of Central America, Louis had recently arrived at the Field Museum after spending many years studying plants there and in South America. He was a knowledgeable contact for anyone interested in those botanically extraordinary areas.

After a call at the University of Michigan herbarium, my father and I drove south to St. Louis, to visit the Missouri Botanical Garden. The Garden was of special interest to me because of its associations with the many systematic botanists—Mildred Mathias, Carl Epling, Dave Gregory, and Louis Williams, among others—whose lives had touched mine. Just before leaving for England, I had read about its newly built Climatron and seen this impressive geodesic dome illustrated in *Life* magazine. The Garden's director, Frits Went, had commissioned architects Murphy and Mackey to design it using the geodesic dome form developed by Buckminster Fuller. It was the first geodesic dome used as a plant conservatory and, at roughly a half acre in size, the second-largest one in the world when it was completed in 1960. My father and I gazed at the Climatron with amazement as we explored the lush plantings in and around it.

I was glad to have the opportunity to meet with some members of the Garden's staff. One of them, Bob Woodson, was a well-known botanist who specialized on the milkweed family. Bob was striving to complete the *Flora of Panama*, which he had initiated in 1943; he asserted during our conversation that he believed the collecting for the project was nearly complete.[1]

As we looked out the third-story windows of the Administration Building, where the Garden's herbarium was then housed, two young staffers, Cal Dodson and Bob Dressler, described some of the inner workings of a botanical garden. They complained that the Garden's research programs had taken a back seat under the directorship of Frits Went, and that the collection of living orchids that they studied was deteriorating as resources were diverted elsewhere. Their perspective certainly interested me, but I had no sense at all that managing such issues might ever become my responsibility.

Continuing our drive west, my father and I experienced our most memorable lodging on the plains of eastern Colorado. In Limon, tired and eager for rest, we misinterpreted a motel's sign displaying the logo "ATA" as being "AAA" and assumed it was American Automobile Association accredited. Alas, it was not to be. The tiny rooms were filthy and the showers had what appeared to be dirt floors—a fine bonding experience for sure, but one which both of us would happily have missed. Leaving there, we swung northward across southern Wyoming and Idaho, collecting samples of a few special plants. Finally, we turned southward toward the Bay Area and a happy reunion with Sally and the children. It had been a wonderful journey, and I was very pleased to have brought the relationship with my father to the level at which we knew each other adult-to-adult.

cro

After pleasant visits with our extended families in San Francisco, Sally and I drove south so that I could take up my long-anticipated position at Rancho Santa Ana Botanic Garden. Reporting to Lee Lenz, who had become director while we were living in England, I was directed to my new office and offered a few suggestions about where we might seek a place to live, a matter of considerable urgency to us. Finding a small apartment nearby in Pomona, Sally, the girls, and I settled in as I began to learn what was expected of me at work. I was not only curator of the herbarium, but also its only paid staffer. We had volunteer helpers for mounting specimens, but it was my responsibility to sort herbarium specimens for loan to other institutions, wrap and mail them, and re-accession those coming back. In other words, everything that went on in the herbarium was to be my responsibility.

As time passed, I took stock of our situation and thought about how I could conduct my duties most efficiently. My predecessor, Phil Munz, was still very active in and around the herbarium. One day I asked him, "Wouldn't it be much more convenient if we could consolidate most of the herbaria around the state to a few central locations?" Munz's response was a hearty and unequivocal "No!" He said with some eloquence that the best hope for botany was to have a robust community of passionate individuals following their own particular interests, each contributing to the body of scientific knowledge with their own local efforts. There would be many fewer specimens to study, he said, if the initiatives of all those individuals weren't encouraged. If the specimens they collected all disappeared into a distant facility, their sense of "ownership" would be negatively affected, and they would not be inspired to collect more specimens.

As I listened to Dr. Munz, I saw that he was making sense. The job of cataloging life on earth was indeed much too large for a few people or institutions to accomplish and could be approached better by many individuals in many places pursuing a variety of specific goals and objectives. One of the best things I could do for botany, I suddenly realized, was to encourage others in their own efforts. This perspective, with its necessary emphasis on collaboration and mutual encouragement, differed from that of a typical research scientist who is completely focused on carrying out his or her own work. It was a powerful insight, and one that began to guide my professional life from that point onward.

Rancho Santa Ana was certainly a nice place to work, but it was a relatively small institution and it didn't feel right for me. In graduate school and then in England, I had been surrounded by dozens of individuals whose pursuits related in some way to my own; visiting botanical gardens and herbaria in Europe and then throughout America, I had broadened my perspective still further. After all that dialogue and collaboration, Rancho Santa Ana seemed isolated and somewhat confining. It might have been the perfect setting for someone with a more introverted temperament and more specialized goals, but I couldn't see myself fitting well there in the long term. I came to realize that I wanted the chance to work with the broader scientific community, to collaborate and communicate with a wide range of people, and to take part in many different projects.

Pomona itself didn't feel quite right, either. Neither Sally nor I felt at home there. It seemed symbolic that the gorgeous mountains beyond our windows were invisible most days, obscured by a thick blanket of yellow-brown smog. Living in yet another unfamiliar place with two small children and little help from me, Sally continued to struggle with depression while the strains on our marriage grew. At home as at work, I had a sense of drifting, treading water, spinning my wheels, with no idea how to get where I knew I wanted to go. On both the personal and professional fronts, I felt frustrated and stuck. As always, however, plant collecting energized me. While my working time at Rancho Santa Ana often seemed routine, the plants that grew wild in the vicinity, happily, did not.

One sunny day in the autumn of 1961, a possible answer for my sense of professional impasse arrived unexpectedly. Professor Paul Ehrlich, whom I had met when I visited Stanford to examine specimens for my dissertation, showed up with his wife Anne at Rancho Santa Ana to visit me and inquire as to whether I might be interested in applying for a position in the university's Department of Biological Sciences. A review committee set up at the suggestion of the National Science Foundation had suggested that Stanford should hire someone who specialized in the field of plant evolution and classification. (At the time, the university housed a substantial herbarium.) My name had appeared on the list of possible candidates, and Paul was reaching out accordingly.

Stanford was that bigger pond, deeper and broader and more challenging, that I had been longing for. And it was in the San Francisco Bay Area, where both Sally and I had deep roots. After consulting with Sally and getting her enthusiastic assent, I readily agreed to apply for the position. We both hoped that the move, seemingly the right opportunity at the right time, would give us a new beginning, and so we anxiously awaited the outcome of the search process.

We didn't have to wait long. Within a couple of weeks of the Ehrlichs' visit, in late October, I received a letter inviting me to interview for a faculty position at Stanford. I was thrilled. On our first night in Palo Alto, Paul and Anne invited Sally and me to dinner at their apartment near the campus. After we finished the meal, Paul put an album on the record player, saying, "You've got to listen to this!" It was a female artist singing songs of the Spanish Civil War. "Who is that?" I asked. "Joan Baez," came the answer. Her voice—pure, strong, and compelling—made a deep impression on me and so did her music. Afterward, I would date the real beginning of my experience of the 1960s to that night in 1961. It was the kind of moment that changes one's sense of what is possible; it promised change, in my own life and in the larger world. My interviews the next day went well, and soon the position was offered to me. By Thanksgiving, less than three months after reporting to duty at Rancho Santa Ana, I had agreed to join the Stanford faculty the following summer.

– 6 –

Stanford

Although our move to Stanford was still months in the future, Sally and I were delighted with the prospect of getting back to the Bay Area. We thought of being closer to family and friends and began to consider what we needed to do in preparation for transferring our activities there. Meanwhile, everyday work went on for me at Rancho Santa Ana and I pursued it diligently.

Eager to take advantage of my remaining time in Southern California, I collected plants locally at a fevered pace. In the spring and early summer of 1962, I visited San Clemente Island, which lies forty-one miles off the California shore. This twenty-one-mile-long island rises to 1,965 feet elevation, plunging down sharply to the ocean on the east side, sloping gently on the west. The U.S. Navy, which owns the island, expedited my studies there, transporting me back and forth, providing barracks housing, and allowing personnel to help me while on the island. I had decided to write an account of all the plants of the island, knowing that they were very interesting and showed various relationships to their relatives on the other islands off the California coast and on the mainland.

San Clemente Island at that time supported hundreds of feral goats that had grazed down all the grassland and scrub habitats that they could reach, leaving remnants of the original vegetation for the most part only on the rocky cliffs of the deep canyons on the west side of the island.[1]

I collected that year elsewhere in several other Southern California localities and worked with graduate students when I could. Before we left for Stanford, I was visited by Dennis Breedlove, then an undergraduate student at the University of California, Santa Barbara, who was only three years my junior. That winter he had driven from the coastal Mexican state of Sinaloa over the Sierra Madre Occidental to Chihuahua, stopping to collect plants along the way. When he showed me the specimens he had collected, I was impressed by how varied and interesting they were; among them were several kinds of Onagraceae that were new to me. It was obvious that Dennis was as passionate about plants as I was and ready to make a difference in the field. After I told him that I was moving to Stanford, he applied to the program there. Confident he would be accepted, I looked forward to working with him as my first graduate student.

After packing up our things, Sally and I headed north at the end of August, settled into a nice house on Middlefield Road in Palo Alto, and set about creating a life for ourselves and the girls. It was finally a place that truly seemed like ours: a freestanding house with lawns and flowers in a very pleasant neighborhood.

I immediately felt very much at home on the gorgeous Stanford campus. The noted landscape architect Frederick Law Olmstead had decided on the placement of the buildings, gardens, and open spaces; the architectural firm of Shepley, Rutan and Coolidge had designed the buildings themselves, choosing a California Mission-inspired Mediterranean style featuring local sandstone and red-tiled roofs. The core buildings surrounded a cloistered quadrangle with north-facing Memorial Church as its focus. The layout seemed perfectly suited to the climate and to the setting.

The university, which had opened for students in 1891, was founded on progressive principles. Nevertheless, when I arrived in the autumn of 1962, it felt almost old-fashioned—female students were forbidden to wear trousers to class, for example, and there were few female professors. At the time, President Wallace Sterling and Provost Fred Terman were intent on increasing the school's size and prominence, in part by expanding its role in developing cutting-edge science and engineering. Terman, who had been a member of the university's engineering faculty from the 1930s, eyed the growing federal funding for science in the 1950s and was determined to get Stanford its share. Terman encouraged his students—who included William Hewlett and David Packard—to stay in the region and go into business right there, thus laying the foundation for what would become Silicon Valley.

Along with other science departments at the university, the Department of Biological Sciences, always vigorous, was actively engaged in strengthening its reputation. Botany, organismic biology, and evolution, traditional fields at Stanford, all benefited greatly from the federal funding that was flowing toward the university. In our diverse and increasingly outstanding department, we had the advantages of being at the vanguard of research, building rapidly as biology diversified. Working with first-rate colleagues, we had the opportunity to participate in interdisciplinary dialogue. My hiring was part of a coordinated effort to build the program in systematic and evolutionary biology, partly in relation to the rich collections that had been assembled at Stanford over the years. I was seen as someone who could take the lead in figuring out how best to use these substantial collections and help make them relevant for university instruction in the contemporary world. This was among the many issues I began to ponder as I settled into my new role.

One of my offices was on the Quad, the big square of buildings that ringed the main landscaped area, and another was in the south wing of the Stanford Art Museum, where the biological collections were housed and a few collection-oriented colleagues had offices. We narrated our correspondence into clunky Dictaphones to have them typed by a department secretary—a relatively cumbersome process, but one that clearly beat typing them out ourselves.

Though research would be a primary focus, I was responsible for shouldering a share of the department's teaching responsibilities as well. Compared with my work on scientific projects, I had had little prior experience in the classroom. Starting with some relatively small graduate-level courses, I was panic-stricken at the idea that the students might not appreciate what I had to say. But I seemed to do all right, and soon graduated to team-teaching in population biology and general biology, classes that might comprise hundreds of students each.

At the age of only twenty-six, I was younger than some of those I was teaching, including a fair number of veterans who were still returning to school to complete their degrees. What if my students walked into my classroom and laughed? What if they walked out? Looking out over the sea of faces (or at least heads), I would see some falling asleep, others engaged in their own earnest conversations with each other. If I wanted to teach anybody anything, I knew I had to get and hold the interest of a good percentage of them. I learned to present myself as a personality, to entertain as well as educate, to find a balance between what was interesting and entertaining and what was perhaps less exciting yet necessary and relevant. It was important to be fully present and aware. I knew that the less-successful teachers were usually those who delivered rote lectures into space rather than engaging with the students actually sitting in the room. Thanks to all this intensive on-the-job training, my skills caught up to the demands of the job before too long. My success in classroom lecturing, particularly for large classes, helped me put to rest any lingering feelings of shyness or anxiety about speaking to groups.

In the meantime, Sally put much energy into the task of making a warm, loving home for the girls. She brought them to church regularly and threw herself into activities like making costumes for Halloween. Our little family made periodic visits to both sets of grandparents, who lived a short distance to the north in San Francisco. Sally's folks were less formal than mine, but both they and my parents doted on Alice and Liz.

I adored my daughters as well, but my workaholic habits continued to limit my time with them, something I would later regret. At the same time, I had little understanding of how my work habits were putting stress on my relationship with Sally and doing nothing to help her cope with her inner struggles. Fortunately, we did have some delightful times together as a family. The summer after my first academic year at Stanford, for example, we stayed for a couple of weeks at The Lair of the Golden Bear, a Cal Berkeley alumni camp in the Sierra Nevada, where there were abundant opportunities for fishing, crafts, and family activities. We also had our meaningful family rituals. Sometimes, to give Liz and Alice a chance to experience my "digs," Sally would bring the girls along when she came to Stanford to pick me up, and we'd leave the museum building by an entrance near some caged turkeys. It was there that I mastered my turkey calls—the finest of all the bird calls—exchanging happy gobbles with the birds and giggles with the girls as we headed for the car.

ꞓꞔ

Soon after arriving at Stanford, I became good friends with two of my
fellow professors, Paul Ehrlich and Dick Holm, both a few years older
than me. Soon we were seen together so often that some of our colleagues
started calling us "Peter, Paul, and Dick" in a nod to the iconic folk group
that was so popular at the time. Over cups of coffee and daily sack lunches
around a heavy round wooden table in the rotunda on the second floor
of Stanford's Art Museum, the three of us spent hours together, talking,
making jokes, and learning a great deal from one another. When we began
co-offering a weekly evening seminar in population biology, we extended
the dialogue to Ming's Chinese Restaurant, where we shared a dinner before each session.

Dick Holm had been a student of Robert Woodson's at the Missouri
Botanical Garden and, like Woodson himself, studied milkweeds. He was
recruited to teach at Stanford in 1949, before he had even finished his
doctorate, and had risen steadily through the ranks. When I arrived at
Stanford in 1962, Dick had recently finished a decade as the curator of the
university's Dudley Herbarium. He had also become director of the biology department's Division of Systematic Botany. Quiet, sincere, and ex-

My closest colleagues at Stanford were Paul Ehrlich (left) and Dick Holm (right), both brilliant and
irrepressible.

tremely intelligent, he was a gifted editor and a highly knowledgeable teacher who taught a wide range of courses.

Paul Ehrlich had grown up in Philadelphia, meeting his lifelong partner and colleague Anne while they were both students at the University of Kansas in the 1950s. There, Paul earned an M.A. and Ph.D. in entomology. I appreciated Paul's biting sense of humor, his refreshingly audacious wit, friendly personality, and his fiercely intelligent mind.

After Paul's arrival at Stanford in 1959, he and Dick had become close friends and collaborators. Stimulating each other's thinking, they had pioneered a new concept: the field of population biology, an integration of the traditionally distinct fields of systematics, evolution, and ecology strongly informed by the behavioral and social sciences. Their keen attention to the implications their concept had for human beings gave their work a special relevance, a usefulness beyond the proverbial ivory tower. Encompassing a variety of fields, population biology was a logical construct for lecturing, writing, and research. It helped form the academic framework for many subsequent efforts, including some of my own.

By the time I arrived at Stanford, Dick and Paul were established friends and partners. I was fortunate to be welcomed fully by both, at a time when everything at the university was new to me and I wasn't sure of myself. Their friendship and generosity helped give me confidence as I solidified my own career, and our dialogue helped change my perspective on my work and myself. I had earlier thought of myself primarily as a student of plant classification (systematics) and evolution. Now, thanks to Paul and Dick, I began to think in terms of the scientific questions that were starting to emerge at the intersection of systematics and evolution and in the fields of ecology and population genetics. It was a significant and exciting shift for me.

Paul, Dick, and I had an early opportunity to cement our friendship in October 1962, when we all attended the annual Systematics Symposium at the Missouri Botanical Garden. Paul and I flew to St. Louis, connecting upon our arrival with Dick, who had spent the preceding months studying milkweed specimens in the Garden's herbarium. The meeting itself was devoted to the then-young subject of numerical taxonomy, and Paul was its moderator. During one session, Professor Richard W. Pohl of Iowa State University, a highly respected international authority on grasses, confronted Paul with a question to the effect of "Are you trying to say that taxonomists could be replaced with computers?" Paul, who was only thirty years old to Pohl's forty-six, brashly replied, "No, excuse me. I wasn't trying to say that at all. Most of them I know could be replaced with an abacus." It was typical Paul—unhesitating, irreverent, direct, and funny. But a sudden hush fell over the crowd as its largely traditional members tried to interpret the exchange and figure out how to respond.

To return home, we rode on one of the Union Pacific Railroad's Dome-liner trains, fitted out with a modern glass-roof lounge and observation cars. Our journey took place at the peak of the Cuban Missile Crisis, an extremely dangerous time for America. School children had been learning to cower under their desks, and the whole country was nervous. All three of us were on tenterhooks the entire way home, jumping off at every station to check newspaper or radio reports. We were relieved to see that all was well in San Francisco when we disembarked, but the actual crisis didn't end for another week.

Another Stanford colleague with whom I began to interact frequently was John Hunter Thomas, who had succeeded Dick Holm as herbarium director. He had completed his M.A. and Ph.D. at Stanford under the supervision of Ira Wiggins, just like my UCLA colleague Harry Thompson. Though only four years separated his thirty and my twenty-six, John seemed almost to come from a different generation. He was meticulous and could sometimes be demanding and impatient, but he was also generous with assistance. John was especially knowledgeable about local plants—his dissertation, *Flora of the Santa Cruz Mountains of California*, concerned the plants of the range that rose up right behind the Stanford campus. Like Paul, Dick, and me, he was interested in population biology and had become passionate about environmental issues. Not surprisingly, given our mutual love of California plants, John became a great help to me in my studies of various species, as well as a good friend.

ᏋᎧ

My appointment at Stanford corresponded with the beginning of a period of explosive growth and fundamental change in the field of biology. One innovation of particular import for my own work was numerical taxonomy, the methodology that had been the focus of the Systematics Symposium at the Missouri Botanical Garden in the fall of 1962. Developed principally by two of Paul Ehrlich's former mentors at the University of Kansas, Professors Robert Sokal and Charles Michener, along with the British scientist Peter H. A. Sneath, numerical taxonomy attempted to determine the relationships between organisms in a new way. Instead of giving a few striking characteristics great importance, it applied numerical methods like algorithms to consider many features at once and see what summing them up would reveal about the organisms' evolutionary relationships. Despite its advantages, numerical taxonomy involved computations that were especially arduous in the days before personal computers. The work behind one of Sokal and Michener's seminal papers was said to have taken two graduate students working an entire year to complete all of the necessary combinations of characteristics between the organisms being studied. We used adding machines, which were becoming increasingly powerful. Our population biology group was elated when we were able to purchase an advanced Wang adding machine, a large and expensive device that we kept in a closet off one of our laboratories.

At the same time, the relatively new field of molecular biology was profoundly altering our understanding of life. As the 1953 hypotheses of Watson and Crick became well established and generally accepted, the number and depth of molecular insights increased exponentially, as did their applications in other fields. Problems that had seemed intractable a few years earlier could now be solved; phenomena that had seemed mysterious became understandable for the first time. Once we understood how genes formed proteins, we could move on to consider the ways in which gene mutations could alter the nature of those proteins and in turn impact the functioning of cells and organisms. Entire new worlds of information and inquiry opened up.

Stanford University was in the vanguard of the shift toward using molecular biology to inform investigations in evolution, systematics, and population genetics. It was invigorating to be part of all the intensive intellectual growth that was going on in developmental biology, in the neurosciences, in molecular biology, and in many other fields. Most of the subjects being explored had no direct connection to my work, but their concepts and questions enriched my thinking. They spurred me to consider the ramifications of what I was learning for other fields, to evaluate its worth in the field of biology as a whole, and to be ready to present these considerations to my colleagues. The sense of excitement, collaboration, and discovery at Stanford banished the feelings of stasis and constraint that had afflicted me at Rancho Santa Ana. Now I was in my professional element, part of an exciting intellectual network, and enjoying every minute.

Work in the department felt serious and intense, but the atmosphere was often playful. When we felt overwhelmed, we might intone "It's all for the glory of God and to hell with it!" or "Let's go down to the garage, idle the motor, and consider the matter further." A line from *Who's Afraid of Virginia Woolf?*—"Jesus Christ, I think I understand this!"—was another favorite phrase. It might not have been the most flattering use of playwright Edward Albee's gift for dialogue, but playful banter helped cut the intensity of the long hours and complex work.

⌘

An important element of my work at Stanford, the development of the concept of coevolution, began with a butterfly and the feeding habits of its caterpillars. Paul Ehrlich and his students were actively studying this butterfly, the Bay checkerspot (*Euphydryas editha bayensis*), an inhabitant of the serpentine grasslands on Jasper Ridge, the biological preserve on campus. The adult checkerspots were conspicuous during their short flight season in spring, and Paul had decided that they would provide a useful lens for examining the factors controlling a population's evolution. Since Paul, trained as an entomologist, was focused on the butterflies and I was focused on plants, our discussions naturally gravitated to the way butterflies and plants interact with each other. One day, Paul raised a puzzle. Why did checkerspot butterfly caterpillars switch every year from feeding on the bright purple-flowered annual *Castilleja densiflora* to feeding on the drab, wind-pollinated, gray-green flowered *Plantago erecta* (another annual, but one that stayed green longer as the slopes dried up) when the two plants seemed so dissimilar? I pointed out that in fact the plants were close relatives, even though they were placed in separate plant families because of a single difference in their pollination systems. We were both so interested in the relationships that we soon began a period of intense study of the food plants that particular kinds of butterflies chose.

Lecturing among the wildflowers on Jasper Ridge.

Searching the literature at Stanford, the California Academy of Sciences, and other Bay Area libraries, Paul and I soon discovered that the caterpillars of some groups of butterflies were very particular about the plants on which they fed, while others were much less discriminating. For example, monarch butterflies and their relatives fed almost exclusively on plants of the milkweed family and cabbage butterflies were equally focused on plants of the closely related caper and mustard families, while the brush-footed butterflies (Nymphalidae) exhibited a much wider taste in food plants. And strikingly, many families of plants (for example, plants of the madder-coffee family, Rubiaceae) had no butterfly larvae feeding on them at all. Because butterfly collectors had so often reared butterflies on their favored food plants in order to obtain perfect specimens, we knew more about the larval feeding habits of butterflies than of almost any other insect group. In short, we had found a wonderfully suitable subject for our investigations.

With those starting points, it wasn't difficult to figure out that some groups of plants had, over the course of their evolutionary histories, evolved certain poisons that deterred herbivores from feeding on them. When a particular group of butterflies or other herbivores evolved in ways that allowed their caterpillars to overcome these chemical barriers, they gained access to a new source of food for which they had relatively few competitors. The plants on which the insects could now feed in turn faced selective pressure to become more poisonous, or poisonous in a different way. In effect, the two species reciprocally affected each other's evolution, each changing in response to the other.

The relationship we were exploring was between insects and their food plants, but the same phenomenon, we realized, could apply to any two kinds of organisms that interacted with each other—predators and their prey, hosts and their parasites, plants and their pollinators. Each would bring selective pressures to bear on the other, and the two types of organisms would evolve in tandem. Paul and I termed this pattern *coevolution*. The new term that we coined came to be used in thousands of studies and experiments thereafter, offering insights far more sophisticated than we had imagined initially. Lots of earlier observations suddenly made sense in light of the concept of coevolution, and this realization suggested many different tests of coevolutionary hypotheses that often turned out to be fruitful.

For me, the close relationship between the lustrous black *Onagrandrena* bees and the species of Onagraceae from which they gathered pollen was a perfect example of coevolution. I also remembered the bright colors of the insects that I had observed feeding on milkweeds when I visited Yosemite Valley with my parents and recognized another product of coevolution. Some insects had developed the ability not only to tolerate the poisons of food plants, but also to retain them in their own bodies and in this way become toxic to predators. With their bright colors, they were "advertising" their toxicity to other animals that might want to eat them, leading those predators to avoid them. Quite possibly, another coevolutionary "arms race" was going on between these colorful herbivorous insects and those predators.

After several months of intensive library review, Paul and I published a paper entitled "Butterflies and Plants: A Study in Coevolution," in the journal *Evolution* in 1965. As is not unusual in science, an idea first raised in casual conversation had led to the discovery of a key scientific concept.

ℰℐ

Dennis Breedlove enrolled in the biology Ph.D. program at the same time I joined the Stanford faculty and, as planned, became my first graduate student. We thought that he might study some group of Onagraceae for his dissertation. Soon, a happy coincidence gave rise to an excellent option for him and an opportunity for me to deepen my knowledge of a new part of the world and of folk taxonomy, the study of vernacular (folk names rather than scientific ones) naming systems for plants and animals.

Dennis had become friends with O. Brent Berlin, a graduate student working with Professor A. Kimball Romney in Stanford's Department of Anthropology. Romney and Berlin wanted to work out the plant-naming systems used by some of the indigenous peoples living in the highlands of Chiapas, the southernmost of Mexico's thirty-two states, and to use this knowledge to better understand the philosophy of naming organisms generally. Professor Harold Conklin of Yale University—a pioneer in ethnoscience, the study of indigenous ways of understanding and knowing the world—had carried out a similar study a few years earlier, with the Hanunoo people in the Philippines. Berlin, with Romney behind him, was anxious to examine the similarities between Conklin's findings and what they themselves would discover in Chiapas. They chose to focus on Tzeltal-speaking people (one of the region's many Mayan ethnic groups) in the municipality of Tenejapa. Historically isolated, these and nearby peoples had come into closer contact with the outside world than ever before after Mexico completed its portion of the Pan-American Highway project in 1950.

Naturally, the anthropologists couldn't carry out the work properly without a botanist, and Dennis Breedlove was their choice. After passing his qualifying examinations at Stanford he moved to the cultural capital of Chiapas, the historic city of San Cristóbal de las Casas. There, over the course of three years, he did much of the project's first-stage work, learning the language with admirable speed and logging many hours of interviews with the Tenejapans, as they are called. It was painstaking work: asking the Tenejapans the names of individual plant samples, making numerous collections of each kind of plant in the designated area as vouchers for individual "naming events," and so on.

Once Dennis got established in Chiapas, a tidal wave of specimens began to flow northward. My job was to determine the scientific names for the different species, even though I knew next to nothing about the flora from this locality when I started. Cases of plant specimens from the project began to accumulate at Stanford, and I parceled them out to specialists as well as I could. Particularly helpful in this enterprise was Dr. Jerzy Rzedowski, one of the leading authorities on Mexican plants. Rzedowski, a Jew, had been born in Poland and imprisoned in a concentration camp during World War II. He had immigrated to Mexico after the camp was liberated by the Allies, earned his bachelor's, master's, and doctoral degrees there, and become a naturalized Mexican citizen in 1955.

All of us working on the project had to be careful not to assume that the Tenejapans gave and used names in the same ways and with the same frequency that we did. Folk taxonomic studies suggested that peoples without a written language could consistently keep no more than a few hundred names for a particular group of organisms, like plants, in mind. Names that are not frequently used are lost over time. Dozens of names might be associated with a plant such as corn that has many useful varieties, while a group of plants with less usefulness might be known under a single name, with distinctions among varieties often recognized but remaining unnamed. More inclusive, hierarchical categories like genera and families were generally not given names by people lacking a written language, even though they were often recognized as groups of similar species.

The short trips I took to Chiapas to work with Dennis were times of pure joy. Dennis and his family lived in a traditional Mexican house in which the dwelling, working, and storage rooms surrounded a central courtyard. The first time I visited, they welcomed my arrival by roasting a goat in a pit dug for the purpose. Waking up early the next morning, I couldn't believe the beauty of the cityscape: the homes of San Cristóbal de las Casas, with their red tile rooftops, sprawled over the hillsides, puffs of smoke issuing

When beginning our ethnobotanical studies in the highlands of the state of Chiapas, Mexico, I was charmed by the venerable town of San Cristóbal de las Casas.

from their chimneys. Market days in San Cristóbal, when Mayan people came from many different settlements in the highlands to sell their wares, were another visual delight. The extraordinary textiles, the vibrantly colored fruits and other wares, and the many scents and sounds came together in a memorable spectacle.

Dennis thoroughly explored and collected in the high-elevation pine-oak forests around San Cristóbal de las Casas, and I was able to join him on occasion. He also often visited lower elevations in other parts of the state to collect plants there, and gradually built up the knowledge of the botany of Chiapas as a whole, ultimately starting to publish a state flora.

*

In the summer after my first academic year at Stanford, I traveled with Bob Ornduff, my close friend from Berkeley days, on a lengthy collecting trip through the American Southeast. I targeted the region because most of the American species of water primroses, *Ludwigia*, were concentrated there, and I wanted to study them in their native ranges. I hoped this fieldwork would allow me to develop my understanding of their evolutionary relationships and fit them better into the context of the rest of the Onagraceae. Bob, who had earned his Ph.D. at Berkeley under Herbert Mason and joined the biology faculty at Duke University, was working at the time on the breeding systems of several other groups of Southern plants.

We each had a list of localities where the particular plants that interested us were growing—*Oxalis* (sorrels), *Pontederia* (pickerel weeds), and

Bob Ornduff, fellow student at Berkeley, and I enjoyed a long road trip around the South in 1963.

Gelsemium (Carolina jasmines) for Bob, roughly two dozen species of *Ludwigia* for me—and we plotted our route so as to include those places. On arriving in North Carolina to meet Bob, I immediately felt assaulted by the unaccustomed heat and sultry humidity of the South in August. As the trip got off to its start, I was also shocked by the poverty of so many African American communities through which we passed and the evident segregation of the region. It became immediately clear that this was a culture very different from that in which I had grown up.

Bob and I took a southwesterly route from North Carolina through South Carolina, Georgia, Alabama, Mississippi, and Louisiana. Although the unaccustomed segregation was obvious and disturbing, there was also much to love about the South, from its haunting landscapes to its tight-knit communities. Every town seemed to have its own little restaurant with delectable local specialties, and the friendly Southern culture of ease and hospitality appealed to me a great deal.

Reaching New Orleans, Bob and I drove directly to Tulane University, where a close friend from our Berkeley days, Don Stone, had become a member of the faculty. We caught up on news and Don took us to a late-night bar for more visiting and drinks. At Tulane we met Joseph Ewan and his wife Nesta. Joe had grown up and spent his early professional career in California and come to Tulane in 1944. He had assembled a private collection of thousands of books and letters written by and about early American naturalists and explorers. Using these materials, Joe Ewan had, through his own publications, played a major role in establishing the history of natural history in America as a subject of scholarship. Though meeting him in person was a delight, its larger significance would not emerge until many years later.

Circling back to the east from New Orleans, Bob and I followed the Gulf coast back across southern Mississippi and Alabama to Florida, where we spent most of our remaining time crisscrossing the peninsula in our search for plants. We passed through miles and miles of orange groves and cattle ranches (some in places where Orlando and its theme parks now sprawl). We especially enjoyed the Everglades, where the waterways and subtropical vegetation seemed to stretch without limit and, at the southern end, crocodiles joined the ubiquitous alligators.

Nearly everywhere we went in Florida, we saw signs over drinking fountains, restrooms, lunch counters, and swimming pools saying "Colored" and "White." I asked Bob if the prevalence of the signage relative to Alabama indicated a greater degree of segregation in Florida. "No, Florida has tourists and many relative newcomers compared to the rest of the South," he said in response. "The signs are for outsiders. In central Alabama, they don't need signs; everybody just knows."

Heading back north, our last major stop was to visit Professor Robert K. Godfrey and his students at Florida State University in Tallahassee. Godfrey was the longtime curator of the Florida State University herbarium (now named in his honor) and an outstanding plant collector with an expertise in the flora of northern Florida and the adjacent areas of Georgia and Alabama. In a pleasant and productive outing, he and his student Andy Clewell drove us to various locations near Tallahassee to find the plants we were seeking. Soon after, I flew back to the Bay Area.

Home on the Stanford campus and settling into the new academic year, I was crossing the street to the Quad on November 22, 1963, when a distraught colleague approached me. "Kennedy's been shot!" he cried out. My first thought was that he was talking about Donald Kennedy, head of the biology department. But of course it was John F. Kennedy, the president, a reality that seemed unthinkable. The weeks that followed were a blur of grief, disbelief, and confusion. America leaned on the comforting presence of Walter Cronkite, a true "anchorman" who helped comfort a mourning nation as President Kennedy was laid to rest. Sally and I were watching live television when Jack Ruby shot Kennedy's alleged assassin, Lee Harvey Oswald. We wondered what would happen to America, not knowing yet that Malcolm X, Martin Luther King Jr., and Robert F. Kennedy would all die by assassination within a few short years. Stanford's first political protest of that era—a quiet vigil protesting the campus's newly stocked fallout shelters—also occurred in 1963, ushering in what would prove to be a decade of often contentious demonstrations.

From late May to early July of 1964, I went on an extended collecting foray with David Gregory to Texas and the Southwest. Dave and I had met at Rancho Santa Ana while he was completing his doctorate in botany under the direction of Phil Munz. Our trip, which covered thousands of miles, was focused on the genus *Gaura* (Onagraceae; then considered a separate group but now included in the genus *Oenothera*). Dave smoked cigars, and our car soon developed its own distinctive stale aroma as we drove on day after day. We crisscrossed the state of Texas, staying in motels, eating delicious local TexMex meals. We listened to the radio constantly, entertained by the top-40 hits, a rich sampling of country and western music, and the Beatles, who had famously made their American debut on the Ed Sullivan show earlier that year. I sent many postcards home to let the children know what we were up to and that I missed them and Sally a lot.

We aimed to measure the features of *Gaura* populations to determine the degree to which their variation patterns overlapped and study the insects visiting and pollinating them in the mornings and evenings. Dave and I gathered copious *Gaura* material for chromosome studies and worked hard to interpret their pollination mechanisms. In all species of the group, each individual flower stays open for less than twenty-four hours. In most of them, the flowers open near sunset and give off a sweet odor. But a few of them, we found, had changed in the course of their evolution to open in the morning. These included *Oenothera (Gaura) lindheimeri*, a plant native to the black-soil prairies of east Texas that has since become a familiar and much-appreciated garden plant. In and around the state of Arkansas we were surprised to find another set of populations with flowers that also open in the morning. I would later name this entity *Gaura demareei* after Delzie Demaree, a Stanford graduate from the 1930s who later moved to Arkansas and became an expert on its plants. The ways in which these entities had evolved were of great interest to us; in the course of studying them, we also took the opportunity to collect and make observations on the other Onagraceae that we encountered.

We found that the evening-opening species of *Gaura* were visited and pollinated by moths and other nocturnal insects, with the soon-wilted flowers that remained the next morning sporadically visited by bees for their residual pollen. For the morning-opening species, in contrast, bees were the predominant pollinators. It seemed likely that in the evolutionary past, in situations where bees had been relatively consistent visitors to the wilted morning flowers, there would have been an advantage to those plants to switch to morning opening. In consequence, wholly morning-opening populations would have developed and then tended, over time, to become distinct in other features from the evening-opening ones. This turned out to resemble, then, the kind of coevolutionary pattern that I had described earlier for *Chylismia*.

As my second academic year at Stanford came to its conclusion, I was busily pursuing research projects concerning various Onagraceae, learning about the folk taxonomy of the Mayan natives of southern Mexico, teaching, and thoroughly enjoying it all. I still hadn't learned to pay nearly enough attention to my family, but for me it was a happy time, one of growth and exciting adventures, learning, and expanding horizons. I awakened happy every day and looked forward to the future with enthusiasm.

– 7 –

Expansion and Changes

AFTER TWO ACADEMIC YEARS at Stanford, it was time to focus on those activities that were most likely to allow me to secure tenure in the coming years: continue to teach and do research, publish additional papers and monographs, win additional grants, and take more graduate students under my wing. Or at least that was the expectation of my colleagues and some of the strongest voices in my head. Other voices (speaking from my heart, perhaps) weren't absolutely convinced. That is not to say that I was disillusioned or unhappy with the academic world. My activities as a Stanford professor brought me great satisfaction, and I was enthusiastic about my research and teaching. But after the 1963–64 academic year, I began to experience some uncertainty about my life and career path. As the 1960s wore on, I increasingly questioned the meaning of the kind of scientific work that I was doing. The feeling that I wanted something bigger kept nudging me, gently but persistently, toward a reimagining of what I wanted to do and accomplish in the time I would have on earth.

A good part of this restlessness, if that's a good word for it, was internal, part of the ambition instilled in me by my mother. Combined with my outgoing nature—and, I would have to admit, my underlying insecurity and need for approval—this restlessness often caused me to long for situations and roles in which I could do more to lead and inspire others, to connect people through networking for common causes, and to exert a recognized influence on something important in the world. But another, probably more significant influence on my state of mind during this period was what was going on in our society and my day-to-day world.

The sixties, as everyone knows, were a time of dramatic social change, unrest, violence, and questioning of fundamental values. After a long period of gradually increasing involvement, in 1965 America entered the war in Vietnam fully with Operation Rolling Thunder. It was traumatic for all Americans to see the daily toll of war dead on the evening news, watch television footage of protests and riots, and continue to live under the threat of nuclear war. For those of us teaching at many of the nation's colleges and universities, the turbulence was magnified. As the sixties progressed, college campuses became hotbeds of protest and questioning of the status quo. Stanford was among the schools at the forefront of it all, with groups of students protesting the Vietnam War, racial inequality, defense research, and the disproportionate representation of young men from working-class backgrounds in the armed forces. Our students were the very harbingers of change and liberation. They burned their bras or draft cards, waved daisies, renounced The Man, and ushered in the Age of Aquarius.

Unrest and counter-culture expression at Stanford grew stronger throughout the sixties. One of the signs of this was the widespread appearance of flowing beards and long hair, with a wildly colorful mix of personal apparel. In the larger classes that I taught, it was not unusual for a dog to wander in and out of the classroom, appearing to listen attentively with eyes bright and ears perked up. It was more than I could say for some of the students. In class, students would read newspapers or secretively nurse joints of marijuana, and all of them felt free to come and go when they wished. A number of the students sported balloons tied to their big toes. Fortunately, *some* of them were seriously interested in biology. Attentively following what I had to say, they took notes carefully and asked informed questions. Since I was sympathetic to the war protests, it was difficult to know what to do, but I tried to err on the side of order. All in all, it was a great time to learn an important life lesson: appearances often tell little or nothing about what a person is really like.

The students who filled the seats in my classrooms disturbed, dazed, and delighted me by turns. When they weren't drugged out and dragging their culture down with their irresponsibility, they were brightening and improving it by questioning everything. They scorned their government but also mourned its leaders as they fell to assassins' bullets one by one. Some thought deeply and searchingly while others, retreating from the frightening intensity of the times, sought oblivion in "weed" or other drugs. The times were definitely "a changing," as Bob Dylan said, and no one knew where it would lead.

During these years, I personally visited the fringes of the world described so engagingly and accurately by Joan Didion, who was a year ahead of me at Berkeley, in her classic essay, "Slouching towards Bethlehem." Not even ten years older than many of my students, I felt that I had a foot in both sides of the generational divide. Although I remained firmly rooted in the world of the academic and continued to teach, con-

The late 1960s were marked by extensive student activism on the Stanford campus, as throughout the country. On April 18, 1969, these students were voting to end their eight-day occupation of the Applied Electronics Laboratory, the campus building where the most military research was being conducted at that time.

duct research, write papers, and move forward with my other scientific work, I also felt kinship with the young people around me, sharing many of their concerns. Their lifestyles, promising relief from stress and hectic schedules, proved very tempting. I didn't take part in any demonstrations, but you don't have to light matches to be affected by a bonfire. As the changes of the 1960s really took hold, the unrest erupting all around me had a disturbing and powerful effect. Watching everything I thought I knew being challenged and even overturned, I truly didn't know where the world was going next or what would happen to me, our country, or our society. Dedicating my intellectual energy to investigating the evolutionary relationships among members of the Onagraceae could be a welcome retreat from current events, but it could also prompt internal questions about relevance and meaning in the world at large.

During this time, inward doubts and confusion combined with a day-to-day experience of societal turmoil and uncertainty to create a fertile seedbed for self-reflection and change. In seeking to resolve inner struggles, some people need dramatic changes in their circumstances and sharp breaks with the past. In my case it was more a matter of gradually re-fashioning my life without major disruption, often without full awareness of the ultimate consequences. That was the case for the scope and scale of my work in botany.

Up to this point in my career, I had been involved particularly and intensively in the study of Onagraceae. At the same time, I had continued to follow Tom Howell's lead and put together floras. Most of them involved just myself or a few others and focused on relatively small areas: San Francisco, the Santa Monica Mountains, Napa County, and San Clemente Island. As the summer of 1964 drew to a close, however, I was attracted by the idea of a flora project covering a much larger geographic area and requiring collaboration among hundreds of botanists from throughout the world. It would turn out to be the first of several such major projects I would eventually help to inspire and carry out, and was therefore truly a harbinger of the future.

The occasion that fired my imagination was the Tenth International Botanical Congress in Edinburgh (1964), my second such meeting. As I prepared to travel to Scotland—my first visit there—I was pursuing several objectives in the family Onagraceae and writing up some of the results of our ethnobotanical project in highland Chiapas. Excited about those projects as well as the concept of coevolution, I was anxious to tell others

It was through Stan Shetler's hard work and organizational ability that the Flora of North America project was launched in 1964.

about them and just as eager to hear what everyone else was doing. Like most professional conventions, the Congress was a great place to connect, meet other botanists, and explore opportunities for future collaboration. Edinburgh was the first international meeting I attended as an independent professional, out on my own and hungry, as always, for fresh inspiration.

At the Congress, the big news was that the first volume of *Flora Europaea*, an encyclopedic account of all the plants of Europe, had been completed. Having prepared the account of Onagraceae for that work while I was in London, I was well aware of the effort, and like other Congress participants, I was excited that its first volume was about to appear. The idea of so many diverse and talented botanists in Europe and beyond joining together to produce this major reference was deeply inspiring. Detailed studies comparing a few species of plants in a particular group with one another, perhaps attempting to reconstruct their evolutionary past, were important, but they were laborious and at best could deal only with a few of the plant groups that occurred in any given region. The appearance of the first volume of the *Flora Europaea* affirmed that the botanists of the world also needed to produce broader works so that the entire world's flora could be understood fully.

Several North American botanists reacted to the news of the first volume's completion with calls to action. Right after the Congress, Áskell Löve, an Icelandic botanist working in Canada, wrote to the American Society of Plant Taxonomists (ASPT), urging that it undertake an effort similar to the *Flora Europaea* for the plants of North America. During the same period, my friend Stan Shetler, a member of the Department of Botany at the Smithsonian Institution, began exploring ways that we might be able to approach such a task.

Shetler and I began talking about the possibilities of a North American flora during the following winter. At its annual meeting in August 1965, the ASPT Council appointed an eleven-person committee to explore the matter. Members came from throughout the U.S., Canada, and Mexico; Bob Thorne, whom I had known in England, served as chair. Our committee held its first open meeting at the Smithsonian in May 1966. We decided that we didn't dare include Mexico, which was home to many more species of plants than the U.S. and Canada combined; its plants were also not nearly as well-known at the time. After the meeting, we shared our recommendations with the ASPT Council. The preparation of a flora of the U.S. and Canada was both desirable and feasible, we reported, and it should follow the style and presentation of the *Flora Europaea* as closely as possible. Several later meetings would further develop the scope of the project and the way in which it would be produced. I knew that I was signing on to a big, long-term project that would take tremendous and sustained effort over many years.

When I returned home to start my third academic year at Stanford, I jumped back in to the many projects I had going on. Primary among these was pursuing the evolutionary and systematic relationships of the main groups of Onagraceae I felt needed to be "solved": various genera of the tribe Onagreae, for which I took field trips up and down the Coast Ranges of California; *Gaura*, as described; *Epilobium*, of which we had a number of samples in cultivation in the greenhouses for hybridization; and *Ludwigia*, which I was cultivating extensively in the greenhouses for chromosome counts and other observations. Revising them systematically, we would have the best possible chances of recognizing the evolutionary units within each of them and the ways in which their individual patterns of variation compared with one another.

The study of *Epilobium* was expedited, and ultimately directed, by graduate students Steven Seavey and Peter Hoch, both of whom became close friends. For *Gaura*, a special passion, I continued to work with David Gregory. The pollination systems in Onagraceae were so diverse that we found surprise after surprise, and then needed additional fieldwork to verify our first observations. In addition to Dennis Breedlove, graduate students who worked with me during this time included Howard Towner, Cliff Schmidt, and Valerie Chase; she studied hybridization and pollination in the columbines of the Sierra Nevada.

My *Flora of San Clemente Island, California* was published in 1963, reflecting many happy days on that goat-infested island. The *Flora of the Santa Monica Mountains, California*, which I had begun with Harry

1. On frequent trips to Golden Gate Park, usually with my father, I got my first taste of nature.

2. My first forays to see insects were to our own sandy backyard, built on the former dunes in San Francisco's Richmond District. There, the four stages of the life cycle of the cabbage butterfly, *Pieris rapae*, were readily available to my inquisitive seven-year-old eyes. Here, the egg, larva, pupa, and adult butterfly.

3. Gorgeous mourning cloaks, *Nymphalis antiopa*, provided real excitement during their rare flights across the back of our garden, where they easily eluded my butterfly net.

4. Jerusalem crickets, *Stenopelmatus*, are large, "squishy," flightless relatives of grasshoppers and other crickets. I could observe them but had no idea how to make them into specimens for my collection.

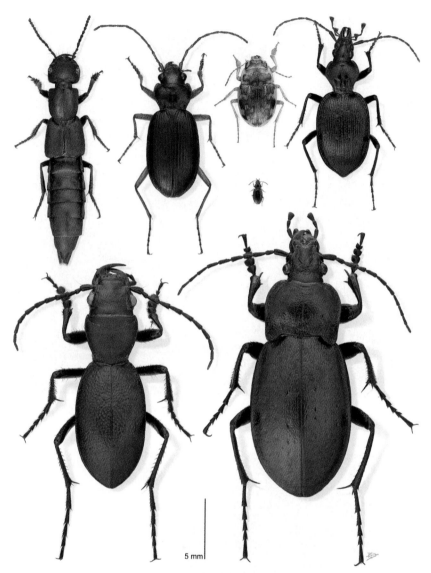

5. With their variety of habitats and characteristics, these beetles fascinated me when I was young. All but *Tasgius ater*, the rove beetle in the upper left corner, are predaceous ground beetles, Carabidae. The rove beetle and *Carabus nemoralis*, in the lower right corner, both introduced from Europe, lived under the pots in a nursery just down the street from our house on 12th Avenue. Next to the rove beetle is *Nebria eschscholtzii*, discovered by the Russians two centuries ago; living under rocks along creeks in the Santa Cruz Mountains, it scampered away to elude capture. *Omophron dentatum* burrowed in sand on the edges of Mountain Lake in the Presidio, whereas the slender-necked *Scaphinotus interruptus* hunted snails on the floor of the redwood forest. Between the two rows is a tiny species of *Tachys*, which finds its minute prey under the bark of dead trees where it lives. At lower left, *Omus californicus* burrowed in sand cliffs near Playland at the Beach, coming out to hunt its prey at night. Karolyn Darrow of the U.S. National Museum of Natural History prepared this painting especially for use here.

6. I vividly remember my excitement at my first sight of the beautiful day-flying moth *Pseudohazis eglanterina*, whizzing low across relatively open hillsides near Zayante.

7. Many organisms fascinated me on my visit to Zayante Creek in the Santa Cruz Mountains. This massive long-horned beetle, *Prionus californicus*, buzzed in to the lights in our cabin and slammed clumsily to the floor, where it was easily captured.

8. A flowering stalk of coltsfoot, *Petasites frigidus* var. *palmatus*, sprouting up from moist soil at the edge of Zayante Creek in the Santa Cruz Mountains of Central California.

9. Vivid blue like the bluing used on clothes, the berries of this lily, *Clintonia andrewsiana*, shone in the undergrowth at the forest's edge.

10. The most beautiful of the lupines, the annual *Lupinus stiversii*, grew on sandy slopes near Angel Falls above Bass Lake.

11. Stream orchid, *Epipactis gigantea*, fringed Angel Falls. I was thrilled with the beauty of one the relatively few species of native orchids in California.

12. The spectacular golden eardrops, *Ehrendorferia chrysantha*, characteristically grows on burned slopes.

13. Many years would pass before I fully understood the significance of the bright red colors of the insects I found feeding on milkweeds at Camp Curry in Yosemite Valley. They included *Tetraopes*, the milkweed beetle, and *Oncopeltus*, the milkweed bug.

14. Another bright-colored insect, with caterpillars that feed on milkweeds, is the monarch butterfly, *Danaus plexippus*. This insect would yield important insights for my later studies.

15. When at the age of twelve I first experienced the unimaginable beauty of one of California's wildflower super blooms, it left a permanent impression on me.

16. The greatest evolutionary botanist of the twentieth century, G. Ledyard Stebbins, whom I met at the age of fourteen, remained an encouraging friend and mentor for the rest of his life—just short of a half century.

17. As was the case for the clarkia, I rediscovered the single clump of this unique manzanita, *Arctostaphylos montana* subsp. *ravenii*, some forty-five years after its only other collection.

18. This cartoon of my re-discovery of the single living plant of *Arctostaphylos montana* subsp. *ravenii* in San Francisco's Presidio was prepared for the Golden Gate National Recreation Area visitor center.

19. I found this species of farewell-to-spring, *Clarkia franciscana*, on serpentine slopes in the Presidio when I was sixteen years old. Learning about it and its relatives helped lead me to a career in botany that focused in part on its family, Onagraceae.

20. When the gorgeous, dark blue-flowered sky pilot, *Polemonium eximium*, occurs with alpine gold (right, #21) they make a beautiful display.

21. Following an invitation by J. T. Howell to collect plants on a Sierra Club Base Camp outing in 1950, six further annual Base Camp experiences introduced me to the beautiful plants that grew at high elevations in the Sierra Nevada. The bright yellow heads of alpine gold, *Hulsea algida*, adorn slopes above timberline along the length of the range.

22. I loved the high mountains and had wonderful times on my Sierra Club outings there from 1950 to 1956. Here I was eighteen years old (1954), teaching on a sandy flat south of Mount Whitney.

23. *Astragalus ravenii,* which I encountered at the top of Sawmill Pass in the high Sierra Nevada, proved to be new to science.

24. Although halted by the flooding Santo Domingo River on our spring trip to northern Baja California in 1958, Harlan Lewis (right), the director of my graduate studies at UCLA, Harry Thompson (center), a member of my Ph.D. committee who was a close friend and colleague for many years, and I had a wonderful time enjoying the plants that beautiful spring.

25. The Bay checkerspot butterfly, *Euphydryas editha bayensis*, which Paul Ehrlich and his group studied in detail for a number of years.

26. A caterpillar of the Bay checkerspot butterfly, whose feeding habits prompted Paul Ehrlich and me to carry out the investigations that eventually led to our formulation of the theory of coevolution.

27. Helen Kennedy (top) and me (center), with the spectacular mountain herb, *Gunnera insignis*, on the OTS course in Costa Rica, August 1967.

Thompson while I was a graduate student, had been completed, and we looked forward to its publication in 1966. A number of papers on the folk taxonomy work in Chiapas were published around this time, with an especially important one in *Science* in 1966.

&

Plant collecting was still a passion for me in the mid-1960s, as well as an essential part of my research, and so I continued to go out into the field whenever possible. The summer of 1965, however, brought an adventure that showed me that fieldwork and collecting could have their challenging moments. In August of 1965, I was on the road, studying pollination and population variation in *Gaura*. I had started the summer's work by myself, sampling populations of *Gaura* in Texas and neighboring states. Later in the trip, I met up with Bob Ornduff, who by that time had returned as a faculty member to Berkeley, the place I had first gotten to know him as a fellow student. We began driving through the Carolinas together, soon finding populations of the plants we were seeking specifically and many others of interest, too.

One of the plant species that I especially wanted to study was then called *Gaura filipes*; it grows on sandy soil in South Carolina and some of the neighboring states and has flowers that open in the evening. Near Orangeburg, South Carolina, we found a population of it growing in narrow sandy fields along the highway, in an area where peanuts were cultivated in small patches. Wanting to study the pollination of the *Gaura* both in the evening and early the next morning, we knocked on the door of a nearby farmhouse. We established that the field did indeed belong to the farmer who opened the door and asked his permission to collect insects on the plants that evening and the following morning, explaining that we would have to be there at 5:30 a.m. when the insects start coming. Not seeming to care very much, he said yes.

After collecting a number of the moths visiting the *Gaura* flowers that evening, we returned to our motel. The next morning, Bob decided to sleep in after dropping me off at the selected site. I wanted to see what insects might be active at the withering *Gaura* flowers. As I collected bees and other insects with my butterfly net, a car drove up and parked. Another car arrived, and then others. Eventually this group of more than a dozen locals—including the sheriff and the same farmer from whom we had obtained permission for our work the night before—walked over and confronted me, forming a semicircle with the sheriff, polished badge shining on his chest and shotgun at the ready, at the center of the group.

As I scanned their stern faces, I felt a hot stab of anxiety. It was just one short year after Freedom Summer, when three young men, two of them white, had been murdered in Mississippi while registering African American voters. Although I was quietly and non-controversially collecting bees, with permission, I couldn't be sure what these folks saw in my presence. Since I was sporting a beard and relatively long hair, I realized that I probably fit very well their image of a hippie. It was a period when the unimaginable could happen, and everyone was on edge. And I had no car.

"What you here for, boy? Where did you spend the night?" demanded the sheriff. My responses didn't satisfy him in the least. Despite showing my ID and appealing to the farmer who had granted me permission to be there—he claimed that he'd never seen me before—I was handcuffed, ushered into a police car, and locked in a cell at the local jail. Fortunately for me, Bob had awakened and started to look for me, becoming increasingly worried; when he didn't find me easily, he feared that I might have been the victim of foul play. Eventually, he and the sheriff encountered each other and Bob was able to clear my name. On releasing me from jail without filing a charge, the sheriff growled, "Don't stop again in this county, boys—I never want to see you again." Needless to say, we took his words very seriously.

Our days collecting together completed, I dropped Bob off at the airport and continued on alone, being *really* sure that I had permission to collect where necessary. Though the Orangeburg incident was seriously unnerving, the trip itself was outstanding for our botanical discoveries. Bob and I collected a number of species in localities from which they hadn't previously been recorded, and we advanced our studies with the new material.

Thereafter, memories of the confrontation in South Carolina often caused me to quake in my boots a bit when I visited remote localities. Back in California, I drove out early one morning to the sand dunes at Bodega Bay, along the coast north of San Francisco. I wanted to see if *Onagrandrena* would be visiting the bright yellow flowers of the dune primrose (*Camissoniopsis cheiranthifolia*) there, since I had observed them visiting flowers of the same species in populations farther south. Driving out among the dunes, I pulled up next to an old house in the predawn darkness. As I collected the bees (which did include a species of *Onagrandrena* as I had hoped), cars began to pull up nearby, a dozen or more in total. Was it a reprise of Orangeburg? Nervously, I went up and asked someone what was going on. The reply came, "Didn't you know? This is the house where Alfred Hitchcock filmed *The Birds*!" If Orangeburg had been a lesson in taking care, this was perhaps a lesson in not taking it all too seriously.

℃

In the middle of the decade, in June of 1966, I turned thirty, passing the "too old to be trusted" benchmark of America's deeply youth-oriented counterculture. As a husband and a father, I was not quite free to jump into the fray, yet I was young enough and in enough emotional turmoil for such a leap to be profoundly tempting. The youth culture that surged around me seemed to promise respite from the hectic press of all the studying, writing papers, discovering new plants, traveling, and giving talks, and I was ever aware that my marriage was crumbling and I was doing little to fix it. One day during this time, Sally and I drove up from Palo Alto into San Francisco to see *Man of La Mancha*, a musical that grew out of the experimental theater of the 1960s. I remember that as one of our last happy evenings amid our marriage's slow decline. We didn't yet openly acknowledge the inevitability of our split, even to each other, but a sense of sadness and uncertainty hung over us.

In the midst of this troubling period came an exciting opportunity to visit the tropics once again. Tom Emmel, one of Paul Ehrlich's graduate students, asked whether I would be interested in teaching part of a course on tropical ecology in Costa Rica the following summer. The course was offered by the Organization for Tropical Studies (OTS), a consortium of American and Costa Rican universities organized three years earlier to offer instruction in tropical biology. Mildred Mathias had been one of the founders, but it didn't take her imprimatur to make the prospect exciting to me. I had enjoyed and learned greatly from my earlier trips to Colombia and Chiapas, continued to sense the importance of the tropics to the planet's health, and was thrilled about the very rich plant life that I would see. Though somewhat smaller than West Virginia in size, Costa Rica was home to about four times as many native plant species, some 10,000 of them.

I made a preliminary visit in March 1967 with Tom Emmel and Roy McDiarmid, a genial student of reptiles and amphibians Emmel had selected to help him organize our course. The visit familiarized me with the field sites to which we would bring students that summer. I collected plants widely and savored the lush and verdant vegetation as well as the beauty of San José, Costa Rica's colonial capital. I loved the exuberance of the people, reflected in the national slogan "¡*pura vida*!" connoting life lived richly and fully.

With Roy McDiarmid, I went to look for an unusual member of the Onagraceae, the very distinctive genus *Hauya*, one of the few trees in the family. It has large white flowers that, as we learned on this occasion, open near sunset. In the beautiful colonial city of Cartago, we were able to watch not only the expected hawkmoths but also bats pollinate *Hauya*. We went on to visit the OTS field site at La Pacífica, where the dry forest was alive with howler monkeys, and ecologist Les Holdridge's boarded-up house at Finca la Selva, in the rainforests on the country's Pacific lowlands. Later, the OTS would acquire La Selva and develop it into one of the world's major tropical field study sites. For me it was remarkable enough even then. Every day brought stunning new sights and ideas, and I savored every moment.

This trip was my first opportunity to witness Dan Janzen in the field. He was totally immersed in the rich world of tropical biology around him, teaching students and constantly looking for new ways to interpret what he was seeing. Still in the early part of his career in 1967, Dan would go on to become perhaps the most original and influential tropical biologist of his time.

Coming home musing on the similarities in the plants of Costa Rica and Chiapas, I began to consider the possibility of treating all the plants from the Isthmus of Tehuantepec in Mexico through Panama in one comprehensive work. Getting involved in the Flora of North America project had definitely set my sights much higher when it came to cataloging plants. What was to become the *Flora Mesoamericana* wouldn't get started until the 1970s, but the seeds were sown right then.

In August, I was back in Costa Rica for my month of teaching. The eight-week "Fundamentals of Tropical Ecology" course was held in July and August for about twenty graduate students. A bright and dedicated group, the students included Gary Hartshorn, who went on to become a noted forester; Tamra Engelhorn from UCLA, of whom Harlan Lewis had spoken highly; and Helen Kennedy. We would assign exercises to the students each morning, to be reported on and graded later that day. I shared the botanical instruction with noted ecologist Rex Daubenmeier of Washington State University, one of the founders of modern ecological studies in the U.S., who took care of the first month of the course before I arrived.

Our first field stop with the class was a twelve-day stay camping near Rincón on the Osa Peninsula. The whole area was covered in dense tropical lowland rainforest, a fairyland of biological diversity and a marvelous place to teach. The forest was flat and open underneath, a bit disorienting once one entered under the trees. One afternoon, Tom and I were alarmed to realize that we were missing two students simultaneously, a worry especially in view of the venomous fer-de-lance and bushmaster snakes that were common in the forest understory. Once we had finally gathered up our flock again, we issued stern warnings about not wandering off alone.

Our course then moved inland to San Vito de Java, a town connected with the rest of the country only when the Pan-American Highway was extended over Cerro de la Muerte, the highest point on the Highway in Costa Rica, in the 1950s. On the ridge above San Vito, an English couple named Robert and Catherine Wilson had constructed a house amid the upland forest. Avid gardeners, they were gradually turning their 360 acres into what would become perhaps the most important botanical garden in the country, a place where precious remnant patches of the original vegetation are still preserved.

The cloud forests above the station were virtually intact, with only some small *milpas*, cultivated areas along the borders of a dirt road that wound its way into the distance. All the slopes running down to the Java River in the valley below were heavily clothed with forests. The fireflies, bright flowers, hummingbirds, lush vegetation: all were extraordinary, a feast for the senses. (Unfortunately, all this exists only in memory today; the whole gloriously beautiful eastern side of the Osa Peninsula and the area around San Vito were both completely deforested before another decade had passed.)

Working with the OTS that summer gave me the opportunity to become the first botanist to collect intensively around Rincón and also the first around San Vito. Back home, I sent off the 600 specimens that I had collected during that one-month trip in 1967 to the Field Museum in Chicago, then the major center for the study of Central American plants.

Waking up every day in gorgeous locations far from my daily routine in California, surrounded by others as passionate about biology as I was, gave me great joy. The special quality of that time deepened in the mutual attraction I experienced with Tamra Engelhorn. Nine years younger than me, Tamra was bright and fearless, a true "flower child" in many ways. She was a student of Hal Mooney's, who was both a good friend and one the first American scientists to develop the field of ecology. Tamra was young enough to seem an enticing link to the freedom of her generation; in her presence my personal and professional stresses seemed to ease. I was still married, but both of us left the course thinking of each other and wondering what, if anything, the future would bring.

Professionally, that month provided a productive break from the intensity of activities at Stanford. It gave my thinking process time and space to unfold and expand, sixties-style, to encompass my constantly evolving understanding of the relationships among the flora and fauna around me, and let it all interweave with my knowledge of conservation, preservation, sustainability, biological diversity, and how humans live on the Earth. My growing awareness of all these interconnections was marinating in the back of my mind as the course ended. Beginning to put the pieces of this great network together, I packed my bags and presses and headed for home.

⁂

In the months after I returned from the OTS course in Costa Rica, the world often seemed to be unraveling. The social mores and standards that we had taken for granted growing up were disappearing, and it was becoming more difficult to know what we could really depend on in our society. During that period I pursued various scientific objectives; made a trip to Chile with Hal Mooney and some of his UCLA graduate students; visited St. Louis for the Systematics Symposium; and pushed on as if by rote. My thoughts were ranging widely, and I was having an increasingly difficult time imagining where things were all headed. Perhaps my Irish heritage kept me on track; in any case, I had adopted a strong emphasis on specific objectives and a focus on bringing them to conclusion. I was eager to be successful in my career no matter what was going on externally.

As I look back on this time, however, I feel a knot in my stomach. I think of my daughters Alice and Liz, for whom I was a very distracted father, and especially of Sally. She deserved better than a marriage that drifted off into nothing, mainly through lack of attention on my part. By late 1967, things had fallen apart for us. We separated, our daughters remaining with Sally.

Even within the confines of my academic work, things were changing, sometimes in bittersweet fashion. In September 1967, at the age of thirty-one, I pressed my plant specimen 22,117 in Louisiana, at the end of a long field trip. Its number signified the total number of collections I had made over the past twenty years. Although I would collect additional plant specimens in the future, that one came to represent the end of my intensive and regular collecting in the field. Collections thereafter would be infrequent, sporadic, and, with one notable exception, not part of any concerted effort to know the plants of a certain area.

What happened to cause an inveterate, compulsive collector of plants to slow down so drastically on an activity that had offered so much pleasure and meaning? In part, my schedule had become too demanding to make collecting a regular part of my life. Another factor was that I had become increasingly interested in theoretical studies, in the ways of tying together observations into useful hypotheses and thus of helping to make sense of the complex network of nature. Although I retained my zest for learning more about the Onagraceae—their past, their present relationships, and the way they differed around the world—and never lost my pleasure in collecting when I could, the direction of my life had been reset without me being fully conscious of it at the time. Operationally and intellectually, I was a different person from that time forward, intent not so much on generating basic scientific knowledge as on using it, generating frameworks for understanding its significance, and encouraging major works that sometimes involved the participation of dozens or even hundreds of other scientists.

As the 1967–68 academic year unfolded, I began spending more time with Tamra, and also with Liz and Alice. I was confused, but looked forward to each visit with Tamra, enjoyed the children, and thought of little else. I still didn't know where the rest of the world was going, politically, socially, or morally, and that bothered me greatly. But I continued simply getting on with life as it was, sometimes almost in a kind of daze, unaware of the shocking events that were soon to occur.

On February 8, 1968, at South Carolina State University, Orangeburg, police opened fire on students protesting segregation, killing three of them and injuring an additional twenty-seven. That such an event occurred in the town where I had been taken to jail three years earlier under suspicion of protesting brought home the violence of the period that the country had entered, and I had no idea of how to deal with it. Later that same month, Tamra and I took a ski trip to Mount Shasta. In short order I took a bad tumble and shattered my leg—a double compound fracture. Tamra and I traveled back home from the mountain, and I entered the hospital to care for my fracture. While in the hospital, my leg in a cast, I received devastating news: Sally had died suddenly, the cause undetermined. This tragedy was made much worse by the fact that it was our two young daughters who had found her collapsed on the kitchen floor. After frantically trying to wake up their mother as she lay dying, eight-year-old Alice summoned resourcefulness beyond her years and called the operator, who sent an ambulance. It was too late. Sally was gone.

I lay mending in the hospital, depressed by these events and worrying about the future. When my leg had healed a bit and I had been fitted with the appropriate cast, I was released from the hospital and gradually resumed my teaching duties. During this period John Cannon cheered me up by writing, "The only thing that I can imagine stranger than you with a broken leg is you on skis!" Unable to handle both work and home life, especially while hobbling about on crutches, I hired a housekeeper, Mrs. Eggerding, to cook and clean and care for Alice and Liz. The girls, of course, were devastated by their mother's death, and it didn't help that well-meaning friends had removed from our house every object that had to do with Sally, thinking it would be easier for the girls not to be reminded of her. They didn't have much use for Mrs. Eggerding nor she for them. It was a very difficult time all around.

As we suffered through our private tragedy, the nation itself continued to be tossed by storms as 1968 unfolded: Vietnam saw the bloody Tet Offensive and the infamous atrocities of the My Lai massacre; civil rights protests and riots surged, with more deadly conflicts between police and demonstrators; the Reverend Martin Luther King Jr. was assassinated, followed two months later by another crushing loss to the country when Bobby Kennedy, running strongly for the Democratic nomination for president, was also cut down by an assassin's bullet. The tension on the Stanford campus grew palpable, as it did across the nation, with rioting taking place in more than a hundred cities.

I continued seeing Tamra through the spring. She wasn't sure what she wanted to do with our relationship and took off for Europe in the summer of 1968 to think things over. She was only twenty-three at the time, and the gap in our ages seemed very large. Further, I had two children who deserved attention and love. As I worried and worried about what would happen while Tamra was in Europe, Judy Collins's version of "Both Sides Now" became a kind of theme for me. There were times when marriage seemed like a way of overcoming the pervasive uncertainty of the times, something on which I would be able to depend amid the growing confusion.

That summer, while Tamra was away, I had a brand-new kind of project on which to focus: developing a college textbook. Earlier, I barely could have imagined myself authoring a text. But after watching my colleagues Paul and Dick do it successfully, I began to wonder whether I might emulate their example. In hindsight, working on a textbook was another manifestation of my gradual movement away from doing basic science, and perhaps yet another way of diverting my attention away from all that was unresolved in my life.

Neil Patterson and Walter Meagher of Worth Publishers had approached me in the fall of 1967, wanting ideas about what the college textbook market might need in the area of the biological sciences. Worth had just published Helena Curtis's *Biology*, and it had been an immediate hit, selling more than 300,000 copies in its first edition. Walter Meagher had met Helena Curtis, who was taking a biology class at Columbia, shortly after Worth Publishers had set itself up for business, and the firm signed her up to do the writing and coordination for that first big textbook. Following upon that success, they were looking for additional promising projects. In the course of that search, they consulted me.

Among other possible projects, I suggested that a new introductory botany text might be in order. The available texts were dated and did not take into account adequately the discoveries that were piling up from molecular biology and other fields. Consequently, there was a growing tendency to abandon introductory botany for an introductory general biology course, a trend that concerned me at the time. Neil and Walter liked the idea of developing such a book and asked if I would be willing to participate in the project.

The model Worth had used so successfully with Helena Curtis involved hiring an excellent writer who was well informed about a given field but not expert in any particular part of it. The writer would then depend on specialized consultants for different areas to provide the details

that they would weave to-
gether into a text. This was the
role that Helena had played in
the development of the biology text. Thus it was natural
that they asked her to play a
similar role for their proposed
botany text, with me as the
principal consultant. In that
function, I would identify the
specialists who would provide
the basic material for the vari-

The remarkable Helena Curtis, with whom I
coauthored a botany textbook that has remained
successful, in subsequent editions, for more than
fifty years.

ous parts of the work. We decided early on that the botany text would be
titled *The Biology of Plants*, intending this title to reflect the integration of
molecular biology with botany.

After preliminary planning, Helena and I decided to begin writing
that upcoming summer. And so it was that she and I rented adjacent
cabins on Cape Cod in the summer of 1968, settling in for a ten-day period of intensive writing. We wrote during the days and visited at night.
I had great fun working with her. She was an extraordinarily lively and
nice woman about twenty years older than me, with curly hair and a twin-
kle in her eye. One night I had lobster for dinner and had to ask Helena
how to take its shell off. She pointed out that I hadn't had the slightest
trouble doing that with my lobster the night before after downing two
martinis. During those ten days and the months and months of further
work that followed, she taught me a great deal, as did the process of textbook writing itself. As time went by, it became clear that I had in effect
become the senior coauthor for the work, and this would be reflected on
the book's cover and title page when it was published in 1970.[1]

~

In the midst of all that was going on for me personally and professionally
in 1968, it is the year that I associate most strongly with my making a
commitment to concentrate on conservation and sustainability. I had
always approached collecting, systematic botany, and pollination studies
as ends in themselves, basic activities of the enterprise of science that had
inherent value because they, like similar research projects being undertaken

by others, contributed to the accumulation of knowledge about the natural world. By 1968, however, I had become convinced that doing this kind of basic research was not enough if the organisms we studied were being threatened. And threatened they were: everywhere I went, natural communities were being destroyed. I knew I had to contribute what I could to the effort to stave off the massive losses of biodiversity we were beginning to see around the world.

This realization was not an immediate conversion. Like many others, I had been deeply influenced by the environmental movement that had taken off about the time I had arrived at Stanford. In 1960, under the guidance of strong leaders including David Brower, the Sierra Club had published *This Is the American Earth*, one of the great conservation statements in American history. When Rachel Carson's book *Silent Spring*, about the deleterious effects of pesticides, appeared in 1962 it greatly raised public awareness about the problem of pollution, causing many citizens and their elected officials to take environmental issues seriously for the first time. As a botanist who frequently went out into the field to collect, I saw abundant evidence that human actions were causing widespread and destructive changes in the environment and that these changes, in turn, were affecting public health and well-being. As I got to know my Stanford colleagues Paul Ehrlich and Dick Holm, my attention was steadily and naturally drawn to problems of the environment and overpopulation in particular. The synthetic discipline of population biology that my colleagues had pioneered led them to be increasingly concerned with human population growth and its consequences on the living world that supports us all.

I enthusiastically joined them in this intellectual development. We pushed back strongly on the continued use of DDT, but soon came to focus on human population growth itself as the principal driver of the environmental changes that we were witnessing. The extremely rapid increases in population that the world was experiencing through the 1960s made it obvious that, just like every other biological population, we would hit our environmental limits. The party could not go on forever. Levels of consumption and waste were exploding upward, driven by population growth, the post-war economic boom, and people's unquenchable desire for new and better possessions. It was no wonder that wild areas were rapidly vanishing, species were disappearing, and our land, water, and air were becoming polluted and even dangerous. By the time I had visited

Costa Rica in 1967, it was obvious that natural vegetation everywhere, particularly in the tropics, was being destroyed rapidly.

So, in 1968, I was already an environmentalist. One catalyst that spurred me toward taking more of an activist stance toward conservation and sustainability came in the form of a letter I received that fall from Norman Myers.

Myers was an Englishman who had gone out to Kenya in 1958 to work for the British Foreign Service and stayed there post-independence (1963) as a photographer and consultant. Observing first-hand the destruction of wildlife habitat in Africa and steep declines in wildlife populations, he was thinking hard about biological extinction. He wrote letters to many of us to see what we thought about the problem of extinction, which he believed would have profound negative effects on us all. Most of the colleagues to whom Norman sent letters let them slide by unanswered, but for me his letter resonated with what I had already been thinking. I read the letter over, and read it again, and it all began to make sense. I began to see the way the world operates from a whole new frame of reference. I was already thinking and talking a lot about the subject of human population growth in relation to environmental damage, but now extinction began to seem a major and closely interrelated issue. I was forced to accept that species extinction was operating on a much more massive scale than I had realized earlier, and my heart sank as I thought about the problem more deeply.

If frightening, these realizations had a galvanizing effect. I began to talk with my colleagues and very soon many of us were spreading the word: the world was changing much more rapidly than most people had realized up to that time. What we hadn't quite understood was the speed with which the species of plants, animals, fungi, and microorganisms that make up the living backbone of the Earth were disappearing. Thus I started to concentrate on the process of extinction, its causes, and what we could do about it. As the problem of deforestation in particular came

Norman Myers, insightful conservationist, during a visit to Duke University in 2004.

more clearly into focus for me, I began thinking hard about what I saw happening to the tropics and what could be done to stop the destruction of tropical habitats and the inevitable loss of species that accompanied their loss. These critical issues would inform much of my work from this time forward, and I would become more strongly and actively outspoken about them as the years moved on.

I began to recognize more fully, too, the complexity of the relationships between human and natural systems. Humans were the obvious causes of ecological decline and the extinction of growing numbers of species, and humans directly felt the effects of the resulting changes; but those were only parts of the equation. Even with our population swelling, using resources at an unsustainable rate and steadily shrinking the areas of wildlands remaining were not inevitable. It mattered *how* we lived on the earth and organized our societies. Adding hugely to the rate of extinction was global inequality and the disempowerment of women. The poor, whom we were exploiting worldwide for commodities that we wanted, were rapidly destroying the ecosystems in which they lived, and women in poorer countries were having many children to work for their families and keep them afloat. The great majority of these women lacked access to education, basic health care, and effective birth control. One thing became ever clearer: the absolute necessity of equitable, sustainable human social and economic systems if there was to be any hope of long-term preservation of the rainforest or indeed of the planet. I realized that human rights, human dignity, and human respect are inextricably intertwined with biological diversity. To save biological diversity, we have first to understand that the fate—and the potential—of all people are one and the same.

꙰

In late August, 1968, I met Tamra at the Los Angeles airport on her return from Europe. Her sojourn across the Atlantic had given her the time and space to reflect on our relationship. She had decided she was ready to make a commitment. We began to talk about how and when we would get married and what that would mean for how we lived our lives. We saw no point in waiting and planned a wedding at her mother's home in San Diego for that autumn. I was overjoyed.

Hal Mooney, Tamra's major professor, was in the process of moving from UCLA to Stanford, which of course was much to our advantage.

Tamra was able to continue her work with Hal without having to live in or frequently visit Los Angeles. (She would complete her graduate studies with a master's degree a couple of years later.) For the children, the time of transition was rough. In October, during Alice's ninth birthday party, Mrs. Eggerding quit in a huff. Soon after, on November 29, Tamra and I were married, and she moved into the home in Mountain View that I had established with the girls. I was certainly happy but kept myself buried in work as much as ever and remained thoroughly confused by the agitated state of the world. Although Tamra had no experience taking care of children, I pretty much left Alice and Liz to her supervision, certainly a difficult assignment.

People across the nation were becoming increasingly disturbed by the social unrest that was gathering around the Vietnam War. In late August, dissension within the Democratic Party about the war had caused about 10,000 anti-war protesters to gather in Chicago near the site of the Democratic Convention. With police intervention, the scene had soon turned bloody and violent; tear gas was used widely to disperse the crowds, and the police assaulted peaceful demonstrators who had broken no laws, disobeyed no orders, and made no threats. American society seemed to be coming apart, riven by irreconcilable views: while great numbers opposed the war, a majority of Americans could not or would not side with the demonstrators and voted to elect Richard Nixon president in November.

From the autumn of 1968 through the spring of 1969, working on the botany textbook took substantial amounts of my time. At the same time I kept up with my students and my teaching responsibilities. I completed and published my large monograph on *Camissonia* (an artificial group of unrelated plants, as it turned out later), and kept several other scientific projects going. Despite the veneer of normality that these activities allowed, I felt that my personal life was verging on disaster. For one thing, my social world had changed. My Stanford colleagues had liked Sally well enough, but didn't quite understand Tamra. They were skeptical and more distant than they had been before. Some of them didn't approve of the fact that I had remarried so soon after Sally's death; others apparently found it difficult to relate to Tamra. Consequently, Tamra and I came to feel somewhat isolated. Tamra had two friends, Mark and Paula, who were genuine hippies, judged by any criterion, and we spent time with them when we could. This entailed more kinds of wild adventures than I would like to admit. It was all very much fun and terribly self-indulgent but did nothing to alleviate the uneasy feelings of uncertainty or help make sense of the chaos that seemed to be exploding throughout the world.

During the first part of 1969, student unrest on campuses nationally was increasing at a fast pace, intensifying the feeling that the world really was falling apart. As resistance to the Vietnam War grew, the situation at Stanford became intense. One day I reached class only by passing through two rows of chanting, protesting students, fists clenched to form a tunnel through which I had to walk. Occupations of buildings and the dismissal of demonstrations with tear gas intensified by the spring of 1969. H. Bruce Franklin—a talented, tenured professor of English who had co-founded the Revolutionary Union, a Maoist organization, and become one of the best known and most-feared critics of the Vietnam War—led a march that resulted in the occupation of the university's computer center. Around the same time, a colleague in the biology department told me that he had prepared space under the home he had built on campus to serve as a hiding place for revolutionaries.

While all this turmoil was going on, I continued to write papers for which I had completed the underlying studies over the past several years. I worked closely with Paul and Dick in various ways: a book of collected readings in evolution by the three of us was published early in 1969, and Paul and I published a paper in *Science*, "Differentiation of Populations," that to my mind pretty well demolished the so-called biological species concept that Ernst Mayr had been pushing for three decades.

During this period, I also tried my hand in a different area, one that had come to my attention while I was gathering information for the botany text. The topic was the symbiotic origin of plastids and mitochondria, microscopic organelles that perform important functions in the cells of all organisms except bacteria. Reviewing the literature and analyzing the situation in her own highly original way, Lynn Margulis, then of Boston University, had strongly advocated a neglected historical theory about the origin of these organelles. The theory proposed they had originated when bacterial cells of different kinds had been incorporated into the cells of the so-called higher organisms and begun to play important roles in their functioning. In the light of this understanding, the reason that the organelles weren't present in bacterial cells became evident: they *were* bacteria.

This symbiotic theory of the origin of mitochondria and chloroplasts was ridiculed when Lynn first started writing and talking about it (today it is universally accepted). I won Lynn's friendship in part by becoming an early adopter of her hypotheses when I was preparing the pertinent section for our botany book.[2]

Throughout this period, my colleagues and I, with the blessing of the American Society of Plant Taxonomists (ASPT), continued planning for the preparation of *Flora of North America*, or FNA, as our project was then designated. It seemed to us that the Smithsonian Institution, where Stan Shetler was based, had the potential to become the most appropriate center for our project. Our colleagues in Europe, led by a small but avid group of British botanists, were by then on track to complete the four-volume *Flora Europaea* in sixteen years, by 1980. That span of time served as our yardstick—a hopelessly optimistic one, as it would turn out.

As I noted earlier, the ASPT had appointed a committee in 1965, charging us with laying out the steps needed to make this large undertaking successful. The American Institute of Biological Sciences had become the FNA sponsor in 1967 because of its fiscal capability and broad representation of the biological community. We lobbied to persuade the National Science Foundation (NSF) to accept the Smithsonian Institution as the coordinating organization for the FNA project because that arrangement seemed so logical to us. The NSF ultimately agreed, even though the Smithsonian was a quasi-federal agency and thus not normally considered eligible for grant support from another federal agency. We received two grants for the project from the NSF in 1969.

Since it had by that time become evident that computers should play a major role in a project of this kind, we were funded in part by the Information Systems Program of the Office of Science Information Service at NSF and strongly encouraged to augment our capabilities in the data-processing area. We secured the services of a computer whiz named Dr. Harriet Meadows and did our best to proceed in a logical way, stressing the importance of maximum compatibility between the written text and the material online. Planning would continue apace for the next several years.

Among all the confusion and tension of that winter and spring, the idea of getting away for the sabbatical year that I had earned began to appeal to me more and more. I began to wonder where I should spend the coming academic year of 1969–70. Academically, it seemed most logical to continue working within the framework of exploring Onagraceae. Within that context, I very much liked the idea of studying the large number of species of *Epilobium* that occurred in New Zealand to see what "made them tick." These species made up nearly a quarter of the global total, and it was obvious that some special factors must have been involved in the evolution of so many of them in such a limited area. I thought that, with Tamra's help, I would be able to manage a study of the whole group in New Zealand together with the fewer, related ones in Australia during the course of a single year—a very ambitious agenda, as I was about to learn.

To make this study possible, I applied for a fellowship at what was then the Department of Scientific and Industrial Research (DSIR) Botany Division in Lincoln, near Christchurch, and at the same time for a Guggenheim Fellowship. I was successful in obtaining both; they would enable Tamra, Alice, Liz, and me to live in New Zealand for nearly a year. The prospect of ranging over the country with Tamra looking for what New Zealanders called "willow weeds" seemed idyllic. So it was that seven months after Tamra and I married, in the middle of one of the most tumultuous periods in the history of the United States, we made plans to fly to New Zealand, the precious islands that the Maori call *Te Aotearoa*, the Land of the Long White Cloud. We would spend a blissful year out from under the dark cloud of social disapproval from my colleagues, remote from the events that were disturbing most of the earth at that time. In Christchurch, over time, some of the confusion about our future would gradually be laid to rest.

– 8 –

New Zealand

THE WEEKS PRIOR to our departure for New Zealand at the end of June were busy and hectic. Some projects underway at the time were wrapped up permanently. I didn't want to come back to much unfinished business, no matter how interesting it seemed. Where possible, I assigned important tasks that had not been completed to others, and during our final push we managed to record twenty-four tapes of necessary letters, objects we were leaving behind, and dictated directions for dealing with various situations that might come up during our absence. Our departure was a scene of utter chaos, with close friends Steve and Anne Seavey helping us pack up the last papers in a frantic evening push the day before we flew off to begin our exciting year abroad.

The preparations were cathartic, as they heralded a new chapter in our lives. Leaving for the year abroad meant that we would leave behind the episodes of hippie self-indulgence and the existential uncertainty. We would land in a new country with a blank slate. The feeling was similar to what I experienced flying to England nine years earlier as a postdoctoral student: I was open, optimistic about the future, and ready for new experiences.

At the airport, we said goodbye to Alice and Liz, who would live with the Seaveys for a couple of weeks while Tamra and I made stops on the way out to New Zealand. Then they would join us in Christchurch. Our first stop was the tropical island of Tahiti, where we checked into a resort hotel and awakened the next day to explore the island. It was filled with Polynesian beauty everywhere we looked. On a quick overnight outing to the neighboring island of Bora Bora, the experiences of the past two years finally caught up with me and, wracked with jet lag and freed from the need to keep it together, I simply broke down and cried, sobbing without really knowing the reason. The next morning I felt much better, and we traveled on to Fiji—very different from Tahiti and fascinating in its own right. Our stopovers allowed us to adjust to different time zones on our way to what would be our host country for the coming months.

We arrived in Christchurch on the fourth of July, 1969, and were met by Eric Godley, the kindly scion of an old English settler family. Tall and well-dressed, with a slightly quizzical expression seemingly permanently impressed on his face, Eric was kind, gentle, modest, and positive. He was head of the government-funded Botany Division in the Department

Eric Godley (1919–2010), director of Botany Division, Department of Scientific and Industrial Research, Lincoln, New Zealand, during the period of my 1969–70 sabbatical visit there.

of Scientific and Industrial Research, our new home institution, which was headquartered in Lincoln, just west of Christchurch. We would learn later from Eric's colleagues that he was considered the father of botany in post-war New Zealand.

Once we had pulled our wits together and found a place to live in Christchurch, Eric took us on a nice drive through some of the neighborhoods of the city and the woods and fields in the near vicinity. A bird that closely resembled the red-tailed hawks back home soared overhead. Eric called it a slow hawk. I asked him if there were vultures in New Zealand, and with his characteristic smile, he responded, "No, when you die in the bush in New Zealand, you die alone." Eric helped us a great deal as we got settled in Christchurch: he proved to be a perfect host and guide for the next phase of our lives. I couldn't help but compare him with Carlos Castaneda's Don Juan, as he guided us into the new country and introduced us to new ways of thinking. He took us under his wing and to some extent we did the same for him, as his recent divorce had brought the usual curse of loneliness.

We collected the girls a few days later, after their long flight from California. We then settled into our flat on Hereford Street, three quarters of a mile from the city's central square, the focal point of a lively neighborhood of theaters, restaurants, and bars and on which stood the Anglican ChristChurch Cathedral.[1] The second school term of four was ending in New Zealand as we arrived. We enrolled the girls in Linwood Avenue School, an Anglican public school within walking distance of our house, where students took turns ringing the bell that signaled the beginning or end of class periods.

Soon after our arrival we bought a used bluish-green Chevy. It served me for getting back and forth to work in Lincoln and was indispensable for our collecting trips, which would eventually amount to more than 15,000 miles of travel all over both islands of New Zealand. Driving on the left side of the road could be tricky when merging onto an empty road without moving cars to tell one which side to take. But we got used to it and survived.

A little over two weeks after we arrived, Neil Armstrong set foot on the moon and America's stock around the world soared, as it certainly did in New Zealand, already and enduringly a friendly place in any case. We had no way to watch the landing; but we quickly heard about it, and Alice and Liz listened to it over the intercom in their school classroom. It was a very exciting and happy time for us all. As the days passed, I started to see a way forward in place of the confusion that I had left behind in California.

Planning the work at hand—unraveling still another set of puzzles presented by the Onagraceae—was a delightfully fresh challenge. When finishing up my graduate studies at UCLA, I had noticed that some forty to fifty species of *Epilobium*, known by the common names willow weed and willow herb, were then listed for New Zealand. Many of those species had features that were unusual in the context of this worldwide genus, which is the largest in the family. Since Onagraceae as a whole are represented more poorly in the Old World than in North and South America, it seemed unusual for so many species of this one genus to occur on a distant pair of islands in the South Pacific. Now I was actually living on those islands, excited to find out what was going on with *Epilobium* there.

Elucidating the deeper past of *Epilobium* species in New Zealand was basic research, and it had a clear role in the much bigger project of understanding how Onagraceae had evolved and formed their pattern of distribution around the world. The overall investigation of Onagraceae was one for which I could find willing colleagues and establish principles relevant to other groups of plants and animals. For example, it meshed in fascinating ways with what geologists were learning about movement of the Earth's land masses over time.

In addition, in New Zealand I certainly was not about to abandon the issues—population growth, biological extinction, and sustainability—that were growing to become my major concerns. In that quiet and peaceful land, I would concentrate on the business at hand while continuing to mull over our common fate and what we might do to affect it. Conversations with Paul and Anne Ehrlich, for example, would continue from a distance, often focused on responses to Paul's influential book, *The Population Bomb*, which had been published in 1968. As we traveled around New Zealand searching for plants, I continued thinking about the future and about the serious impact of what we were beginning to view as a serious wave of biological extinction.

The task in New Zealand, which I undertook as a joint project with Tamra, was to sort out the dozens of species of *Epilobium* there, clarifying their limits and their relationships. Some of them were widespread, others narrowly distributed, and we studied their features comparatively in natural populations and in the herbarium. Many of the species had unique features compared with those of *Epilobium* found elsewhere. We grew many populations together in the greenhouses at Lincoln so that we could compare them directly with one another. To determine their chromosome numbers, we depended on John Hair, an excellent cytologist and staff member at Lincoln, to work with the populations we were growing. All of them turned out to have the same number of chromosomes in their somatic cells—$2n = 36$—as did most of their relatives in other parts of the world.

Tamra and I had to reserve some time for collecting in Australia as well—to be able to compare the Australian populations accurately with those in New Zealand—and we had one field season of several months to see them all in their native habitats, with a few additional months to study them in the herbarium. The first order of business, which we commenced without delay, was to go through the numerous *Epilobium* specimens in the herbarium at Lincoln, start to become familiar with the individual species and the observable characteristics that separated them from each other, and learn where they had been collected in nature.

Earlier classifications of the group had been uneven in quality, with poorly delimited species put on the list along with others that were clearly distinct. We easily recognized the species that obviously stood apart from the others, mostly growing in special habitats like the huge scree slopes on the mountains. Others, like the ones in the *Epilobium glabellum* complex, varied greatly over extensive ranges and had been treated in a number of different ways taxonomically in the past. For these, we had to decide how to subdivide them and which should be given names marking them as distinct. Finding the distinctive species while studying as many wild populations as possible of the widespread, variable ones required the careful planning of excursions. We were especially interested to find places where more than one species grew together so that we could see whether they intergraded or formed hybrids with one another. Such patterns were of critical importance in deciding whether to maintain different names for the entities or not.

Many people, generous with their time and advice, helped us with the process of planning our trips into the field. This was not a simple matter, because there were so many places we wanted to visit and the overall time was short. Naturally, our future schedules kept getting revised as we went along.

It was July and we were in the Southern Hemisphere, so *Epilobium* populations would not start to flower until the arrival of spring near the end of the year. This gave us about twelve weeks to visit the herbaria in other centers and talk with the botanists with relevant knowledge. Two with particularly wide knowledge were W. B. Brockie of the Otari Open Air Plant Museum in Wellington, who had spent twenty years studying the genus, and A. P. Druce, of Botany Division on North Island, who had critically evaluated the entities in the genus during his many years of insightful fieldwork. These specialists described problems in delimiting particular species and told us about interesting places that we would want to visit.

Our first outing consisted of several exciting days spent on Stephens Island, in Cook Straits between North and South Islands, which we visited with a party from the New Zealand Wildlife Division. There we helped our colleagues plant small shrubs to help re-vegetate slopes that had been grazed bare by sheep over many decades. We also helped band individual fairy prions, small seabirds that flew back to their clustered burrows each evening. Tuataras, unique lizard-like reptiles, lived in the same burrows and fed on the prions; we were helping band them also. Seeing them in nature was a treat, because tuataras are the only living members of a group of reptiles that flourished 200 million years ago; today, they survive only on about a dozen islands scattered around the coasts of New Zealand. One tuatara scratched me as I was banding it; the scratch became a badge of honor that I was able to display for about a decade before it faded to the point at which it was no longer visible.

Returning to Lincoln, we turned to the other native genus of Onagraceae that occurred in New Zealand, *Fuchsia*, which began blooming around the end of September. We studied the two *Fuchsia* species and their frequent hybrids up and down both coasts of South Island.[2] The two species, both common around Christchurch, were the tree *Fuchsia excorticata* and the straggling shrub *Fuchsia perscandens*. Introduced Australian possums (very different from our North American ones) love to eat the leaves and branches of the tree species, *Fuchsia excorticata*, and could reduce a whole tree to a skeleton in one or two nights. (Tens of millions of these exotic possums exist today in New Zealand, where no natural predators can act as a check on their population growth. Fortunately, many are harvested for their fur each year.)

During this period we took a trip of some weeks to North Island, where we could study additional populations of *Fuchsia* and where *Epilobium* comes into flower earlier than in the south. We traveled widely along the coasts and into the interior, with its huge flat volcanoes, beautiful lakes, and small patches of archaic Southern Hemisphere temperate rainforest dominated by huge conifers and species of southern beech, *Nothofagus*. Eric Godley flew up from Christchurch to show us some of the populations of *Fuchsia* that he had studied earlier. The third new species of *Fuchsia*, *Fuchsia procumbens*, occurs in only about a few populations in the relatively flat lands north of Auckland. The wiry stems of this unique plant creep along the ground, with their large, erect flowers standing above the leaves. For our visit to them, Eric had summoned a

high school chum of his to join us, and we enjoyed a magical night together on a warm summer evening camping on the beach north of Auckland. We started a fire, pried oysters off the rocks, roasted them, and
shared views of botany, life, and the world. It simply doesn't get any better
than that.

Since the species of *Epilobium* that occur on North Island all occur
on South Island also, it was time to head back south once we had evaluated the northern populations that we were able to locate. On the 29th
of November—our first wedding anniversary, just to put the time warp
we were experiencing into perspective—we were driving up over Arthur's
Pass, which leads west from Christchurch over the central ranges to the
West Coast, beginning our detailed study of the more numerous South
Island species and populations of *Epilobium*.

Most species of *Epilobium* reach their heaviest bloom on South Island
in December, January, and February—summer in the Southern Hemisphere. This caused us to map out the most productive schedule for December and January, refining what we had developed in our first months
in the country. South Island is large, about the size of New York State,
and dominated by a rugged range of mountains with peaks over 12,000
feet tall running north to south. Only a few roads cross over the mountains to link the east and west coasts. We ended up doing some 500 miles
of hiking along with our extensive driving to reach all the wonderful
places we visited during those months.

One of our earliest trips was to Molesworth Station in south Marlbor-
ough, the largest upland sheep station in New Zealand, comprising
nearly 700 square miles and a fascinating locality for a number of very
special kinds of plants. Many different habitats occur on its dry slopes.
Here, as throughout our travels, we studied the plants in their natural
habitats, wrote up our notes about them, and brought many specimens
back alive to Lincoln, where
they were established in the
greenhouses. We were fortunate to be accompanied on
this trip by the eminent botanist Lucy Moore, who was
sixty-three years old and on the
verge of retirement from an
outstanding career. She was a

Epilobium pycnostachyum, showing its long woody
stems growing out under the moving scree, with
the end popping up, as in color image #30, into the
light.

short, wiry person, unfailingly cheerful and helpful, who freely shared her extensive knowledge about the plants and habitats we were seeing. With Lucy we climbed high on the slopes of Mount Murphy, where colonies of two really unusual species of *Epilobium* grew on the huge expanse of shifting scree. If their branches were broken off by the moving rocks, new ones would regrow from the base, which was anchored by the roots to the more stable bedrock below. Scree is widespread on South Island, a product of geologically rapid uplift followed by the erosion of rocks on the rising slopes, and species in several different plant genera— distinctive products of local evolution—have adapted to grow in this unusual habitat.

We then headed down to the ranges in the interior of the south part of the island, in Otago, where additional species of *Epilobium* occurred. Here there were pairs of entities that we worked to tell apart all summer, trying to decide whether to recognize them with distinct names ("split them") or merge them into one ("lump them"). Ultimately this kind of decision is an arbitrary one, but it needs to be made carefully and supported by evidence. My friend Harry Thompson, one of my professors at UCLA, had wisely warned me years earlier against regarding such taxonomic decisions as the ultimate be-all and end-all of studies of plant populations. He cautioned me to keep track of the data on which all taxonomic decisions were based because these were really the important findings, the real knowledge that was useful for the future.

The biggest adventure of our fieldwork in New Zealand occurred on a hike I undertook without Tamra in mid-January, to study an unusual species that had been collected and named *Epilobium purpuratum* in the mid-nineteenth century. It was discovered by James Hector, the energetic director of the Geological Survey of Otago, during the course of his pioneering explorations in 1863. Although it seemed to resemble other species pretty closely, the herbarium specimens looked distinctive enough that we wanted badly to see living plants in order to understand them as well as possible. The best place to look for it seemed to be Hector Col, a 4,900-foot saddle on the crest of the Southern Alps, almost certainly the place it had first been encountered over a century earlier. Hector Col would clearly be a tough place to reach, and we asked Alan Mark, a professor at the University of Otago in Dunedin, a city down the coast from Christchurch, to lead the climb. Alan was conducting important and useful studies of the tussock grassland and lakes in the interior of South

Island and was extremely knowledgeable about all aspects of New Zealand botany. Besides that, he was physically fit and tough.

To reach Hector Col, we hiked along a trail up the south bank of the West Fork of the Matukituki River, carrying our sleeping bags and other gear for an overnight stay. The trail in was fairly easy, and toward evening we reached the flat basin at the head of the river, where we had a stunning view of the col and the surrounding rugged mountains of the Southern Alps. Rolling out our sleeping bags in a persistent rain, we spent a damp and fitful night, but awakened the next day to bright sunshine. After breakfast, we began to climb up the steep wall of a glacial cirque a hundred or so feet high at the northeast foot of the col—tough going but just possible for me. On the climb, I was startled when a stone about the size of a loaf of bread *blew* off a ledge beside me, and in that way got an idea of the strength of the wind gusts and the possibility that they might just blow me off too. Fortunately, that didn't happen, and we reached the flat rocky slopes leading up toward the col. As we neared the top of the saddle, we were delighted to find that our target *Epilobium* was locally abundant. At the very top of the col we had the thrill of looking down nearly 5,000 feet straight onto the breakers of the Tasman Sea, crashing wildly on the rocks below. We gathered some living plants of the very special *Epilobium* into plastic bags and headed back down the treacherous slopes.

When we got back to the Matukituki River, however, we found it had risen to nearly waist height, greatly swollen by the rain that had fallen during the night. We were on the north side of the river, and the trail we had taken up led up the much gentler and flatter south side. We wanted to get back to the south-side trail to make our way out to the road. Since Alan was very fit and near the peak of his physical condition, he was the one who roped up and tried to cross the surging waters with a large pack of our camping gear on his back. The strong current immediately knocked him over, ripped him out of the rope, and tore away the pack; he was barely able to fight his way out of the raging waters and scramble up on shore 150 feet or so downstream, fishing out his pack as he emerged. We had no option: we would have to find our way out along the north side of the river, where the steep slope, covered with dense thickets of southern beeches, dropped almost directly into the river. We did this in pouring rain and had to spend an extra night sleepless in the rain among the dense thickets, eagerly awaiting the next day's light so we could resume our scramble downward. We finally made it out, to the relief of Tamra and the waiting party at the cars, worn out and tested severely, but with the living plants of *Epilobium purpuratum*—which we would later confirm as a distinct species—securely stowed in our backpacks.

Much too soon in the course of our mad scramble to learn about the regional variation in species of *Epilobium* in New Zealand, we headed off for Australia to see how closely the species there resembled those of New Zealand, and how many different kinds there were on the continent. Alice and Liz stayed home in New Zealand, on a sheep farm with Hugh and Marge Wilson, a couple we met soon after we came to Christchurch. They enjoyed a fine time there, watching sheep shearing, attending a country school, and riding their bikes along dirt roads to buy comic books at a nearby shop.

We had done our homework about Australian *Epilobium* in advance so that we'd know where to go when we landed. We flew to Melbourne, that very attractive European-style city, rented a car, and drove first west and then east up the coast through dry grassland and scattered *Eucalyptus* trees. The lowland species of *Epilobium*, as we knew in advance, were the same as the ones we had come to know in New Zealand, but we collected them avidly for comparative purposes. On our third day collecting around Melbourne, in early February, we headed north up Mount Wellington, through beautiful groves of gigantic *Eucalyptus* trees. At about 3,900 feet elevation, around springs in patches of moist grassland, we encountered *Epilobium gunnianum*, a beautiful species with large purplish flowers, growing with two other species that were widespread in the lowlands, together with an introduced species from North America that was also common in New Zealand.

One of the creatures that we encountered along our way in the ranges of southeastern Australia, a wombat, was coaxed into my arms, a great treat. Once there, it seemed to enjoy the experience—I certainly did!

Dropping down into Canberra after another day of collecting in the high ranges, we had the great pleasure of meeting Dr. Nancy Burbidge, a wonderfully knowledgeable and extremely helpful botanist who directed us up various roads into the surrounding ranges to localities where species of *Epilobium* abounded. Nancy, then sixty-eight years old and retired, had directed the Commonwealth herbarium since its establishment and built a sturdy foundation for the study of Australian plants, which played a major role in the initiation of the *Flora of Australia* series. We drove to the summit of Mount Kosciuszko, the highest point in Australia at 7,310 feet, through patches of low *Eucalyptus* trees (snow gums), and bare slopes, collecting the same set of species of *Epilobium* repeatedly along the way, while enjoying the distinctive countryside and its inhabitants. A few days later, we drove north to join two extraordinary botanists, L. A. S. "Lawrie" Johnson and Barbara Briggs, in the Northern

Lawrie Johnson and Barbara Briggs, leading Australian botanists, who assisted us with our fieldwork in Australia in early 1970, became lifelong friends and colleagues.

Tablelands of New South Wales, camping with them in the well-watered meadows at Barrington Tops, at around 4,900 feet elevation. Lawrie and Barbara were two of the finest botanists in Australia, both from the Royal Botanic Garden Sydney.[3] Stopping to make more observations and collections in the mountains around New England National Park, we headed together back to Sydney.

By mid-February, it was time to search for *Epilobium* populations on Tasmania, the large island that lies south of southeastern Australia. In Tasmania's capitol, Hobart, we hooked up with Winifred Curtis, the doyen of Tasmanian botany, who was wonderfully kind in guiding us all around the island. She and a colleague accompanied us throughout almost all of nine days of fieldwork on the island, helping us efficiently to find the species we wanted. Tasmania was very important to our study, being located at a higher latitude and thus home to more species and populations of *Epilobium* than is mainland Australia. It was important that we get to know them well for comparisons with their relatives on the Australian mainland and in New Zealand. Visiting much of the more accessible parts of Tasmania, we had especially fine collecting above timberline on Ben Lomond and Mount Field near the center of the island.

There we found an undescribed species of *Epilobium* that is entirely restricted to Tasmania, which we later named *Epilobium fugitivum*.

Back in New Zealand by early March, we did a little more fieldwork, but I spent most of my time in the herbarium and library. To be able to understand the distributions of the species we had been studying, it was necessary to map the localities where they had been collected. That required weeks of work with atlases and gazetteers. We worked with Keith West, a wonderfully talented artist who beautifully illustrated our works. By that time, New Zealand had begun to feel like home.

In the matter of classification, our observations and studies led us to recognize thirty-seven distinct native species of *Epilobium* in New Zealand.[4] In 1961, H. H. Allan's *Flora of New Zealand* had recognized fifty species from New Zealand. The differences between our treatments were accounted for by the ways we looked at variation and the amount of material and observations on which we based our conclusions. In addition, Tamra and I made many discoveries about the species involved and reported them. Twelve native species occur in Australia, half of them also in New Zealand. Five species of *Epilobium* in Australasia turned out to be introduced from the Northern Hemisphere. In addition to these, there were four additional native ones found only in the mountains of New Guinea, which we were unable to visit.

The results of our studies led me to ask questions about how the ancestors of *Epilobium* had reached the places where they now occur. These questions represented a field of inquiry distinct from finding the most appropriate taxonomy for these plants. Eventually, we were able to demonstrate that *Epilobium* entered the region from Eurasia, and that a tremendous evolutionary burst followed for the genus in New Zealand, mainly associated with the geologically recent uplift of the Southern Alps and the development of a number of unique habitats there into which new species of the genus evolved. Once we understood the pattern of distribution and evolution in *Epilobium*, it became obvious that other groups of plants had followed a similar pattern: dispersal and subsequent speciation. In that way, our investigation of *Epilobium* enabled us to add depth to the overall understanding of the biogeography of Australasia at large.

In working to solve the biogeographic riddle of *Epilobium*, we benefitted greatly from what geologists had learned in just the previous decade or so about the makeup and movements of the Earth's crust. The theory of plate tectonics had been proposed and validated, and many of the specifics of plate movement worked out, mostly in the decade since I had graduated from college. By 1970, geologists had come to understand that the lowland evergreen forests of southeastern Australia and Tasmania, New Zealand, New Caledonia, and southern South America were connected directly by land about eighty million years ago, before the constituent land masses began to separate. This relationship explained the close similarities between their rich assemblages of often-archaic organisms. On the other hand, the higher elevation habitats of New Zealand appeared as the mountains were uplifted only a few million years ago. The ancestors of the relatively few groups of plants and animals that occur in these high habitats could only have been derived from lowland populations, arrived directly by long-distance wind dispersal, or been carried there by birds. The prevailing westerly winds that occur at about 35°–45° latitude in both the Southern and Northern Hemispheres are much stronger in the south, where much less land, and hence less friction, slows them down. In view of the strength and trajectory of these winds, it is not surprising that plants with light, airborne seeds, like *Epilobium*, have been easily transported between the southern lands in the past and show many similarities in the present.

It was an exciting time to be considering the present-day distributions of plants and animals and how they were determined, in part, by movements of the Earth's surface in the distant past. The explanations that we worked out for the patterns of distribution of *Epilobium* would have not seemed even remotely plausible only a decade earlier. The ability to use geologists' rapidly expanding knowledge of the past to understand patterns in the places where plants grow today gripped my imagination. After our return to the States, I would continue this line of work in collaboration with my friend and former professor Dan Axelrod.

Near the end of our time in New Zealand, we spent an evening with Eric Godley at his home, barbecuing our meal in the garden. Godley had encouraged us to consider staying in New Zealand, but he knew by then that it wasn't to be. It had become clear to me that to be most effective in the world and *for* the world, I needed to be back on American soil. Taking that into account, he predicted that evening that I might become a preeminent botanical organizer like the famous German Adolf Engler (1844–1930) and go on to write or edit major floras from broad areas in various parts of the world. It wasn't quite like St. Paul falling off his donkey and being struck with the thunder of the Lord, but Godley's words went directly into my soul. They had a significant impact on my self-conception and ambitions from that point forward. One might say with hindsight that Godley was extraordinarily prescient, but it was really a matter of the workings of his self-fulfilling prophecy being set into motion.

℘

Then it came time to head back home. Five months remained before the start of the 1970–71 academic year, so we didn't have to return directly. Our time abroad had been magical, very productive, and exciting—a kind of prolonged honeymoon, in retrospect—and we thought it would be wonderful to continue it in a somewhat different form. Wanting to learn something about the rest of the world on the way home, we planned an itinerary that would take us across Asia in April and then through much of Europe during the summer. The girls would stay with the Wilsons on their farm during April and then fly to join us in Istanbul for our four-month exploration in Europe.

We flew back to Australia to snorkel at the Great Barrier Reef; to Port Moresby, New Guinea, and its surrounding eucalyptus woods; and then on to Hong Kong and later Cambodia. In Phnom Penh we dined in a riverboat restaurant with only about two other couples; the country was almost deserted by tourists. We had no idea that only two weeks hence the first American bombing of Cambodia would begin. A guide showed us a boat being towed down the Mekong River, explaining its passengers were Vietnamese fishermen being moved to better fishing grounds. Later we would find out that the Vietnamese actually were being taken away to be massacred.

After enjoying the impressive ruins at Angkor Wat, we traveled to Bangkok and toured the city, marveling at its great reclining Buddha. From Thailand, we flew northwest to Kolkata, India, where we visited the Indian Botanical Garden and admired its enormous banyan tree, estimated to be about 1,250 years old and spreading with many trunks over an area of about four acres. We then traveled on westward to Kathmandu, Nepal, for a short visit, marveling at the large fruit bats hanging on the bushes in town and then driving up along a Chinese-built road for a fine view of the Himalaya.

New Delhi was special, both in itself and because we were able to visit the Taj Mahal on a day trip, traveling there by train through largely empty countryside. On the return trip, we shared a compartment with an Indian army officer, a very cultivated Hindu gentleman, along with one of the men reporting to him, a Sikh. The officer told us many interesting things; in particular, he pointed out that life was a voyage and if we push too hard for our own advancement the voyage would be a tumultuous one indeed. On a piece of newspaper, he drew a picture of a couple in a small boat, saying that if we overreached in self-promotion, the rocks at the bottom of the ocean would move, causing the water to roil and the boat to pitch dangerously. I saved this diagram, thinking it clearly represented good advice, and have always tried to take its message to heart.

Flying on westward from India, we stopped briefly in Tehran, saw the sights, and then flew on to Cairo. I loved seeing the Great Pyramids in person! At the Cairo Museum, the displays were heavily protected with sand bags, reminding us of the state of war with Israel and the general condition of the world. From there, we were off to the sparkling city of Istanbul, where we would reconnect with the children.

When we collected Alice and Liz at the airport after their adventurous flight from New Zealand, it was time to celebrate Liz's ninth birthday. We did this by going sightseeing in Istanbul and visiting the Grand Bazaar. Among the stalls, we all had a good laugh watching Liz teetering on a stool while trying on a leather jacket she was about to purchase with money my parents had sent her for her birthday. Flying from Istanbul to Vienna, we picked up a new VW camper that we had ordered in advance. With our own transportation, we were set for the next four months of wandering through Europe.

While we were driving through the countryside in Austria, Tamra fell seriously ill. She was diagnosed with paratyphoid, and so we paused in Zurich to see her nursed back to health in the Kantonsspital. After her recovery, we spent almost two months traveling through Italy and Greece, visiting as many ancient sites as we could. Our camper gave us great flexibility in where to stop, and we occasionally stayed in hotels to wash up more thoroughly than we could in the camper. Then moving slowly through Spain, we visited Barcelona, Sevilla, Toledo, the cave at Altamira with its enthralling paleolithic paintings, Madrid, and Cádiz, the port from which Columbus had sailed on his first voyage to the New World. It was a great learning experience for the children, and a huge pleasure for us, too.

☙

Returning to the United States at the end of the summer, everything felt different from the way it had when we left for my sabbatical. The interlude abroad—a total of fourteen months—had given me time and space to reflect about many aspects of science and life, and I had come to view myself and the world from a new perspective. Moreover, the world itself had continued to change rapidly. Even though we had remained aware of events in the U.S. and elsewhere while we were away, the feeling of having entered a new era was palpable. Riots and demonstrations had become a way of life, the Charles Manson trial was underway, the U.S. had been bombing in Cambodia, and the draft had gone into effect. At Stanford, students had stepped up their demonstrations and sit-ins, loudly objecting to the university's research role in U.S. military defense and decrying CIA recruitment on campus. Significantly for me, the first Earth Day had been held on April 22, 1970. This inaugurated a full-fledged environmental movement that over the next few years would result in the passage of the Endangered Species Act, the National Environment Policy Act, and other basic pillars of our environmental legislation.[5]

Against this background of social change, I realized that my values had slowly changed as well. This was most evident for me when I considered them in relation to those of my parents. My mother and father were both socially conservative, naturally content with the period of prosperity following the war and now generally anxious about the transformations underway. I, on the other hand, increasingly understood the riots and unrest as the growing pains of a society striving to make forward progress. Our differing reactions to Bob Dylan exemplified the divide: my mother had a fit whenever she heard his raspy voice, while I and other people of my generation loved his poetry, prophetic words, and what they said about our emerging values.

After a short period of adjustment, I plunged back in to my required teaching duties and the ever-present committee meetings. There was also some work to do in connection with the imminent appearance of the botany textbook. Much of my time during the first months back at Stanford, however, was spent intensively working to complete the monograph on *Epilobium* in Australasia, which would be published as a book in New Zealand a few years later. I also concentrated a good deal on thinking about the connections between continental movements and the distribution of organisms, beginning my work with Dan Axelrod on the analysis of such patterns. Periodically throughout that winter and spring I was preoccupied with the Flora of North America project, working, in the context of committee meetings, to lay the groundwork for important administrative changes and funding successes that would occur in the next couple of years.

During the winter of 1970–71, amid everything else that was going on, I learned that the director of the Missouri Botanical Garden, David M. Gates, was leaving, after a successful tenure of six years. He had accepted a position at the University of Michigan Biological Station, which his father had founded and where he had spent a great deal of happy time as a boy. He was to depart from St. Louis in the summer of 1971. By that time, I had some knowledge of the Garden, having visited several times, and my overall impression was favorable. (Gates was at that time the chairman of the Flora of North America Steering Committee, and Dr. Walter Lewis, the very able curator of the herbarium at the Garden, was a fellow FNA editor; and so several FNA meetings had taken place in St. Louis.) With a degree of negative social pressure at Stanford and the promise of a fresh start in St. Louis, I began to wonder whether the position of director of the Garden might be a good fit for me and a move to St. Louis a wise choice for our family.

This idea was at first just idle fantasy. Before learning of Gates's coming departure, the notion of directing that or any other botanical garden was the furthest thing from my mind. I wasn't an administrator; I was an academic—one who, having embraced the spirit of the 1960s, had grown rather long hair and a beard. I was committed to studying the Onagraceae and other plants, teaching, and publishing papers, and valued the freedom to travel and immerse myself in botanical research. I knew very little about botanical gardens, much less how to run one.

The matter suddenly became real and concrete for me, however, when later that winter an unexpected visitor made an appointment to see me in my office at Stanford. The visitor was George Pake, who had recently resigned as provost of Washington University in St. Louis to join Xerox Corporation. In doing so, he had moved to Silicon Valley to establish the research campus known as Xerox PARC, which under his guidance would turn out to be a highly successful venture. As provost, George had represented Washington University on the Board of the Missouri Botanical Garden and so was well known to the other board members. Gangly, tall, slim, and of pleasant demeanor, he sat in a chair in my office, put his feet up on the central table, and began to talk. I learned that he had been asked to see if I might be interested in applying for the director position at the Garden, and specifically if I would be willing to come out to St. Louis for an interview. I thought about the matter for a few moments and agreed to the visit, thus putting the possibility of a move directly into play.

Over the next few weeks, I pondered the Garden in a new and more focused way. I had never thought much about it as an institution, but I knew that it had a major herbarium and library, the Climatron, and some outdoor displays. Two of my mentors, Mildred Mathias and Carl Epling, had spoken warmly about the graduate program the Garden co-sponsored with Washington University and about their times there in the 1920s. I knew that a number of other prominent botanists had graduated from the program, which dated back to 1885. Dick Holm, one of my closest friends at Stanford and a St. Louis native, was a graduate who shared many tales about his times as a student and the eccentricities of the people with whom he had worked. Dave Gregory, a recent graduate and close friend, had regaled me with tales of his times at the Garden in the 1950s, and Ledyard Stebbins had introduced me to Edgar Anderson, the great Garden geneticist who had visited him in California. Lee Lenz, the director who had welcomed me at Rancho Santa Ana a decade earlier, was also a Garden graduate. All in all, it was clearly an important botanical center.

As I looked into the institution's history, I came to appreciate the fact that it was the oldest continuously operating botanical garden in the country, opened to the public in 1859 by an English merchant named Henry Shaw (1800–1889). Shaw had come to town forty years earlier as an agent for his uncle, a manufacturer of iron goods in Sheffield, England. Broadening his inventory to hardware in general, Shaw became a very wealthy man by the standards of the time. Retiring after two decades of active work in St. Louis, Shaw turned his attention to the idea of doing good things for his adopted community. In addition to the Garden, he had established the adjacent Tower Grove Park for public enjoyment. In 1885, he had established what became known as the Henry Shaw School of Botany, permanently linked with the Garden, at Washington University in St. Louis.

A series of directors had served the Garden since Mr. Shaw's death in 1889, the most recent of whom was David Gates. Since his arrival at the institution in 1965, Gates had made a solid start on reviving the Garden's fortunes, which had lagged since the end of World War II despite the construction of the Climatron in 1959–60. Recognizing that the herbarium and library had become greatly overcrowded in the Garden's Administration Building, David had persuaded the trustees to help him raise the funds necessary for the construction of a new building to house them properly. The capital fund campaign had been successful, with most of the funds coming from generous Garden patrons John and Anne Lehmann. Construction of the new building, to be named for John S. Lehmann, had begun only a few months previously, in late 1970.

Tamra and I spent some time discussing the possibility of a move to St. Louis. We both agreed that it could be a positive change. It had been difficult for us as a couple to integrate into the professional and personal social milieu at Stanford in which Sally and I had been well established. The move could mean a new start for our family. Tamra had completed her master's degree work and was ready for a change. I disliked the idea of leaving California, to which I was tied so closely in so many ways. Obviously, though, an excellent opportunity would require that sacrifice.

As the date for my interview in St. Louis approached, I realized I did want the job—passionately, in fact. It was a chance to grow, to test myself, and perhaps to have a broader impact in the world. I was already moving in a direction counter to that of many academics. Whereas most university professors become increasingly specialized during their careers, my interests were widening rapidly. Academia didn't warmly encourage, much less reward, that kind of general interest. At the Missouri Botanical Garden, I would have the scope and resources to do and to support broader work. And I would have the ability to help get talented people involved in a variety of projects with real impact.

Eric Godley's prediction took center stage in my mind. Obviously, I couldn't begin to undertake an effort even remotely on the scale of Adolf Engler's as a university professor with a full teaching schedule and roster of specialized research projects. In contrast, the Missouri Botanical Garden was clearly a place where I might be able to fulfill the destiny that Eric had laid out.

Another enticing element was the international scope that would come with my assuming the helm at the Garden. I was already interested in Latin America, and the Garden was operating a program in Panama. The Garden's herbarium and, even more, its library were global in scope. Since my earliest days I had been interested in countries all over the world, and it was an exciting prospect to combine that interest with my love for botany and to be in a position to encourage others to learn about, describe, and catalog their plants. The potential to partner the Garden with botanists and scientific institutions in many countries and lead the expansion of its program into one that was truly global held a great deal of appeal.

To be honest, I also was attracted by the opportunity to be a leader. By nature, I loved to take charge, to make things happen. It wasn't about power so much as the drive and urgency to get things done. The Missouri Botanical Garden directorship, I realized, could be the perfect fit for that element in my personality. All of this swirled through my mind as I discussed the position with Tamra and with my friends, including Dick Holm, who was an invaluable sounding board.

When I flew out to St. Louis, my meeting with the trustees seemed to go well, as did the seminar I presented. After the seminar, I gossiped avidly with all comers to try to see what was going on and gauge whether I would have a good chance to be selected.

I knew as well as anyone that my curriculum vitae presented challenges. I had never hired anyone, fired anyone, run an organization of any size, or done institutional fundraising. Even so, I had wide connections in the world of botany and science, an international awareness and perspective, a real passion for building teams and supporting others' work, and a genuine liking for people. And, of course, I had a strong background in botanical research, a qualification that was not insignificant. I knew there were other strong candidates but had no idea what my own chances might be.

Soon, however, I learned that the committee was seriously considering the possibility of hiring me when, in April, I was invited back to St. Louis and instructed to bring Tamra along. After an interview with the board's search committee, we were to attend a party given by Warren Shapleigh, chair of the search committee and head of the Consumer Products Group at Ralston Purina, and his wife Jane, at their home outside of St. Louis. We were driven to the event by Powell Whitehead, retired CEO of General Steel Industries and then chairman of the Garden's board of trustees, and his wife Georgia. Powell, an old-style businessman, had a difficult time navigating that far out into the country, and nothing happened to put us at ease along the way.

The principal purpose of the party, I surmised, was to present us to Mrs. Lehmann, who with her late husband John had been by far the most important patron of the Garden since its founder, Henry Shaw. Though she had no official veto power, her approval was crucial. Wiry and with an old-fashioned hair style, she propped herself against the wall of the Shapleighs' back porch and studied us carefully. I still had long hair and a full beard, since that was the way I was and really hadn't thought of any other possibility. Mrs. Lehmann might well have taken exception to me on my appearances alone. I must have passed muster with her and the trustees who were in attendance, however, because within a few days of returning to Palo Alto, I received a call offering me the job. I came out once again for discussions and accepted the position in late spring of that year.

On that same visit, we were able to enjoy the lovely spring flora of the Midwestern and Eastern United States for the first time. We were taken to visit the Garden's natural reserve, now called the Shaw Nature Reserve, about thirty miles west in the foothills of the Ozark Mountains. Redbuds and dogwoods were in full bloom at the time of our visit, and we were excited to discover a beautiful luna moth resting on the trunk of one of the trees. They were a harbinger of many unusual botanical "treats" we found over the following years, including butterfly weed (*Asclepias tuberosa*), which had been studied in depth by Bob Woodson years earlier. Our own evening primroses were well represented also, with the Missouri evening primrose forming beautiful colonies in glades, open rocky areas among the woods. It was a lovely day in beautiful surroundings, and helped cement our desire to move to St. Louis.

In the months to come, I learned that Tom K. Smith Jr., who was to become our close friend and advisor in St. Louis, had made the motion to hire me, stressing my youth, my qualifications, and his perception that I really wanted the job. He was certainly right about our eagerness to accept the new job with all of its attendant promise and challenges.

So there we were, ready to move to St. Louis and take up my new duties on August 1, 1971. At Stanford, the general reaction was one of mild disbelief. Don Kennedy, who was the chairman of the biology department at that point, kindly let my promotion to Associate Professor with tenure go through "in case you ever decide to come back." Don even offered to step aside and make me chair of the department if it was an administrative role that I was seeking, so I left feeling honored and appreciated. Naturally, my closest friends in the department, Paul Ehrlich, Dick Holm, and John Thomas, subjected me to endless kidding about how unwise my choice had been. Dick Holm jovially advised me that a major day of celebration in St. Louis was "The Melting of the Asphalt," which occurred in late spring each year. It was a sad time but also an exciting one, since I knew we would be able to start a new life on our own terms and I would have the opportunity to stretch and grow.

– 9 –

The Missouri Botanical Garden

Tamra, Alice, Liz, and I arrived in St. Louis on the first of August, 1971, tired from the long drive from California and wilting in the humid summer swelter, but excited to have reached our future home. We parked in front of the residence provided for us, located within the Garden's historical limestone walls. It was an attractive Arts-and-Crafts-style brick mansion with external Tudoresque half-timbering that had been completed in 1914 for then Garden director George Moore and his family. We got out to look over the residence, delighted by the prospect of living in such a lovely place. We admired the rich oak paneling, high ceilings, ornamental moldings, large fireplace, and bay windows looking out on the grounds. The girls told us that they loved the lily pond behind the house; at the hour we arrived it was being tended by a genial volunteer named John Brown, who said that it would be his delight to make us feel special.

The house had not seen much repair work over the years, and although its renovation had been begun by David Gates, it still needed a good deal of attention. That didn't bother any of us at the time: it was basically the centerpiece of a fairyland and only steps away from the main public spaces of the overall garden. My commute to the office would now be a pleasant five-minute walk.

After moving in, we attended to the many practical matters that are part of adopting a new community and establishing new lives. We enrolled the girls in Wade School, which had a special program for gifted students. Since it was located just west of the Garden, the girls could walk there across grassy fields and through the abandoned plant nurseries that lay beyond the developed part of the Garden. And we went about getting to know the neighborhood. Our main driveway was across the street from the Tower Grove Baptist Church and School. Henry Shaw's lovely Tower Grove Park, 300 acres in size, adjoined the Garden to the south. More stores and other services were conveniently located along main streets a few blocks away to the east and the west. Not far away was The Hill, the predominantly Italian district just west of the Garden, which had been founded as men moved from the Milan region in the late nineteenth cen-

The director's residence, located in the southeastern corner of the Missouri Botanical Garden.

tury to work the economically important clay mines and brick kilns in the district. Filled with excellent restaurants, The Hill could lay claim to having raised such luminaries as Yogi Berra and Joe Garagiola.

As coastal Californians, we weren't used to the midsummer heat and humidity we found in our new city, but we liked St. Louis from the start. We were soon made to feel at home. A couple of weeks after we arrived, Nora Stern, then president of the Members' Board, invited us to dinner with her husband Walt at the University Club, on the top floor of the University Club Tower west of the city proper. From that high perch I looked out westward and southward, impressed that St. Louis appeared as one great forest. Amongst all the trees, the widely spaced homes were scarcely visible. It certainly was not the kind of cityscape one would see from such a vantage point in my native San Francisco.

Other invitations from generous and friendly St. Louisans followed. Over the next few months we would make many new and lasting friendships. Among those who welcomed us warmly into their lives and community were Tom and Cindy Woolsey, Debbie and Alexander "Pompey" Bakewell, Walter and Mary Randolph Ballinger, Peter and Marie Louise Pastreich, Buster and Marge May, Johns Hopkins, and Tom and Mary Hall. At the Garden itself, Tom Smith, Howard Baer, and Frank Wolff were quick to offer their friendship and helpful advice, and many of the other trustees helped in various ways, making our new home feel like a very welcoming one indeed. Bob Hermann was a bright star for us, enlivening our lives and helping to show us the way in St. Louis. Engrossed in exploring all the opportunities that lay before us, we could easily ignore the political and social turmoil that was still swirling around the country.

My predecessor, David Gates, kindly stayed in St. Louis for a few days after our arrival to help me get oriented to the Garden. David and his assistant Mark Paddock offered sage advice based on their experiences, and Mark went over the list of personnel with me so I'd have an idea of the people with whom I would be dealing. I found these sessions, and the knowledge they provided about the opportunities and difficulties that awaited me, to be quite helpful. I certainly appreciated having some realistic suggestions as I eased into my new job.

In its basic outlines, that job involved supervising about 150 full- and part-time staff members, attending to the needs of about 2,500 members and perhaps 150 volunteers, managing a budget with a total income of $850,000 (about $5.4 million in today's dollars), and overseeing 79 acres of grounds and facilities that included a herbarium housing about two million specimens. The institution wasn't small, but it proved to be just small enough that I could learn on the job.

In addition to serving as the Garden's director, I had duties as a tenured full professor at Washington University, an arrangement that founder Henry Shaw had initiated in 1885. I would attend faculty meetings, teach occasional courses, and supervise or serve on the committees of some doctoral candidates—all activities that were familiar to me from my earlier experiences, and ones that I enjoyed. At the Garden, I continued research on some of the projects that I had begun while at Stanford.

The new John S. Lehmann Building was still under construction, and so I settled into the traditional Garden Director's office in Henry Shaw's former study on the second floor of his townhouse. That building had been constructed in 1849 at the corner of 7th and Locust in downtown St. Louis, moved to the Garden in 1891, as Henry Shaw had stipulated in his will, and enlarged subsequently. Sitting at my desk for the first time, I was well aware that as a Stanford professor, I hadn't had many opportunities to exercise a leadership role, much less one involving more than one hundred people and a somewhat complex organization. Fortunately, I turned out to have good instincts and plenty of fine help. With the support of a dedicated, talented Garden staff, I did my best to hit the ground running.

Needless to say, the role of director of a large botanical garden involves much more than science. I discovered very quickly that the director has to be a politician, a fundraiser, a long-view strategist, a tactician, an administrator, a change agent, a peacemaker, a provocateur, and of course a leader, among other roles. With so much on the plate, delegation and teamwork were absolute necessities.

My mother's philosophy was that you should know how to do everything yourself; my father, in contrast, believed in paying people with expertise to do things for you. At the Garden, it soon became clear that my father's approach was the only one that had any chance of working. No one person could do it all, or even keep his fingers on the pulse of everything that was happening in the organization. If I wanted to accomplish the ambitious goals I had set out for myself, I would have no choice but to delegate well and then rely on the results. I had learned in my previous jobs that I could trust others to do high-quality work, and—even more importantly—that *showing* your trust in them would encourage them to do their very best for you and for the institution. For me, this would turn out to be an effective approach to leadership.

It was a matter of first importance to become familiar with the Garden's plantings and horticultural facilities. Relatively little of the Garden's area was developed or presented as a display at the time we arrived in town; much of the grounds remained rough lawns, weedy fields, abandoned earlier plantings, and overgrown woods. I strolled through the Gladney Rose Garden and examined the large brick 1882 greenhouse called the Linnean House, which had originally been Henry Shaw's orangery. As I had a decade earlier, I marveled at the Climatron, the world's first geodesic dome greenhouse. I walked along the old north wall of the Garden, which projected east and west from the Linnean House, and visited the gazebo at the south end of what is now the Anne L. Lehmann Rose Garden, erected in honor of Anne's late husband John, who had passed away four years earlier. I examined the small rose garden that extended north from the Lehmann Gazebo; delighted in the herb garden built by the St. Louis Herb Society on the south side of Tower Grove House; and circumnavigated the small swampy lake in the southwestern corner of the Garden that had been spruced up a bit during David Gates's tenure. I paid a visit to the statue of Victory, carved in 1885 by Carlo Nicoli to replicate a statue in the Pitti Palace, Florence, and noticed the first line of the verse the female figure is writing on her shield: "The Victory of Science Over Ignorance. . . ."

But there wasn't much time to savor the Garden's beauties. As its new director, I had pressing tasks. My examination of the grounds led me to conclude that a great deal of work needed to be done. The physical plant needed repair and upgrading and the planted area of the garden was small and much of it run-down. Among the full-time employees there were only three research scientists. To do any of the needed work, hire new staff members, and begin to realize the Garden's obvious potential for growth, we needed more funding immediately. Despite the vigorous capital campaign of my predecessor, the Garden's finances remained too precarious to allow for stability, much less growth. Dr. Gates had obtained some large grants for his important research studies on plant biophysics, which carried some overhead that helped the Garden overall. But, especially with him leaving, the need for new funds was both large and urgent.

Among the other issues facing me was a rather weighty one involving the Garden's properties. In addition to the grounds on Tower Grove Avenue, the Garden owned nearly 2,000 acres of mostly undeveloped land at the Shaw Nature Reserve, which I had visited briefly while considering the move to St. Louis. It was located near Gray Summit in Franklin County, about thirty-five miles from the Garden proper. Much of this land had been acquired in 1925 when the trustees, increasingly concerned about the heavy, toxic smog that blanketed St. Louis and damaged the Garden's living plants, purchased three run-down, contiguous farms that became the nucleus of their new facility. They built greenhouses and began to develop the property as a place they could grow orchids in the fresher country air. This was especially important, since selling orchid blossoms for the then mandatory corsages was an important source of revenue for the Garden in the 1920s. Later, some of the trustees, enjoying cleaner air in St. Louis as the burning of coal for heat was phased out, began to wonder what they were doing with such a large swath of land well outside of the city. Although David Gates had begun the process of convincing the trustees of the land's growing value in the face of urban sprawl, the problem of defining its future remained, and many trustees were fretting about the usefulness of the Gray Summit holding to the Garden. With some urgency, I had either to find a logical rationale for keeping the "Gray Summit Extension of the Missouri Botanical Garden," as the facility was then called, or let it be sold to help meet the Garden's financial needs.

Overall, and from the start, I realized that the only potential source for sustained major support for the Garden was the St. Louis community. The Garden was a well-established, much-beloved institution. But the more attractive we could become as a place for both residents and out-of-towners to visit and enjoy—and the greater the number of satisfied members we had involved in our affairs—the stronger our financial base would become. With that, our ability to grow and do all the wonderful things of which I knew we were capable would strengthen as well. I was convinced that with good community support we would be able to do an outstanding job in horticultural display, education, and research; without it we would be able to do very little.

In order to gain such support, it was clearly important for Tamra and me to get to know St. Louis's social terrain. In forming a social network, we would become familiar with the community, including people and institutions who might eventually wish to make major contributions to the Garden or help in other ways. In that regard, moving to St. Louis was a bit like jumping onto the shoot-the-Chutes carnival ride. Soon after I arrived, with the special help of trustee Tom Smith, I was put up for membership in, and joined, the University Club, the St. Louis Club, and, eventually, the Veiled Prophet Organization, all of which helped in getting to know prominent people and their places in the community.

As time went by, we were gradually immersed in the world of movers, shakers, philanthropists, and wealth. It proved to be surprisingly easy to make friends with people of many different backgrounds. I learned what to say, when to say it, and how to convince people with very different political values about the importance of conservation and the environment, while also, where possible, illuminating the beauty and usefulness of plants. The joke-telling skills I had cultivated since second grade paid off. Get someone to laugh and you are halfway to a new friendship. And it always helped to remember that wealthy and prominent people are, after all, just people.

At parties, Tamra and I were often the only representatives of any cultural institution, which made us both novel and popular. We were also, typically, more politically progressive than almost anyone else there. It took careful explanations and the unemotional presentation of facts to keep matters on an even keel, but we usually succeeded. Fortunately, I am genuinely interested in people and what they care about, which also proved to be an advantage. The various causes toward which I worked—growing the Missouri Botanical Garden, improving life in St. Louis, preserving plant life, advancing science—were goals with which almost everyone agreed and was willing to support. I generally avoided political topics unless they directly pertained to the botanical cause at hand or to the future of the Garden itself. I adopted this approach instinctively, only later recognizing it as an important unwritten rule for how the head of an organization must conduct him or herself. Some of the people we met probably saw Tamra and me primarily as curiosities, but in these early days we developed warm rela-

The Chrysanthemum Ball, September 10, 1976, including Trustees Tom Smith (center) and Bob Hermann (upper left). Between them is Gloria Vanderbilt. Such events, new to us, helped to provide funds for the Garden's growing program of activities.

tionships with a number of people, some of whom would go on to give generously to the Garden.

It complicated matters a bit that while I was learning my new job, socializing with potential donors, and getting to know my colleagues and community, I was also busy finishing up and publishing several projects that had followed me from my Stanford days. My work on Onagraceae—now involving a number of students and other collaborators, including Peter Hoch, Maggie Sharp, Steve Seavey, Bill Tai, and Dave Gregory—continued strongly, and dozens of papers were published. Maggie Sharp, who completed her Ph.D. at Stanford with Paul Ehrlich, actually joined us in St. Louis as a postdoc that first year and began studies of pollination in the natural communities at the Gray Summit Extension property.

The Flora of North America project had been a major preoccupation during my last year at Stanford and continued to be so after my move to St. Louis. After discussion at committee meetings, the management of the project was transferred from the American Institute of Biological Sciences to the Smithsonian Institution in July 1971. We began to proceed with blessings from the Smithsonian and the understanding from Dillon Ripley, the Secretary, that if we were successful in obtaining National Science Foundation funding, the Smithsonian would gradually absorb the project into its budget. That year, we submitted a proposal for some $8 million to the NSF, with additional requests to Canadian governmental bodies. We envisioned seven regional centers in the U.S. and Canada, one of which was to be located at the Missouri Botanical Garden. The principals at each center would not receive financial compensation for their own time but would each have an editorial assistant paid for by the grant, as well as some funds for meetings and other expenses. Stan Shetler would coordinate the project from the Smithsonian. I was excited that this long-anticipated project looked like it might actually get off the ground.

As if my day-to-day responsibilities as director and my ongoing research activities weren't enough to keep me busy during my first year at the Garden, I also traveled a lot. Despite having become the director of a botanical garden, I didn't really know much about the way they operated in many fundamentally important areas. It seemed that one good way to learn what I needed to know was to visit others and learn from my new colleagues. What were other gardens like? What kinds of horticulture and display did they feature? What kind of research was being done? How did they interface with the community? I knew I had a great deal to learn.

My visit to Longwood Gardens, founded by Pierre DuPont on Du-Pont land near Philadelphia, was one of the more memorable. I was overwhelmed by Longwood's size, beauty, and budget. The grounds were immaculate, without a single weed or scrap of litter. Trying to figure out how that was possible, I asked how many horticulturists the garden employed. The answer was "two hundred." Two hundred! I then understood how they could keep the grounds so perfect. I also turned green with envy: at our Garden we had substantially fewer employees on the total roster!

Of particular use in learning about botanical gardens and what made them tick was being elected to the council of what was then called the American Association of Botanical Gardens and Arboreta (it's now called the American Public Gardens Association). Some of the other large gardens were not particularly active in the association, but I was very happy to support it, since I considered it to be the principal, and most important, professional organization for these institutions and their staff. I learned a great deal about gardens and horticulture from the eminent horticulturists and botanical garden leaders who served with me on the council: Betty Scholtz, Mike Dirr, John Creech, Bill Gambill, Fred Widmoyer, Francis Ching, Mai Arbegast, Les Laking, Dick Lighty, Roy Taylor (a classmate from Berkeley), and Joe Oppe.

After less than a year at the Garden's helm, I had a hard time when people began coming up to me and saying what a wonderful job I was doing. At that point in the game, their praise seemed ridiculous. I followed my natural instincts in passing the praise off to the staff members and volunteers, but soon recognized how much people want heroes. Individuals are taken to personalize a cause or an event, and although one can be modest about it, that turns out to be the way that people like it to be, to an unrealistic degree. It was one of many lessons that I had to learn, and something I had to accept.

∾

In the spring of 1972, almost a year after my arrival as director, the gleaming new Lehmann Building was ready to occupy. The facility had been designed to house the Garden's overcrowded herbarium and world-class library, both then jammed into inadequate space in the Administration Building. The architect was Gyo Obata, an internationally noted architect and St. Louis native whose parents I had gotten to know on the Sierra

Club Base Camp outing of 1951.[1] Gyo had made the decision not to connect the new facility directly with the existing Administration Building, but rather to erect a bold new independent structure, its one-way mirrored glass allowing one to appreciate the historical beauty of the garden from the inside and to see it artistically reflected from the outside.

Wanting to think in terms of advancing the Garden forward into the future, I chose to move my office to the Lehmann Building as soon as it was completed. Its clean, modern design appealed to me; its sleek lines and reflective surfaces felt conducive to research and progress. I had arrived in time to design my own office there, on the building's upper level and near the library, looking out at the beautiful row of tall dawn redwoods (*Metasequoia*) that had been grown from seeds transported from China in 1947.

To supervise the process of moving the herbarium and library from their crowded quarters in the Administration Building across to their new home, I recruited Tom Croat, a research scientist who preceded me at the Garden. Tom ingeniously rigged conveyor belts on which the specimens could be moved readily from the different floors of the Administration Building to ground level in front of the Lehmann Building. An exciting feature of the newly designed herbarium was its space-saving shelf compactors, which allowed the aisles between metal storage shelving to glide open and close as needed. David Gates had observed the use of compactors in herbaria while visiting Australia and had wisely integrated them into our new building's design on his return. We also moved our extensive library into the Lehmann Building, where it formed an integral part of our new center for research and study.

Along with the opening of the new facility came some changes and additions in the staff. My old friend Dick Holm, himself a graduate of Washington University's Missouri Botanical Garden program, had given me advice about hiring people when I left Stanford to take the director position. He said that the most important thing was to hire sane, self-possessed people, arguing that no matter how talented people are, they are very unlikely to be good, helpful team players if they are not reasonably balanced in their approach to life. That advice seemed wise to me, complementing my sense that I should trust absolutely the people to whom I delegated responsibility. So, as it became possible to expand the Garden's staff and fill open positions, I used an approach similar to what Steve Jobs has been quoted as saying: "It doesn't make sense to hire smart people and tell them what to do; we hire smart people so they can tell us what to do."

I had already made one major personnel decision. Paul Brockmann, a graduate of the University of Missouri's Forestry School, was an assistant in grounds maintenance at the Garden when I arrived. With Mark Paddock's strong endorsement, I had promoted him to be in charge of our whole maintenance program, leaving lawn cutting and trimming with our horticulturists. Paul quickly showed himself to be well suited for this key leadership role, one he has succeeded in for nearly fifty years.

Around the time we moved in to the Lehmann Building in 1972, I was able to persuade Dr. William Klein, whom I had met as a graduate student at Rancho Santa Ana when I was at UCLA, to come to the Garden as assistant director. Bill, a wonderful person with a mischievous smile, was actually on leave from the University of Colorado and agreed to remain in St. Louis only for the one-year duration of his leave. (To my delight, however, he would end up staying with us for a very productive five years.) Bill's friendship, humor, ingenuity, and problem-solving abilities were hugely helpful in many of our projects. He would succeed in raising the Garden's first grants from the state and help a lot in fundraising from other institutions and individuals as well. Besides, he was lots of fun!

With the move to the new building completed, Bill Klein installed as assistant director, and the initial challenges of learning the director role successfully met, I was ready by the end of 1972 to jump with both feet directly into the task of planning for the Garden's future. At this point, I had broad goals and some specific objectives, but little sense of how to go about systematically working toward accomplishing it all. I wanted to get the institution on secure financial footing, improve the grounds and facilities, transform the Garden into an even more popular destination for St. Louisans and visitors, greatly expand the research programs, and raise the institution's profile and reputation internationally. It was an ambitious set of interlocking goals, all requiring a great deal of work and substantial additional funding.

My natural tendency was to pursue a few objectives simultaneously and then add more as the first ones were realized. I had not come to my position with a grand scheme in mind, but rather would set off in directions that seemed both needed and promising—or, perhaps, in a number of directions at once—and be guided by events as they unfolded. As matters developed, they could be woven into a functional whole, and other goals added.

As I began working on plans for the Garden's future, I was fortunate to have a crucial set of knowledgeable allies and advisors in the form of the Garden's board of trustees. Particularly notable for his guidance was St. Louis philanthropist and civic leader Howard Baer. As president of the St. Louis Zoo board for many years, he had started the Children's Zoo and Zooline Railroad and, with Judge Thomas McGuire, had led the efforts to pass the Zoo-Museum Tax District, which united the shrinking city of St. Louis with St. Louis County in providing much-needed support for institutions critical to the health of our community. Howard Baer was rightly called "a guardian of civic institutions," and his was just the kind of vision, experience, and support I needed as the Garden entered this new chapter.[2]

A second important mentor was Tom K. Smith Jr. A longtime executive at Monsanto Company, Tom had been a member of the Garden's board of trustees since 1967. Tom helped me in many ways with the fine points of administration and budgeting, fields that were new to me when I took up my position at the Garden. He pointed out to me, for example, that it was not necessary to know a lot about a field when supervising someone else working in that field, but it *was* necessary to ask that person good questions and let them teach you by answering those questions well.

One key element of planning fell into place in a rather natural way. I was confronted with a decision as to whether to spend available funds to enlarge the gate building at the original main entrance on Tower Grove Avenue to provide more space for our gift shop. In considering this issue, it immediately became clear to me that a decision to put a structure in one place would end up precluding future options for other structures or features in that same place and nearby. With more than three quarters of the Garden's seventy-nine acres essentially undeveloped, there was plenty of room to add features and buildings, but short-sighted decisions could come back to haunt us later. We needed to consider the total space available and decide what we really wanted it to contain for the long-term future.

I discussed the need for planning with my closest advisors, Tom Smith and Howard Baer, and asked for their advice on how to proceed. It was Howard who brought full shape to what was nascent in the idea of planning: we needed a comprehensive master plan for developing the Garden.

We couldn't identify a suitable planner in St. Louis and so moved on to consider the national picture. Using the connections I had forged with other botanical gardens, we were able to get their advice on the matter of commissioning a master plan. We found that the Environmental Planning and Design Partnership in Pittsburgh had designed the Chicago Botanic Garden and was working on a plan for the Cary Arboretum in Dutchess County, New York, then part of The New York Botanical Garden. The firm clearly had the experience we needed. We hired them in 1972. One of the firm's partners, Geoff Rausch, a thoughtful man with a perpetually quizzical smile, became our principal, soon joined by his energetic partner Missy Marshall.

The planning team began its work by collecting our ideas on what features we would like to see in the Garden. They incorporated these ideas into a draft Master Plan, which we then studied carefully. Early on, we came together with the planners to the radical discovery that the historical entrance to the Garden on Tower Grove Avenue was not in the best place to introduce the Garden's diverse features and programs to the public. The Climatron dominated the first view of the Garden there too completely. There was no room for parking near that entrance, and no room to develop the facilities that we wanted to build, including classrooms and other facilities, a lecture hall, a restaurant, and a shop. A broad area at the north end of the Garden, lying beyond the historical internal limestone wall that projected from the Linnean House, still had room for the new developments that we envisioned and so seemed a good location for a new entrance. Developing the area in this way would involve tearing down the historical wall, replacing the antiquated maintenance buildings that lay just beyond that wall, and putting up a new entrance facility facing large parking lots to the north. At first I was shocked when Geoff and Missy suggested that we take this drastic step. But I soon came to believe that it was the right way to go. Despite our early fears, all interested parties readily approved the proposed shift in orientation, and the new entrance facilities became a major focus of our planning from that time forward.

In the early part of 1973, while the Master Plan was still being developed, it was fortuitous that a small delegation representing the St. Louis Chapter of the Japanese American Citizens League (JACL) came to pay me a visit. They had a question: would we consider placing some Japanese features in our Garden? I was immediately supportive of the idea because, quite familiar with the popularity of the Japanese Garden in San Francisco's Golden Gate Park, I knew such features would add interest for our visitors. Behind the Climatron there was a small pool that we had decorated with a lantern that one of our trustees had purchased when the Japanese exhibit at the St. Louis World's Fair in 1904 closed. I thought that the surroundings of that pool might be a perfect place for a small exhibition of the Japanese gardening style. As soon as Geoff and Missy heard of this idea, however, they basically asked, "Are you nuts?" The scope and placement I had envisioned didn't please them at all; it would be more suitable, they thought, to develop a much larger Japanese garden near the improved lake that they had planned for the southwest corner of the Garden. Some trustees were skeptical for different reasons; they wondered if it would be appropriate to include such a starkly different kind of garden in one that had fundamentally English roots. It seemed to me, though, that we had the scope to display different sorts of garden design, and that doing so would enrich the experience of our visitors. I also saw the Japanese Garden as a way of strengthening the global outlook that was indispensable for building a sound regional and national future. Eventually my view prevailed, and the Japanese Garden was incorporated into the Master Plan. (It would rapidly become the Garden's most popular display feature after its completion in 1977.)

Once the draft Master Plan had been accepted by the Garden's board of trustees, we were ready to begin to implement some of its elements. One of the first projects was the construction of the A. Wessell Shapleigh Fountain. Dedicated to the memory of Warren Shapleigh's father, a former trustee of Washington University, it featured the pink Missouri granite that is so characteristic of that campus. When it was finished, it sat alone in the woods among grassy fields, away from any existing Garden feature, but eventually the completion of other elements of the Master Plan would make it a focal point for visitors. We envisioned its rhythmically rising and falling jets of water being a lasting source of fun for generations of children visiting the Garden on hot summer days.

It seemed that things were going very well. My fundraising was beginning to bear fruit, my desire to expand the Garden's research programs was being realized, and we were taking the first steps in transforming the exciting vision of the Master Plan into reality. I felt very optimistic about the future. Although we would continue to move forward overall, the inevitable bumps in the road did occur. In 1972 we had received our grant award for the Flora of North America project from the National Science Foundation, and I had stepped up my efforts to organize an editorial center for the project at the Garden. Not long after we had celebrated what seemed to be the launch of the Flora of North American project, however, it came to a screeching halt. The Office of Management of the Budget ruled that the Smithsonian could not approach Congress for incremental funding then or in the years to come. The Smithsonian being unwilling or unable to appeal that decision, the NSF withdrew the funding for our program in early 1973, forcing us to suspend all the operational aspects of the embryonic FNA program. We began looking for new ways to keep our project moving forward, but it was many years before the program became operational once again, with the first two volumes not published until 1993.

❧

While all of this was taking place, my job as director continued to be heavily consuming, but it was certainly not the only dimension of my life. Alice and Liz were growing up rapidly. They both did well at school, ultimately completing their studies at Wade and entering John Burroughs, a high-quality private school that was recommended by several of the Garden's trustees and other friends.

My parents had entered their retirement years, and I made a point of visiting them several times a year. They loved it when I brought along Tamra, Alice, and Liz. My father, a kid at heart, enjoyed playing with the girls. Their relationships were warm and mutually loving. With Social Security, retirement savings for each of them, income from the Breen family ranch, and the proceeds from the sale of the house in Sea Cliff, my parents had enough money to live comfortably for the rest of their lives. When they decided to move out of San Francisco in 1974, they chose Rossmoor, a large retirement complex near Walnut Creek that offered a combination of independence and support. It was fun to think that they chose the same community where my father's ancestors the Bakers had farmed in the 1850s, at a place that was also near where my great-grandmother Amelia had settled with her aunt when she traveled from Australia as a teenager in that same decade.

At Rossmoor, my parents purchased a nice duplex up on a lovely hillside near the community's clubhouse. Dad got into hiking and volunteered for Meals on Wheels, and they joined a church nearby. My parents were well-liked people who loved parties and socializing, and Rossmoor was a good place for them to live. They both appreciated their daily cocktail hour, taking the time to catch up on the day with other residents while sipping vodka martinis. Around their attractive home, my mother pursued her love of gardening and kept her surroundings as pristine. I was happy to see that their marriage had continued to grow and deepen over time and that they were a stable, loving, mutually supportive team.

ex

Through the mid- and late-1970s, development of the Garden's displays and facilities continued apace. One of the first and largest pieces was the design and construction of the Japanese Garden. Since Geoff and Missy lacked the background it would have taken to design a proper Japanese Garden, we had to search for someone who could lay out the space properly. George Hasegawa, head of the local Japanese American Citizens League chapter, put out a call to other chapters around the country to see whether they might have anyone to suggest, and indeed they did. Professor Koichi Kawana, of the Department of Art at UCLA, was in charge of the Japanese gardens then owned by UCLA. He had recently designed some original Japanese elements for the plantings around the Los Angeles Waterworks. It seemed that Professor Kawana and his partner Carol Parish might be the ones to fill the bill for us.

Following his first visit to St. Louis, Koichi produced a painting of how he thought the eventual Japanese garden might look, and we knew that any idea of having a small Japanese garden "near" the lake was history.

In Koichi's bold concept, *Seiwa-en* (the garden of clear, pure harmony and peace) was to encompass the whole lake, in the style of a strolling garden that might have been built by a wealthy individual in Japan in the late nineteenth century.

Presented with such an exciting vision, Bill Klein and I began seeking funds to make our Japanese garden a reality. With the help of state House of Representatives leaders Ken Rothman and Richard Rabbitt, both from St. Louis, we secured state appropriations of $300,000 in 1975 for excavating the greatly enlarged lake and then $75,000 in 1976 for additional work. Bill Klein worked with the legislature and the governor's office tirelessly to persuade them to provide these appropriations for our project.

In order to be eligible as a private institution to receive these funds, we granted easements to the state in perpetuity for the land where the Japanese Garden was to be located. We also received grants from the Missouri State Arts Council, and some of the design elements were funded by the National Endowment for the Arts. In addition, several of the trustees and other enthusiastic patrons of the Garden supported the project generously. After a groundbreaking ceremony in the fall of 1974, the project was underway.

Actual construction began in the summer of 1975, starting with a year of excavating for the future lake. While the digging was proceeding, Koichi visited for a few days every two or three months and began to place stones—which we brought in from selected quarries in Pennsylvania, Colorado, and elsewhere by the truckload—in specific orientations and at specific places around the shore and on the Tea House Island and other islands in the lake. He would stand and direct the placement of the individual stones to a man with a crane, and once that was done, the stones would be two thirds buried in the soil, in keeping with Japanese custom. During the times when Koichi was unable to be on site, Karl Pettit III supervised the placement of stones and features, expertly carrying out Koichi's plan. A great deal of triangulation was necessary to be sure that the various stones and lanterns would appear at exactly the right position on the lake's edge when it was refilled. In addition, wooden rangui posts had to be placed around part of the lake's margin, the two waterfalls built and their stones anchored, and bridges constructed. Karl either supervised these tasks or carried them out himself; he also designed and supervised the construction of the Plum Viewing Arbor and the restrooms, in consultation with Koichi, who remained in charge of the overall project management.

The project was large and demanding, so it was exhilarating to see the new garden gradually take shape. The stone lanterns, water basins, waterfalls, and eventually a teahouse—a gift from Missouri's sister state in Japan, Nagano Prefecture—slowly came together to make an aesthetically pleasing whole. As the work progressed, we built out the individual gardens, planting them primarily with traditional Japanese plants including oversize bonsai pines, flowering cherries, peonies, azaleas, water lilies, and lotus. For the selection of these plants, our outdoor horticulturist John Elsley, newly arrived from England, suggested sources and choices to Koichi and secured the plants. He, Karl, and Koichi were a formidable trio, and we owe the great beauty of the resulting garden largely to them.

The completed garden occupied more than 14 acres, almost a fifth of the entire Missouri Botanical Garden, with its features laid out around the irregular shore of a 5.5-acre lake. In keeping with the traditional Japanese idea of a garden as a symbolic microcosm of nature, the garden was simultaneously a horticultural display, a work of art, and a place inspiring contemplation and peace. Every aspect was imbued with meaning and purpose. Koichi told me that if I ever allowed anything in it to be painted red (a later design feature that originated in China) he would surely come back and haunt me forever.

The Japanese Garden was formally dedicated on Thursday, May 5, 1977. During the ceremony, a group of Shinto priests held traditional ceremonies by the lake in somewhat cloudy weather. Their leader had assured me that it would not rain, and sure enough it did not. My parents were able to be there for the ceremonies. My father and I stood on the hill at the north end of the garden, looked out over the crowd, and watched everything that was going on. Dad was quiet, but I could tell how pleased he

In 1977, we celebrated beginning to fill the lake in *Seiwa-en*, the Japanese Garden. From the left are pictured Roy Taylor (University of British Columbia Botanical Garden, visiting), me, Geoff Rausch, Karl Pettit, Koichi Kawana, Jim Casper (representing our contractors, Harold Casper & Sons), and John Elsley.

was. In return, I felt warmly about all he had done through his life to encourage and support me.

On the following Sunday we held a Japanese Festival in the Garden—the first edition of what would turn out to be an extremely popular annual event. That day more than 18,000 visitors crowded into the Garden—really too many for the space, but a wonderful testament to the general excitement that the new garden had generated in the community.

Other developments soon followed, some of them directly related to the Japanese Garden. John Elsley conceived the idea of developing a woodland garden in the English style using the overgrown, garbage-filled block of woods between the Japanese Garden and the Lehmann Building. He saw to the implementation of his dream while completing the Japanese Garden. A lovely stream featuring Karl's stonework flowed through the woodland garden, entering the lake as a sparkling, three-step waterfall.

Another readjustment to the developments that were taking place at the Garden was caused by an important physical loss that we suffered in 1978. Walking near the Organization for Tropical Studies' La Selva Field Station in Costa Rica early that year, I was intercepted by people bringing me bad news. A fire had started in greasy pipes above the stove in our Snack Bar on the upper level of the Floral Display House, jumped to the plastic roof of the structure, and burned it to the ground in a matter of minutes. For many years, we had depended on that structure to house our popular seasonal flower shows. Its loss was a major blow. Bill Klein and I put on our most creative thinking caps, deciding that in the short term we would move our orchid shows to the Climatron, focus on the rose gardens, and exhibit chrysanthemums outside in order to maintain the seasonal rhythm of special shows to which our members and visitors had become accustomed.

At the same time, it proved possible to increase the size of our land-holding in Gray Summit. In this connection, we learned that the Butler family had decided to sell their property, a parcel of 223 acres at the east end of the Garden's property. They lived in a lovely house that St. Louis philanthropist David P. Wohl had built in 1932 to use as a hunting lodge. With a state easement and a generous contribution from Adlyne Freund, another St. Louis philanthropist, we were able to acquire the property. Combined with another large parcel adjacent to the west side of the property that the Garden had received from the estate of Stratford and Elise Morton by way of the Nature Conservancy, the Gray Summit landhold-

THE MISSOURI BOTANICAL GARDEN

ing had grown to 2,532 acres—nearly four square miles—and its future was essentially guaranteed. Bill and I decided to streamline its name and call it the Shaw Arboretum.[3]

Once the Japanese Garden was completed, it came time for us to turn our attention to a capital fund drive that would make possible the construction of the new visitor reception facilities at the north end of the Garden. We were able to secure generous major funding for what would be called the Ridgway Visitor Center from Dolly and Elmer Kiefer. Dolly's father, Edmund Ridgway, had been one of the founders of the 7-UP Company, and had benefited from its sale in 1978 to Phillip Morris Company.

Planning and additional fundraising for the Ridgway Visitor Center began in earnest in 1978. In the light of Gyo Obata's success with the Lehmann Building, we once again secured his services as its architect. We designed the new facility to house our reception and orientation areas, a floral display hall to replace the one lost to fire, a spacious auditorium, a gift shop, a café, a gallery space, and educational facilities, with a large parking lot between the building and the north end of the Garden. Gyo came up with plans for an iconic building that would blend the best of old and new, its striking glass barrel-vaulted ceiling inspired by Joseph Paxton's Crystal Palace, the site of the 1851 Great Exhibition in London. This was particularly appropriate in view of the role that Paxton's design for the Duke of Devonshire's greenhouses at his estate at Chatsworth had played in Shaw's original decision to build a botanical garden.

When it came time to build our visitor center, however, the U.S. was being racked by a period of runaway inflation, peaking at nearly 15% annually in April 1980. We had not foreseen the resulting rapid increases in building costs; the financial goals we had set were based on much more modest annual increases. The situation became so bad that the board of trustees, under the leadership of President Tom Smith, held a meeting to decide whether to continue with our plans to construct the facility or to return the money to the donors and wait for a more suitable time. At the meeting, Howard Baer told a story of standing with his wife Isabelle on the steps of the 700-year-old Ducal Palace at Mantua in northern Italy at twilight and hearing an elderly man nearby comment, "No one asks how much it cost any more, do they?" Inspiration overcame fiscal conservatism, and we proceeded with our new entrance building.

We had to cut some corners to realize our overall plans. Even though we projected an attendance that would climb to an estimated million vis-

itors annually, we had to narrow the barrel vault, cut out the
viewing platform overlooking
the floral display hall, and elim-
inate a few other features. We
were also very careful in choosing the materials that we would
use in construction. Fortunately, I had been able to hire
a very talented woman named
Pat Rich to supervise the construction. She proved particu-

Our construction manager Pat Rich showing the
Garden's trustees the state of construction of the
iconic Ridgway Center some months before its
opening in 1982.

larly well suited for the job, her first after raising her children. We got
through the experience, and the beautiful new building was opened to
the public in 1982.

⁊

While the Garden's St. Louis facilities expanded and drew increasing
numbers of visitors, and our research programs overseas multiplied and
grew, my family experienced a number of significant milestones. In the fall
of 1976, Alice, sixteen years old, left home to study in France for a year.
Shortly thereafter, Tamra announced that we were going to have a child.
Liz, who had turned fifteen earlier that year, was suitably amazed and
delighted at the prospect of having a little brother. Francis Clark Raven,
named for my father, Walter Francis Raven, and Tamra's mother, Marjory
Clark Engelhorn, was born at Barnes Hospital on July 3, 1977. Times had
changed since Alice and Liz were born, and I was present in the delivery
room for his birth and not in a smoky waiting room somewhere down
the hall. Life at home changed dramatically with a baby in the house; I
was happy to have a son and once again be a father to a little one.

Sadly, in 1980, my own father was diagnosed with liver cancer. It was
clear he didn't have long to live—probably less than a year. I made visits
an even greater priority. My father and I would take walks together during
which he talked about his life and the world. I got a much fuller picture
of his early days during that time and gained the impression that, all in all,
he was content with his lot. Despite the many disappointments he had
suffered he remained a sweet man, entirely without anger or bitterness. In

June 1981, the night before he died, I visited him at the hospice where he was staying. Looking up at me, he remarked, "You've been a very good boy." I replied, "You've been a very good daddy," and I leaned over and kissed him. What a gift to have had that final exchange! In the morning, staying with my mother, I learned that he had died quietly during the night.

After the funeral, according to his wishes, we had his ashes scattered over Mount Diablo. He had spent many happy days hiking with Rossmoor friends along the trails on that beautiful mountain, such a landmark in the East Bay, and his final wish showed how much he felt at home in the area. Returning to St. Louis, I endowed a pair of benches at the Garden in his memory, placing them near where we had stood together for the opening of the Japanese Garden four years earlier. At that time, I had had no idea that he would be gone so soon.

After my father passed away, Mom stayed for a while in their house, before moving to an apartment in a large community building nearby at Rossmoor. Though a tall woman, she had grown stooped and begun to suffer from rheumatoid arthritis, especially in her fingers. She also had cataract surgery and trouble with her eyes, but she remained convivial and social. As a testament to her likability, her ophthalmologist, Dr. Ronald Shigio, enjoyed seeing her so much that he and his wife virtually adopted her and stopped charging her for visits. Life in the community building was perfect for her, since she could cook on her own or go downstairs for prepared meals, and in either case visit with friends. Under those circumstances, she maintained an active social life that she enjoyed greatly over the next several years. She had a spare room, and I visited as often as I could.

Another delightful family addition came a couple of years after my father's death. Kathryn Amelia Raven was born September 11, 1982. Kathryn is the namesake of Tamra's paternal grandmother, who was an opera singer. Her middle name, Amelia, honors my maternal great-grandmother, the one who had traveled alone from Australia to California as a teenager in the 1850s. When we brought Katie home, I held her on my arm in our garden and said to her, "Oh, dear baby, you have much better opportunities ahead of you than practically any woman born in earlier generations." I remember that moment well, and Katie—who soon became known as Kate outside our family—has achieved a great deal working diligently for liberal values over the course of her life.

With two young children in the house, it almost seemed as if I had been transported back in time. On Sundays we made a great production of cooking waffles together. Francis would "shave" with me on many mornings, using shaving cream and an old toothbrush. We also had many enjoyable family evenings. A favorite destination was a Thai restaurant on Grand Avenue called "The King and I," run by a Thai gentleman named Suchin Prapaisilp and his family. When little Kate got sleepy, one of the women from Suchin's family would take her in her arms and walk with her, rocking her gently. Then we'd put Kate to sleep on a folded tablecloth right next to the table.

On weekends, we frequently drove out to the Shaw Arboretum and stayed at the original elegant brick home that has since been renovated and renamed the Bascom House. We took the two younger kids on summer vacations to the West every summer, camping on visits to the Rockies, Yellowstone, parts of California, and ultimately Alaska. Fortunately, there were no cell phones or email accounts then, so it was possible to really get away. I collected a few plant specimens, mostly Onagraceae, and we thoroughly enjoyed the places we saw and the time we spent together. By this time, Alice and Liz had grown up into delightful young women—bright, capable, and curious.

გა

Throughout the Garden's period of expansion and renewal during the 1970s and early 1980s, its finances remained a constant source of concern. I was very successful in convincing a number of wealthy St. Louis citizens to make generous donations, usually for specific purposes, and the constantly expanding numbers of visitors and members brought in an ever-growing income. But what we really needed for stability and security and sustainable growth was a significant amount of permanent, annual public funding. Although we had some temporary successes in obtaining some public funding year to year during the 1970s, we didn't achieve this long-term goal until 1983. When it was finally realized, the steady supply of funds was a great boon to the Garden's fortunes and future, and in many ways marked the beginning of a new era for the institution.

In April 1971, just before my arrival as director of the Garden, St. Louis voters had passed a ballot initiative creating the Zoo-Museum District, which provided tax-funded support for the St. Louis Zoo, the St.

Louis Art Museum, and the very young St. Louis Science Center. The funds came from set millage amounts added to the real estate tax bills of properties in both the City of St. Louis and St. Louis County. As mentioned earlier, our trustee, Howard Baer, had played a major role in conceiving that plan and nursing it into reality. The Zoo, Art Museum, and Science Center were all publicly funded institutions, so providing them with tax dollars through the District was consistent with the existing legal framework. Although the Missouri Botanical Garden was very much like the other three institutions, it was private and so fell outside that legal framework—or so we believed when we first began considering the problem of its possible admission to the District.

After a series of frustrating failures to obtain some modest appropriations from the state, Bill Klein and I began looking for alternative ways to obtain tax support. One we day called on Ken Rothman, who had helped us so much in obtaining state funds to build the Japanese Garden. Ken suggested that even though we were a private institution, there must be some way to retain that status and still gain regular public support. Our piecemeal approach wasn't working, and we felt that Ken's challenge to find a regular source of funds from the state was definitely one that we ought to accept. We considered that this would probably involve somehow gaining support from the Zoo-Museum District.

Challenged by the concept, we set to work trying to find a way to accomplish this goal. The Garden's attorney Frank Wolff stepped in and designed a model: a public Botanical Garden Subdistrict would be set up at the same level as the governing bodies of the other institutions; the tax funds would be appropriated to the Subdistrict and granted to the Garden on the basis of a successful annual application. The model passed legal muster, but we needed enabling legislation from the state legislature in order to go to the city and county voters for approval of our tax rate. Garden Assistant to the Director Rick Daley, who had recently succeeded Bill Klein in the post, began working full time on the matter in the autumn of 1979, getting to know the state legislators and assessing their potential for supporting our effort. When the 1980 legislative session opened, it soon became clear that we had the support we needed in the House, but the Senate was quite another matter. In the end, the Senate failed even to consider our bill during a welter of last-minute horse-trading.

After another dismal failure in 1981, we staked our hopes on the 1982 state legislative session. The issue now was that we charged admission, while the original members of the District did not. We proposed opening free on Wednesday and Saturday mornings, and that satisfied the opposition. The enabling legislation finally passed both houses of the legislature. But that was only the first step; we still had to win voter approval on the ballot.

For the measure to pass, separate majority votes for each institution were required in both city and county. Hiring advisors, we worked with the three existing members of the cultural tax district, all of whom requested increases in their own funding at the same time. Together, we appealed to the public in August 1982, not being sure whether our efforts to gain wider public support had been sufficient to win approval. In any case, *all* the institutions were defeated in that election. As we glumly pondered the results at a post-election gathering in the Carpenter's Hall on Hampton Avenue, we wondered whether any path to success was open to us.

Within a few weeks, Bob Hyland, a strong civic leader, stepped forward with an offer to lead the four institutions in a new attempt to win voter support. Like Howard Baer, who with Judge Thomas Maguire had led the initial drive a decade earlier, Bob had been a long-term president of the Zoo: he badly wanted to increase funding for his favorite institution. With some misgivings about the expense of doing so, we joined the group, participating in a campaign marked by Bob Hyland's strong leadership. He simply refused to take no for an answer. Not everybody liked that about him, but he got things done, and it was great to have him on our side.

After a second, more energetic campaign, our measure came up to Election Day on April 5, 1983—almost exactly twelve years from the day that the original legislation forming the District had passed. When the results came in, we found that we had won easily in the city. But the county results were confused, and the recount would take several more days to accomplish. Listening to the radio as I was driving to a National Academy of Sciences regional meeting at the University of Illinois in Urbana a few days later, I finally heard the verdict: the initiative had passed in the county as well.

We had won our long, hard-fought battle. From that time on, the Garden could operate from a greatly strengthened financial base. Of all the changes at the Garden during my tenure, this was, perhaps, the most meaningful and enduring.

෴

After twelve years as director of the Garden, I had a sense of accomplishment, of having taken a big risk and successfully met the challenge. Over those years, we had implemented a master plan, built the English Woodland and Japanese Gardens, built the Ridgway Center, greatly expanded our research program, enlarged the Shaw Arboretum by a third, and gained long-term tax funding from the Zoo-Museum District. With the Garden on a new, successful course, it was a natural time to consider finding a new personal challenge. That possibility arose, right on cue, when I was invited to apply for the position of Secretary of the Smithsonian Institution in 1983.

I had previously received several inquiries about positions at other institutions. With only one exception, I had considered those positions not sufficiently different from the one I already held at the Garden to warrant even thinking about moving. I agreed to a very few interviews as a matter of courtesy and tried to keep an open mind, but the alternatives never seemed right for me. My position at the Garden was energizing, enjoyable, and well paid, and it allowed me to concentrate on plants. It would take a very special opportunity—and an enticing new challenge—to induce me to consider moving on. Leading the Smithsonian certainly fit the bill.

The opportunity to be considered as a candidate for the Secretary of the Smithsonian came to me through my colleague Don Duckworth, a friend from OTS meetings in Costa Rica. Don worked in the office of the institution's Assistant Secretary for Museum Programs, Paul Perrot. In this capacity, he had learned that Dillon Ripley would be retiring as Secretary in 1984 after twenty wonderfully productive years in the position.

Encouraging me to allow myself to be considered as Dillon Ripley's successor, Don guided me through the process. It was a plum job, much noted in the press, and one in which I could definitely see an expanded (if not overwhelming!) role for myself. In preparation for a possible move, I studied the culture and daily life of Washington, D.C., where we would live if I were to be selected for the position. I quickly came to the conclusion that I would have to learn to dance and to stay up late: the social life of upscale St. Louis was quiet compared to that of the nation's capital. I was interviewed by the Regents collectively and by some individually. I called on the Congressional Regents one by one and did everything that I could think of to present myself as a suitable candidate. Bill Bowen, president of Princeton University and head of the Regents' search committee, interviewed me several times and visited St. Louis to interview members of the Garden's board and other civic leaders. I couldn't help wondering what they were telling him!

For much of the time the search was being conducted, from 1983 into early 1984, I was considered the front-runner, so much so that Barry Goldwater, who was a Regent from the Senate, stated publicly that he thought that the next Secretary for the Smithsonian "is going to be a botanist from Missouri." Fairly far along in the process, matters seemed to be developing in my favor.

In 1983, I had been elected a member of the Council of the National Academy of Sciences, a role in which I got to know many distinguished scientists, including Harry Gray, Millie Dresselhaus, Frank Press, Maxine Singer, and Alex Rich. One of my new acquaintances was Robert McCormick Adams, a distinguished archeologist who was then serving as provost of the University of Chicago. In the spring of 1984, at a Council meeting in the board room at NAS headquarters, I was sitting next to Bob, my mind doubtless wandering over various aspects of the Smithsonian, when he turned to me and asked, "What do you know about that strange institution down on the Mall, the Smithsonian, and their Natural History Museum?" Seeing the surprise run over my face, Bob explained that Bill Bowen had been attempting to persuade him to apply for the position, and that he was considering the possibility.

Bob Adams, at that time fifty-eight years old, had at first declined to apply for the position but changed his mind in the summer of 1984. Ten years my elder, and with extensive administrative experience, he must have seemed the more attractive candidate. Perhaps Dillon Ripley's twenty-year term had made the Regents wary of younger candidates. Well before the day when the appointment was to be announced, I knew that I had not been selected, since if I had been, I would have been summoned to Washington. On the morning of the chosen day, September 17, 1984, Bill Bowen called and gave me the news. Bob Adams had been selected. I had had a seemingly charmed history of getting the opportunities I wanted, so losing the post was something of a surprise as well as a disappointment. I loved the Garden, however, and was not the least bit unhappy to recommit myself to my work there.

– 10 –

Global Reach

WHEN I ARRIVED at the Missouri Botanical Garden in 1971, it was very much like other medium-scale botanical gardens around the world, with a varied collection of living plants assembled for display and education. When it came to research, however, the Garden had a few characteristics that set it apart from its peers. Its extensive herbarium and library were of worldwide scope and therefore more like those of larger botanical gardens such as Kew, New York, and Geneva. The Garden also had a strong and productive partnership with Washington University, dating back to 1885, that had enabled the training of hundreds of graduate students over the years. Finally, although its own field activities had mostly been confined to the U.S., there was one long-standing foreign program concerned with the flora of Panama.

To the extent that I had thought about the matter, I was sure that an international program could be built at the Garden. My personal outlook— programmed into my personality from infancy and fostered by living in London studying largely African and Asian plants, extensively exploring New Zealand and Australia, and teaching and collecting in Costa Rica—

was naturally directed abroad, and with the collapse of the Flora of North America project there were more resources and time to devote to other areas.

Because of its age, important collections, and close connection with Washington University, the Garden had a solid foundation as a research and educational institution. During the first half of the twentieth century, many prominent botanists matriculated from the Henry Shaw School of Botany at Washington University and became faculty both in the U.S. and abroad. Like those of many comparable institutions, the Garden's activities slowed following World War II, but they were accelerating by the time I arrived as director in 1971. Particularly important in the revival was Walter Lewis, who had joined the Garden's staff in 1965, a short time before David Gates. Lewis, who was a participant in the Flora of North America project, had revived the Garden's scientific publication, the *Annals of the Missouri Botanical Garden*, and pushed along the *Flora of Panama*. He played an especially important role in revitalizing the Garden's graduate program, forging new links with St. Louis University, the University of Missouri–St. Louis, and Southern Illinois University–Edwardsville that would prove fruitful over the years. His contributions, like those of David Gates, were key to the growth of the Garden overall; they gave the Garden's research program a new impetus and a way forward that it otherwise would have lacked.

When it became my turn to think about the future, it seemed logical because of our work in Panama to get started in South America, where the rich forests and other ecosystems badly needed additional exploration, as soon as we could. In this endeavor, I had the assistance and inspiration of Tom Croat, a rising expert on Latin American plants, who, as I've mentioned, was the imaginative mover of our herbarium into its new quarters at the Lehmann Building. Tom earned his Ph.D. at the University of Kansas and initially moved to St. Louis to prepare an account of the plants of Barro Colorado Island. The island had been formed by the flooding of the Panama Canal and then operated as a research site by the Smithsonian Institution, who wanted a solid account of its plants as a basis for other biological studies there. The flora was published in 1978, and along the way Tom chose to specialize on the large and poorly known arum family, Araceae, which, it turned out, had thousands of still undescribed species in Latin America. Continuing with his interest in other plants, Tom eventually became one of the most prodigious plant collectors of all time. In

Tom Croat, prodigious collector of tropical plants, is pictured with a huge leaf from his 100,000th collection, the type (original) specimen of *Anthurium centimillesimum*, from the moist forests of western Ecuador.

a half-century at the Garden, he would collect well over 100,000 individual plant specimens, contributing especially strongly to our success in Latin America. Only one or two individuals in the entire history of botany have collected more.

In addition to Tom, we needed someone to direct the Garden's research program and form a platform for extending our reach. The person who seemed right for this position—Marshall Crosby—was already, like Tom, a member of the Garden staff. Marshall had been hired by Walter Lewis upon his graduation from Duke University as a curator and the editor of the *Annals of the Missouri Botanical Garden*. After I appointed Marshall as research director in 1972, we worked together to build the Garden's research staff.

To keep the *Flora of Panama* moving toward completion, we hired Bill D'Arcy, who was already working on the project with grant support. The project was twenty-nine years old, and Bill pursued it to completion and put together a very useful checklist of the plants of Panama along the way. In addition, he pursued studies of the potato family, Solanaceae.

We were then successful in obtaining a grant from the National Science Foundation for the curation of our newly housed herbarium. Al Gentry was just at the point of graduation from our program in 1972, and being fully aware of his capabilities, we hired him at the first opportunity, using funds from that grant. With an energetic disposition, a facility for languages, and a fervent, irrepressible devotion to collecting and learning about plants, Al proved to be perfectly suited for the rigors of fieldwork. At my suggestion, he continued his research in the incredibly rich and poorly known western region of Colombia, the Chocó. Al partnered with Colombian botanist Enrique Forero, and their joint project exploring the Chocó pulled us directly into South America for the first time. Al soon proved himself to be not only a prodigious collector but also an innovative analyst of tropical plant communities, pioneering the technique of sampling along transects as a way of assessing the structure and composition of tropical forests—far superior to the results that could be gained by collecting individual specimens wherever they were encountered in a given area. Leveraging a remarkable ability to identify species from their leaves and branches only, not just the ones that had flowers or fruits attached, he was able to discover patterns of distribution that others had missed. His outstanding *Field Guide to the Families and Genera of Woody Plants of Northwest South America* was published in 1993. Tragically, only a few months later, the small plane in which he was riding between field sites in Ecuador crashed, killing Al in his prime at the age of forty-eight. He had given his life for his love of plants.

The same NSF grant with which we hired Al also enabled us obtain the services of Gerrit Davidse, who was just at the point of graduating from Iowa State University. A student of grasses, Gerrit set about to explore several areas of South America to gain an understanding of their regional differences. Gerrit would go on to become an internationally recognized expert on New World grasses, a pillar of the Garden's research efforts, and eventually the coordinator of *Flora Mesoamericana*.

Another significant addition to the Garden's staff occurred in 1972, when we were able to recruit Ron Liesner from the Field Museum. As a member of the herbarium staff, Ron spent most of his time sorting and identifying the large piles of specimens that now were streaming in from all parts of the globe. By organizing the specimens, he made them accessible for further study and insured that they were filed properly in the herbarium, findable by specialists examining our material for plants in the groups in which they were particularly interested. We would simply have been unable to accomplish so much as an institution in so many places in Latin America and elsewhere if Ron had not been a key member of our program.

At about the same time I was bringing Al, Gerrit, and Ron to the Garden, another opportunity presented itself. James Duke, of the U.S. De-

Al Gentry pursued his studies of tropical plants tirelessly until his premature death at the age of forty-eight, the result of a 1993 airplane crash in a small plane in Ecuador.

partment of Agriculture, had indicated the possibility of funding research on the relatives of opium poppy, which were then poorly understood.[1] My close friend Bob Ornduff suggested that a young South African botanist named Peter Goldblatt, whom he had met on a sabbatical visit, might be an ideal candidate for carrying out the study. We hired Peter, who proceeded to conduct extensive fieldwork on the poppies in Turkey and Iran, followed up by laboratory studies in St. Louis. Over two years, Peter clarified the relationships among the group of poppies that he was analyzing. Fortunately for him and certainly for the Garden, it would become possible to hire Peter on a permanent basis after he completed the poppy project, a development that facilitated the expansion of our field-based programs into Africa.

⌘

With its extraordinarily diverse—and, to North American eyes, unusual— flora, Africa had long attracted my interest as a botanist. I was not alone in this sentiment; based on our relatively strong holdings of historical herbarium specimens from the continent and with a large measure of hope, Walter Lewis had, just before my arrival, suggested that the Garden become the New World center for the study of African plants. As of 1973, however, no one from the Garden's staff had ever specialized on African plants or collected there, even though we already had accumulated more

Gerrit Davidse, the John S. Lehmann Curator of Grasses at the Garden, was hired at the same time as Al Gentry in 1972.

No person has played a more important role in the growth of the Garden's herbarium from about two million specimens in 1971 to about seven million today than Ron Liesner. Ron possesses an uncanny ability to identify plant specimens from all over the world.

African herbarium specimens than any other New World institution. But that was about to change.

My enthusiasm for doing research in Africa was kicked up a notch with a visit to South Africa and Kenya in the summer of 1973. Tamra and I had been invited to South Africa by the newly appointed director of the country's Botanical Research Institute, Bernard de Winter. The visit was intended to help celebrate the formal opening of the Institute's new main building with a symposium jointly sponsored with the Botanical Society of South Africa. After we landed in Johannesburg, Bernard and his wife Mayda greeted us and escorted us to their home in the South African capitol, Pretoria.

After festivities in Pretoria, we were whisked off to see something of the country. Ours was a distinguished group, including Jack Heslop-Harrison, the director of the Royal Botanic Gardens, Kew, and his wife Yolande, a distinguished botanist in her own right; Hermann Merxmüller, head of the Botanische Staatssammlung and of the botanical garden in Munich; and Bob (R. B.) Drummond, head of the Botanical Research Institute in what was then Salisbury, Rhodesia, who had an amazing stock of knowledge about the plants we would see.

Heading over the mountains northeast of Pretoria, we stopped in the open forest on a ridge to look at the plants and scenery in late autumn. In the drier and sub-desert areas we saw some wonderful aloes and other desert plants; learned about the characteristic dry forests of the region, dominated by widely spaced Mopani trees; and held intense discussions of all we were seeing. Overall, it was a great treat for us all to be in such good company as we crossed the Eastern Transvaal.

Dropping to a lower elevation, we entered the kingdom of the Rain Queen, Makobo Modjadji VI, in the Greater Letaba Municipality. An amazing forest of tree cycads (*Encephalartos transvenosus*) covered large areas of the hills—most of us had never seen its like before. Cycads are gymnosperms, relatives of pines and the ginkgo, ancient plants that have survived in widely scattered places around the world, but the hills we were viewing certainly seemed to be their metropolis.

We drove on and camped at Lake Sibhayi, on the northeast coast of South Africa, an important resting stop on one of the world's great flyways for migratory birds. Bedding down for the night in the open subtropical forest near the lakeshore, we had a wonderful time observing the

Near Modjadji in South Africa's Transvaal we visited this stunning forest of cycad trees, *Encephalartos transvenosus*.

plants and wildlife around us that evening and on the following morning.

Returning to Pretoria, Tamra and I flew north to Nairobi to begin an exploration of the human and biological riches of Kenya, a former British colony that had become independent at the end of 1963. We were anxious to meet Norman Myers, who had been so important in the late 1960s in helping us all to recognize the major biodiversity crisis the world was facing. Norman had gone out to Kenya as a colonial administrator in 1958, at the age of twenty-four. When Kenya became an independent country in 1963, Norman, who loved the country and its people, became a Kenyan citizen, taking up teaching, wildlife photography, and tour-guiding. We visited him, his wife Dorothy, and their three-year-old daughter Malindi in Nairobi, starting a lifelong friendship. From Nairobi, we set off to see the country. It was different in those days: we simply rented a car and headed out, without reservations, looking for places to stay day to day as we traveled.

First we headed northwest thorough open country with occasional clumps of biologically richer moist forest. As we reached the crest of some hills and caught sight of Lake Nakuru, we saw a broad band of pink rimming the entire east side of the lake. At first I thought this band must have been caused by masses of pink, single-celled algae like those that colored the drying salt beds I had known around San Francisco Bay. Instead, it turned out to be an enormous flock of flamingoes wading together in the shallow water around the edges of the lake. Hippos stood here and there around the lake, antelope hid in the bushes and thin groves of trees, and multi-colored birds were everywhere. Passing by the lake, we headed up the Aberdare mountains and reached our destination, the overnight animal-watching resort called The Ark. Elephants and Cape buffalo could be seen in the surrounding meadows, and rhinos appeared by the lodge's windows after we turned in to our bunk beds.

Moving down to the plains of Masai Mara and Amboseli parks with their flat-topped acacias, we made the acquaintance of more of East Africa's rich wildlife: giraffes, zebras, elephants, several kinds of antelopes, occasional lions and cheetahs, hyaenas, secretary birds, vultures, kingfishers, and ground hornbills. The animals were so close, and so easy to see, and the whole scene so wondrous and evocative of the Pleistocene, that from that day forward I began to strongly recommend to anyone who might care that they should see Africa's magnificent animals while they survive.

After returning from Africa and beginning to think about how the Garden might work more actively on that biologically rich continent, I made the acquaintance of a Russian émigré scientist named Boris Krukoff. A colorful and forceful character who took to calling himself Bob because so many people mispronounced his given name Boris, Krukoff would engage a good deal of my energy for the remaining decade of his life.[2] At the time of the Russian Revolution, Krukoff, born in 1898, had fought his way across Siberia to Vladivostok, where he and his companions commandeered a ship and sailed it to Manila. Claiming asylum, he made his way to Canada and then to New York. There, based on his botanical knowledge, he became a consultant with Merck & Co., helping them in their search for natural products from which to extract useful drugs. Based at The New York Botanical Garden, Bob had organized eight very successful plant-hunting expeditions to the Amazon basin of Brazil from 1928 to 1955. During World War II, he operated a cinchona plantation for Merck in Guatemala to help provide quinine for the war effort. Once the war was over, Merck sold the plantation to him for a very favorable price, and he began to grow cardamom there. Given the increased popularity of that spice as a dietary supplement, Krukoff's eventual sale of the plantation to Dutch interests earned him a fortune, which he had used primarily to endow botanical projects at The New York Botanical Garden, Kew, and Leiden. Bob once told me that he wished a large statue of him could be erected on the hill overlooking the plantation. Thus the workers could know he was still glowering down at them lest they even think about getting out of line. He may have been joking; but knowing Bob, I'd say not.

After some discussion, Bob agreed to endow our nascent program in African botany. With the income generated by his gift, I was able to offer Peter Goldblatt a permanent position as B.A. Krukoff Curator of African Botany. We also put a portion of the endowment income to good use in

Boris Krukoff, botanical explorer of the Amazon and important patron of botanical studies here and at several other large herbaria, generously endowed our program in African botany by providing funds for the purpose.

purchasing available duplicate specimens of African plants, building up our stock of research material in that way.

In 1973, as our Africa program was just starting to get off the ground, I turned my attention to Central America and the organization of an effort of regional significance that could usefully bring together the information being gathered in several regional projects. I proposed that this project would produce a flora that accounted for all the plants that occurred in Middle America, that wonderfully varied block of land that reaches from Panama to the Isthmus of Tehuantepec in southern Mexico. Because of its broad scope, such a flora could not be undertaken by the Garden alone; indeed, it would have to involve a large international team of scientists representing several institutions.

Participation in this effort would be a natural fit for the Garden, with its work in Panama and our impending completion of the *Flora of Panama*. And it excited me personally, having delighted in studying the plants of the highland parts of Chiapas a decade earlier. Further, I knew that the region—about 10% larger than the state of Texas, but home to about four times as many kinds of plants—was particularly important as a meeting place of plants originally from North America with those from South America; the continents had drifted together to form a land bridge only about 2.7 million years ago. I thought that a comprehensive account of these plants might be appealing to various institutions, particularly the Field Museum in Chicago, which had been active in Honduras and Guatemala and had a flora of Costa Rica underway. However, when I gathered a group of scientists together to discuss the project in 1974, their interest in it was tepid at best. The project was too big, they thought: destined to fail or, at best, to overwhelm their existing projects. The lack of enthusiasm came as a surprise to me. I would have to protect that little seed of an idea until conditions became conducive to its germination.

ↄﬞ

While still at Stanford, I had become fascinated by the field of biogeography—the study of the distribution of different kinds of organisms over the surface of the Earth. By the time of my sabbatical studies in New Zealand, it had come to be understood that the continents, until then thought always to have occupied their current positions, had in fact moved large distances in the past, and were continuing to move. Their movements had obviously had major effects on the distributions of organisms—New Zealand being a case in point—and what we knew about past patterns of dispersal needed to be re-examined in the light of the new findings. Through various techniques, it had become possible to date the positions of lands in the past and to consider the ages of different groups of plants and animals against this background.

I carried my interest in biogeography with me to St. Louis, and before long I was able to delve further into its complexities. Long-time friend Dan Axelrod, who had moved from UCLA to the University of California, Davis, proposed that we collaborate on some papers. As a paleobotanist, he knew the fossil record well. He and I began with a paper on the derivation of the biogeographic patterns of the Australasian area, published in *Science* magazine in 1972. We then started to examine the distribution of all seed-forming plants, adding background information from vertebrate animals and other groups of organisms. This work resulted in a 134-page paper entitled "Angiosperm Biogeography and Past Continental Movements," which was published in the *Annals of the Missouri Botanical Garden* in 1974. This paper proved generally useful and attracted a good deal of attention from people considering such problems. For Dan and me, working out all these important and newly understood relationships made us feel giddy, like children in a candy store!

Writing that summary paper inevitably spurred my thinking about new research sites for the Garden. Learning more about the plants of any region is inherently valuable for science and for conservation, but botanists and other scientists are often attracted most strongly to areas that are especially rich in numbers of species, particularly endemic ones—those found nowhere else. The history of some of these areas is tied up with the ancient movement of pieces of continental crust, made up of relatively light rocks, that split off from larger landmasses and then drifted away to become islands or in some cases separate continents. The islands were originally populated by the plants and animals of the area from which they had separated. For this reason, they have often proved to be intriguing playgrounds for evolution after their isolation from continents. This is true, for example, of the island of New Caledonia, home to the most distinctive group of endemic plants found anywhere. Located on an arc extending north from New Zealand off the eastern coast of Australia, the island's flora could be understood only in terms of tectonic movements over very long periods of time and subsequent evolution. New Caledonia, New Zealand, and the chain of associated islands were, in fact, pieces of the ancient megacontinent of Australia and Antarctica that had split off and started moving toward their present positions about eighty million years ago. The plants and animals of their parent regions had started out the journeys as passengers and in some cases had survived down to the present, while others had reached them by being blown or drifted over the intervening water gaps. Then the passengers had evolved to varying degrees in evolution to produce the species seen there today.

Intrigued by the implications of these movements for plant evolution, I encouraged Pete Lowry, an incoming graduate student at Washington University, to take up the study of the diverse and spectacular members of the aralia family that are found on New Caledonia. Pete's project was very successful, and it would eventually lead him to undertake a lifetime of important studies of the plants of that family. In addition, the project initiated what would become the Garden's long-term research interest in the island.

Another place that stood out vividly in our study was Madagascar, an island about the size of California and Arizona combined, which lies roughly 250 miles off the east coast of Africa. Although there are not as many plant families restricted to Madagascar as there are to New Caledonia, many of its subfamilies and genera and about 95% of its flowering plant species (then considered to number about 8,500) are endemics. In addition, it was clear at the time that a high proportion, perhaps more than three quarters, of the original forests and other kinds of vegetation of Madagascar had been destroyed as a result of grazing and other human activities. Both because of the unique nature of the flora and the severe threat of extinction facing the plants there, I was determined that Garden scientists begin to study the plants of Madagascar, locate and catalog those still there, and evaluate their conservation status as soon as possible. Finding funding for the purpose, I sent Tom Croat, Al Gentry, and Marshall Crosby to Madagascar on individual collecting trips in the mid-1970s. Each of them returned richly laden with specimens, their efforts representing the Garden's first fieldwork in Africa. I was anxious to place a resident botanist on the island, but Madagascar in the mid-1970s turned on to a pro-communist path, formed alliances with China and Russia, and made it difficult for foreigners from other countries to visit, much less to conduct scientific studies. We were not ready to give up our promising program without a struggle, however, and resolved to wait until the political situation improved.

✧

Planting cooperative research projects around the world was not my sole objective in widening the Garden's global reach. It was important also to forge partnerships with botanists in other countries where we could be helpful to them, but would not be playing such an active role. In the 1970s, after a few years at the Garden's helm, I made several visits abroad that did not result in Garden research programs per se, but nevertheless powerfully influenced my thinking about plants and the world that we inhabit with them and helped to broaden the Garden's network of botanical connections.

One of the most memorable of these trips was to Israel in 1975. I had been invited to the country by Professor Uzi Plitmann of the Hebrew University of Jerusalem, who had visited Stanford and worked with me there for a few months in the 1960s. We had grown to like Uzi and his family very much. He was keen for me to visit his university for a term and give lectures to the students, but I felt that I couldn't spare that much time away from St. Louis. We settled on a visit of three weeks. Tamra and I arrived in Tel Aviv in mid-January, a mere two years after the Yom Kippur War that Israel had fought against Egypt and Syria. The country was confident in the extent of territory that it occupied but at the same time insecure, fearing the possibility of renewed armed conflict on all sides.

In preparation for our visit, I read several books on the history of Israel, of Zionism, and the quarter century that had elapsed since the modern state of Israel was established in 1948. Nothing could have prepared me for the view of Jerusalem that opened up as we drove over the hill from the Tel Aviv Airport. As soon as I saw the city, I understood that it was a sacred place, so important to so many people over the millennia.

After recovering from the flu that I arrived with, I lectured for four days each week. On the weekends we traveled all over the country: from Beersheva, Engedi, and Masada to Jericho, the Dead Sea, Bethlehem, Nazareth and the Sea of Galilee, the Hula Valley, Golan Heights, Mount Meron, and Haifa, meeting people and seeing sights along the way. Avinoam Danin, a young lecturer at the university, became a close friend and was often our guide. He explained the ecological function of the crusts of lichens and algae that formed on the desert soils; the irrigated agriculture of the lower Jordan Valley; and the northernmost occurrence of the African umbrella thorn acacia, *Vachellia tortilis*, as a single tree near the Dead Sea. In this extraordinary setting, it was simple to understand why prophets had gone into the desert to meditate.

A second trip later that year, to the Soviet Union, would have an even larger consequence for my future work. The occasion was the Twelfth International Botanical Congress, held in Leningrad in the summer of 1975. The Soviet Union had been closed to most foreign visitors for a very long time. The idea of actually being able to go there and see the country for ourselves was both unexpected and attractive, especially since Tamra and I knew there would be outings and chances to see other parts of the Soviet Union in the period around the Congress.

When Tamra and I arrived in Leningrad, we went by taxi to the Hotel Europa and immediately learned first-hand about the drab décor and poor service that were so characteristic of Soviet hotels at that time. The city, however, was filled with magnificent sights, including the great gray battleship Aurora on the Neva River, the equestrian statue of Peter the Great, the unparalleled Hermitage Museum, the numerous and elegant royal palaces, and the beautiful high-domed cathedrals.

The sessions of the Congress itself were mostly held in buildings belonging to the Soviet Academy of Sciences, some in halls with very uncomfortable wooden seats. I had been invited by the famous academician Armen Takhtajan to speak at a symposium on the symbiotic origin of certain cell organelles by the incorporation of bacteria. As noted earlier, I had written a paper on the subject, entitled "A Multiple Origin for Plastids and Mitochondria," while at Stanford; and although the hypothesis I proposed in the paper would eventually turn out to be incorrect, my original contributions to the topic had attracted Takhtajan's attention. The symposium was held in a grand hall in which Takhtajan, as moderator, sat on a sort of throne high above those presenting the papers—a memorable scene indeed.

In addition to attending sessions at the Congress, I was able to visit the famous Komarov Botanical Institute, founded by Peter the Great in 1714 as a medicinal herb garden. The 1913 building that housed the Institute's herbarium and library had significantly deteriorated. The building had ineffective heating and ventilation, making it impossible to work there on cold winter days. The corroding metal roof was supported by massive, decaying slabs of wood, and the plaster walls were slowly melting off in the rain, with the window sills slowly falling apart. To have one of the finest libraries and herbaria in the world, with its unmatched collections of plant specimens from the Soviet Union and throughout Asia, as well as the Americas and Africa, sitting in such a building dismayed me and the other visiting botanists. After all, the type specimen of the California poppy (*Eschscholzia californica*), collected in the Presidio of San Francisco by Russian scientists in 1816, was preserved in that herbarium!

As part of the Congress-related activities, we left Leningrad and traveled to Moscow, staying in another hotel with remarkably poor service. Here we were more or less on our own, visiting the Pushkin Museum, viewing the portraits and icons of the Tretyakov Gallery, admiring the high red walls and soaring golden towers of the Kremlin, and touring the Tsytsin Main Botanical Garden, in which special plantings, representing the different vegetation zones of the Soviet Union, had been organized for Congress delegates.

At the Main Botanical Garden I visited Alexey Skvortsov, an outstanding botanist who, like me, was a student of *Epilobium*. He was a very nice man and someone with whom I would stay in contact for the rest of his life. I was astonished when he told me that if I had written him about *Epilobium* or any other subject in 1960, or even worse before Stalin died in 1953, he would almost certainly have been imprisoned, no matter what the letter said. Even in 1975, most conversations with Russian botanists were held discreetly, safe from the eyes and ears of the ubiquitous censors—for example, after they invited you to board a bus with them and ride around. Several manuscripts were secretly entrusted to me to submit to journals in the West, but I was told under no circumstances to have proofs sent back to their authors in the Soviet Union.

After Moscow, we took a field trip to the Caucasus Mountains and the Republic of Georgia. Among the notable botanists with us were Vernon Heywood, Ledyard Stebbins, and Reed Rollins, the Asa Gray Professor of Botany at Harvard. We were all excited to be visiting the Caucasus, a biologically rich area that had been essentially closed to foreign visitors for many years.

Naturally we botanists visited the Georgian Botanical Garden (now the National Botanical Garden of Georgia), which occupies a picturesque canyon on the outskirts of Tbilisi. There we were warmly received, holding interesting discussions with our Georgian colleagues. We traveled on to visit the Lagodekhi Protected Areas, a forested region that features a beautiful lake and waterfall. It was particularly important for us to see this forest because the Caucasian forests are the richest botanical community in western Eurasia. They are filled with remnant populations of plants that grew in the diverse forests that covered Eurasia up to about fifteen million years ago, with the kinds of plants that were once widespread now surviving most abundantly in China. Making our way back to Moscow, we flew home filled with rich and varied memories of the Soviet Union, its plants and people. I did not begin to imagine at that time how extensive my future interactions with Soviet and then Russian botanists would prove to be.

☙

Closer to home, I continued my strong interest in filling out our knowledge of the plants of all parts of North and South America. In the early 1970s, I had read an account by Frank Seymour of the University of Vermont about bringing groups of students and others to collect plants in Nicaragua. Through Seymour's account, I learned that the plants of Nicaragua were far less well known than those of any other Central American country. With this intriguing bit of knowledge tucked in the back of my head, I realized a few years later that the time may have come for the Garden to do something about that gap in botanical knowledge. I knew from our experience in Chiapas that it would not be possible to survey a country's plants properly without having a botanist actually living there, and this meant that the effort would require special funding. Fortunately, Bob Krukoff's deepening involvement with the Missouri Botanical Garden presented a means of making it happen. Eventually, I asked Krukoff if he would be willing to make a grant of $100,000 to get the work in Nicaragua started. Bob liked the idea, on the basis that the plants of Nicaragua were so poorly known; he dreamt of the wonderful and varied trees that he had known in the Amazon ranging that far up the Isthmus (they actually don't) and of the wonderful botanical novelties that he hoped would be discovered there.

With Krukoff's grant, I hired Doug Stevens, a recent graduate of the University of Michigan, to be our resident botanist in Nicaragua. I introduced him to Krukoff and sent him off to begin the project in 1977. I also looked around for additional support for the Nicaraguan research. Roberto Incer Barquero, the head of the Central Bank of Nicaragua, was keenly interested in our program and provided funds to help us get it launched. The following year, we recognized his assistance by presenting him with our Henry Shaw Medal, the highest award at the Garden. He accepted the medal in St. Louis, giving a moving speech about the need for the preservation of nature in his native land. The following year, however, the corrupt dictatorship of Nicaragua that the U.S. had been supporting for many years fell; Incer fled, and we feared that the future of our program might be in doubt. Doug Stevens left the country for a couple of months. On his return, he found that soldiers from the new Sandinista government had seized our project's vehicle, but that the Nicaraguans working with Doug had bravely seized it right back, convincing the authorities that the project was in the interest of the country and should not be disrupted. And so Doug and his colleagues were able to continue their important research efforts with no lasting interruption.

Doug would reside for more than a decade in Nicaragua, training Nicaraguans to conduct botanical surveys all over the country, seeing to the gathering of some 85,000 collections of Nicaraguan plants, and helping to establish a national herbarium and training programs in two universities. By the time Doug left the country, the plants of Nicaragua were relatively well known. Fortunately for us, Doug had met and married an exceptional Nicaraguan, Olga Martha Montiel, who accompanied him back to St. Louis and would subsequently figure prominently in many Garden activities.[3] Doug and Olga Martha have continued to advance the overall knowledge of the plants of Nicaragua and their conservation ever since.

∽

At the same time that our activities in Latin America were expanding, I began dreaming of the possibility of working with botanists in China. It was after all, the country of my birth. Because of my parents' residence there, the indelible influence of my Chinese amah, my relationship with Dr. Sun Ko-Chi, and a childhood spent among Asian artifacts that my parents had brought back with them when they sailed away from Shanghai in June 1937, I had always maintained a special interest in China and Chinese culture. With the closing of the country to Westerners after the success of the Communist Revolution in 1949, my chances of ever visiting the country of my birth had always seemed remote. The virtual end to communication with the outside world that began in 1966 with the Cultural Revolution seemed to push the country even farther out of reach.

In the late 1970s, however, the walls separating the U.S. and China appeared, if not crumbling, to be showing some hopeful cracks. Official rapprochement had begun in 1971, when U.S. Secretary of State Henry Kissinger secretly visited Beijing and began working with President Nixon to find a way to normalize relations, leading to Nixon's highly publicized visit to China early in 1972.

While these official diplomatic efforts garnered the headlines, scientists working in the background helped greatly to forge what eventually became a solid working relationship between the two countries. In 1971, Art Galston of Yale and Ethan Signer of MIT applied for visas at the Chinese Embassy in Hanoi, Vietnam; received them; and became the first American scientists to visit the People's Republic of China since 1949. A few years earlier, in 1966, the Committee on Scholarly Communication with the People's Republic of China (CSCPRC; later the Committee on Scholarly Communication with China) had been formed with the goal of opening dialogue between Chinese and American scholars. It should be no surprise that science was used to rebuild communication between our otherwise antagonistic countries. Science is a universal language, one that can bridge long-established chasms of culture and politics in very genuine ways. Its reach, its ideas, and its progress are all intrinsically global.

In any case, the gradual relaxation in tensions between China and the U.S., along with the precedent of Galston's and Signer's visit and a very few additional delegations, seemed to point the way to further opportunities. In 1975, when I was serving as president of the Botanical Society of America (BSA), the CSCPRC suggested that, as BSA president, I might write to my Chinese counterpart, the president of the Botanical Society of China, and suggest whatever joint activities I thought might be appropriate. Following up eagerly, I started to organize exchange visits, enlisting

Art Galston's assistance. I also had important help in this effort from Bruce Bartholomew, who had begun his graduate program with me just before I left Stanford University in 1971. Bruce had become increasingly interested in China, studying the language and, like me, dreaming of possible connections. It took several years to make the arrangements and receive official sanction, but we finally managed to set up an official trip to China. Under the auspices of the Botanical Society of America, Bruce, Art, and I selected a group of ten scientists who would visit botanical centers in China. The trip was funded by the U.S. National Science Foundation. Our delegation traveled to China in June 1978. In Beijing, Tang Pei-sung, a leading botanist then in his seventies, welcomed the members of the delegation in impeccable English.

A year later, in 1979, Tang led a delegation of Chinese botanists that visited us in return and attended the Botanical Society of America's annual meeting in Stillwater, Oklahoma. Memorably, Tang requested hamburgers and a dish of vanilla ice cream upon arrival, presumably favorites from his earlier days in the U.S. decades earlier. At a symposium at the conference organized to honor the botanists from China, Tang likened the visit of the American delegation the year before to "opening windows and bringing fresh air to a house that had been closed too long."

At the end of the Chinese botanists' visit, we met with them at the University of California, Berkeley, to discuss future opportunities and direction. In the Women's Lounge of the Faculty Club, we discussed many ideas to improve the quality and breadth of our future joint efforts, including reciprocal visits and fieldwork. We also broached a project that would turn into an important part of my life as it became concrete over the following years. Two of the Chinese members of the visiting delegation, Yü Te-Tsun of Beijing and Wu Zhengyi of Kunming, suggested that we work together on a revision of the *Flora Reipublicae Popularis Sinicae* (FRPS), the monumental 80-volume account of all the plants of China, begun in 1959, of which Dr. Yü was then chair, and publish it in English as the *Flora of China*. The original Chinese-language work was finally completed in 1992, and the first volume of the English flora was published in 1994.

Their reasoning was simple and very appealing to me. The naming of plants and the organizing of those names according to their relationships—taxonomy—are governed by global convention and depend on reference to the specimens and published materials on which the names were initially based. The Chinese scientists who were working on accounts of individual groups of plants did not have access to the thousands of specimens of Chinese plants that had been collected by Europeans and Americans before the development of botany and other fields of science in modern China. Particularly important among these were the original specimens of the many Chinese plants that were new to science and had been assigned names by foreign botanists, thus becoming the "types" for those names—standards by which other members of the same species could be recognized. It was especially important to examine these types in the context of all the other specimens of their respective plant groups that were available. Moreover, to understand properly the many species of Chinese plants whose ranges extend outside of China (about half of them), it was necessary to be able to compare the Chinese examples with the extensive samples of these plants from throughout the world that are preserved in foreign herbaria. And, of course, because of the censorship that severely limited the entry of printed material from abroad, the world literature on botany, especially that published since the end of World War II, was poorly represented in Chinese libraries. For all these reasons, our Chinese colleagues, while proud of their efforts in bringing together the knowledge of all Chinese plants for the first time, understood that it would take international collaboration to update treatments of Chinese plants properly and integrate them accurately into the global system of plant names.

From our subsequent discussions, I learned that our Chinese colleagues were thinking mainly about preparing and publishing a translation of the original flora, with the names and references brought up to date. Although it turned out that actually the new work would, in due course, evolve into a complete and thorough revision of all Chinese plants, I didn't think much about the details at this early stage. I did understand that a joint project of this magnitude would be an excellent way to bring Chinese, American, and other foreign botanists together, a concept that I strongly supported. Although it would take years for the English-language revision to get underway, its seeds, slow-growing as they would turn out to be, were sown that day on the Berkeley campus.

The exchanges of 1978 and 1979 in general, and the idea for the *Flora of China* collaboration in particular, presaged a future of expanding opportunity for Chinese–American collaboration in the biological sciences. For me, this was thrilling. For a biodiversity-oriented scientist, the importance of understanding China and its plants, animals, fungi, and microorganisms is obvious. A warm-temperate forest that occupied the whole of what is now the North Temperate region of Eurasia and North America and included the only surviving populations of such plants as dawn redwood and ginkgo has survived in its richest form in China, the only Northern Hemisphere country that includes unbroken transition zones connecting tropical, subtropical, temperate, and boreal forests. To understand any group of plants or animals with a North Temperate distribution well, it is necessary to be able to consider its Chinese representatives. The prospect of having new access to this botanical wealth and the botanists who studied it was dazzling, like suddenly being able to visit the moon; it gave me the opportunity to be involved in a productive new round of botanical bridge-building between China and the U.S. It was clear that a great deal of exciting collaboration lay ahead.

✧

On the other side of the world, in Central America, another opportunity for building up botanical knowledge presented itself in the late 1970s, in the context of my involvement with the Organization for Tropical Studies (OTS). I had come to St. Louis in 1971 with rich and pleasant memories of working in Costa Rica in 1966 and 1967 under the auspices of the OTS. Given that background, I had urged Washington University to become a member of this organization as a way of providing a place for our students to learn about the tropics. The university formally joined the organization in 1977, at which time I had become a representative to the OTS board of directors (I would serve in that capacity until 1991). This set the stage for many enjoyable experiences in Costa Rica, especially in connection with attending the semi-annual meetings of the OTS board.

The OTS had acquired land and buildings on Costa Rica's Caribbean plain at Finca la Selva in 1968 and there established a permanent field station. From the beginning, those developing the station were concerned that active deforestation would soon isolate it as an island of vegetation, with a consequent loss of species and diversity. With the establishment of the magnificent Braulio Carrillo National Park along the summit ridge of the mountains above Finca la Selva, however, a solution to this problem came into sight. We could protect our station and the diverse vegetation zones that lay between it and the tops of the mountains by permanently acquiring what would essentially be an arm of the national park extending down from the top of the range to the lowlands at La Selva.

With the active collaboration of Bill Burley, then of The Nature Conservancy, and the encouragement of the OTS board, I set out in 1979 to raise the necessary $1.7 million to make the acquisition of this vitally important additional land possible. Eventually we were successful, and the connecting acreage was purchased parcel by parcel over the next couple of years. Within the new lands, animals and plants could migrate freely up and down the mountainside, and La Selva no longer faced the danger of being cut off as an island of vegetation in which there would have been a much greater danger of local extinction. For me personally, raising funds for the extension brought me into contact with private foundation personnel like Bill Robertson and Dan Martin, whose acquaintance would prove to be of great importance for me in the years to come.

Also during the late 1970s, I was able to take some important steps in realizing my long-dormant dream of launching the comprehensive flora of Middle America. Having never given up on this vision, I began to think seriously once more about how I might renew efforts to convince people to come together and get the project off the ground. I was determined to place the northern boundary of the regional flora at its natural break, the Isthmus of Tehuantepec, and this meant that a significant piece of Mexican territory would be included in it. But Mexico, I knew, was the location of much of the opposition to the project. Along with some others, my good friend Arturo Gómez Pompa, an influential Mexican botanist, was dead set *against* Mexico's involvement. He felt that studies of Mexican plants ought to be carried out predominantly by Mexican botanists so that their knowledge of Mexican plants would be enhanced, while this was clearly to be an international project.

At the meeting of the American Institute of Biological Sciences at Oklahoma State University in Stillwater, Oklahoma, in the summer of 1979, I discussed this problem with my colleague and friend Jerzy Rzedowski, the leading Mexican botanist who was mentioned earlier in connection with his help for our project in Chiapas. When I asked him how I would be able to involve Mexican botanists in the project, he suggested that I invite José Sarukhán and Mario Sousa, both of the Universidad Nacional Autonóma de México (UNAM), to join the project. If they accepted my invitation, he suggested, other Mexican botanists would come to accept the idea.

I followed Rzedowski's suggestion up with a trip to Mexico City a couple of months later, and found that both José and Mario were willing to join us in making the flora a reality. Greatly encouraged, I followed up on the possibility of involving the Natural History Museum in London in the projected flora. John Cannon, in whose department I had worked during my postdoctoral year in London in 1960–61, had become the head of the museum's botany department and remained a close friend. Under his direction, the department was looking for a major new project, and I thought the Middle America flora project could be something they might like. The great encyclopedic work *Biologia Centrali-Americana*, still the fundamental source for information about many groups of plants and animals of Mexico and Central America, had been produced at the museum from 1879 to 1915, and much of the material on which it had been based was preserved there. While John Cannon and his wife Margaret were vacationing in the U.S. during the torrid summer of 1980, I invited them and José Sarukhán to join us in St. Louis for a meeting on the proposed flora. Being sensible botanists, we abated some of the awful summer heat by carrying out some of our discussions almost fully immersed in the waterlily pool in the back of our residence at the Garden, with the temperature in the shade hovering just below 100°F!

Our discussions were productive. John agreed that the Natural History Museum would participate, and José persuaded us that the work should be published in Spanish and by UNAM's distinguished press. He also suggested that we call the work *Flora Mesoamericana*, a felicitous title that would serve us well over the years. We knew that a great deal of work lay ahead, and that publication of the first volume was years in the future, but the project had been launched! We were excited about the exploration and collecting that putting together the flora would entail, and the prospect of new botanical discoveries that would enlarge our knowledge of the plants of this fascinating area.[4]

ℰℐ

After the successful exchange visits of Chinese and American botanists in 1978 and 1979, I anticipated a future chance to visit China myself. Our invitation came in 1980, but, worried about the amount of time a proper visit would take, I was hesitant to accept. I consulted with Bill Tai, a close friend born in China who had worked with me as a postdoctoral fellow at Stanford. He suggested that in order to make the trip fully worthwhile, I should simply request to visit as many places as I wished, and then go if my requests were granted. This seemed solid advice, and I followed it. I was still studying the genus *Epilobium* avidly and trying to understand it throughout its worldwide range. From the many Chinese specimens of *Epilobium* that I had reviewed in the Harvard University Herbaria and elsewhere, I knew that several interesting species occurred on Mount Emei in Sichuan, a classical locality that had been visited by several earlier foreign botanists. So I put Mount Emei on my list, half expecting to be refused permission. I also explained that if possible I'd like to visit Shanghai to see some of the places associated with my family and meet the children of Dr. Sun, the doctor who had delivered me forty-four years earlier, if they could be located. The Chinese authorities agreed to both ideas in short order, and we were off to the proverbial races. Somewhat surprised that our seemingly audacious plan had been accepted, we enthusiastically began to make the arrangements for our travel. That same summer, the Chinese also approved a joint field trip to Shennongjia in Hubei province; the group on that trip, which included Bruce Bartholemew and another former student, Dave Boufford of Harvard University, would be leaving Beijing at around the same time Tamra and I would be arriving there. With this auspicious boost, Dave was to become a leading global expert on the plants of eastern Asia for many decades into the future. Botanical relations between China and the U.S. were clearly starting to warm up.

The chance to return to China was a dream come true, a long-desired fantasy adventure. It was with great anticipation, then, that Tamra and I took off for China in August 1980. After landing at the Beijing airport and passing through customs, we were collected by our hosts for the short drive to the Friendship Hotel, where virtually all foreigners visiting Beijing in those days were accommodated. The sun had already set and I drank in the sights as we drove through the dark Chinese countryside: people playing cards and clacking mahjong tiles under the streetlights along the edges of the road, interspersed with street vendors with aromatic charcoal grills selling everything from rice balls to duck blood to glass noodle soup.

In the morning, our first destination was the Institute of Botany of the Chinese Academy of Sciences (IBCAS). Founded in 1928, the Institute is one of the oldest in China and one of the few institutions authorized to confer master's and doctoral degrees in the plant sciences. Its acting head was our friend Tang Pei-sung. The Institute would become a critical part of all my future dealings with China.

Bill Tai was in Beijing at the same time, working as a visiting lecturer at the Academy of Sciences Institute of Genetics. He accompanied us as we took in the sights in Beijing. Tamra visited Tiananmen Square first and requested that I be allowed to visit there also. An official car was ordered, and I was amazed at the sight of that huge square and of the Imperial Palace that stood at one end, adorned by a large portrait of Mao Tse Tung. Later, we climbed a restored section of the Great Wall. With great delight, I was able collect one sample of *Epilobium*, my first from China, in a low-lying field along the way.

A couple of days later, accompanied by Wang Siyü from the Institute of Botany's foreign office, we traveled southward from Beijing by train, visiting the Botanical Garden in Nanjing on our way to Shanghai. In Shanghai, representatives of the Chinese Academy of Sciences took me to see an apartment house in which my parents had lived during part of their time there. The tall building was still in good condition and fully occupied. Of course, I had left Shanghai too young to have any memories of it, but it was thrilling to know I was walking in my family's footsteps.

A high point of the Shanghai visit was meeting Yi and Yobonne, the daughter and son of Dr. Sun: the Academy of Sciences had succeeded in locating them for me. They were middle-aged, bright, and evidently as happy to see me as I was them. The first thing they said was, "Ah, Peter Raven! Our mother told us that your mother was good with chop sticks, while your father never really learned to use them properly."

What I learned from Dr. Sun's children surprised me. Far from being executed during the dark times in China, as some had told us, Dr. Sun had continued to run the maternity hospital he had built even after it was taken over by the state; he remained at the hospital's helm until his peaceful death from liver failure years later. Both Yi and Yobonne had been able to attend college before being sent away to the countryside for "re-education" during the Cultural Revolution, but returned to become teachers in the 1970s. So while their father's hopes that they would attend college in America had not come to pass, his wishes for their education and success were fully realized.

I had a surprise for Yi and Yobonne. By then the original $5,000 Dr. Sun had entrusted to my parents in 1948 had grown to over $20,000, a substantial sum in China in 1980. I was delighted to be able finally to

When we visited the Shanghai hospital where I was delivered by Dr. Sun Ko-Chi, we were pleased to find a portrait of the architect Jun Zhuang, who designed it for Dr. Sun.

give them the proceeds, and they were touched to receive them. Since its original purpose of education had been fulfilled, they each used their share of the money to buy a house. I was very glad to close the loop of the story for my parents after so many years of worry, and I knew that the news of the money's delivery would bring smiles to their faces. This closure would prove timely: it was soon after I returned from China that my father was diagnosed with liver cancer.

We continued our travels, heading westward now to Sichuan Province, the "rice bowl" of China. As we drove through the countryside toward Sichuan's capital, Chengdu, my mind returned to my family's history and their years in China before I was born. I recalled stories about my great-uncle, the banker Frank Raven, finding ways to get funds through this rough land to missionaries in the interior, an important element in his banking success in the first decades of the century. In his day, it was a long and dangerous overland trip to Chengdu, with warlords sometimes intervening and demanding tribute along the way.

Our main objective in Sichuan was to explore and collect plants on Mount Emei, which is a very special place, exceedingly rich in species of plants and other organisms. Among those accompanying us on this trip were Chen Jia-rui, with whom I had begun to collaborate in studies of Chinese *Epilobium*, and Professor Wu Zhengyi from Kunming, one of those who had suggested the Flora of China project in Berkeley a year earlier. Professor Wu was the greatest expert on Chinese plants, the country's leading botanist in terms of plants discovered and named, and so a wonderful person with whom to enjoy this botanical paradise.

Mount Emei is one of China's Four Sacred Buddhist Mountains, the site of a temple dating to the first century, soon after the time when Buddhism was introduced to China from India. Today a road snakes halfway up the mountain and a cable car covers the rest of the distance to the monastery on top. But in 1980 neither road nor cable car had yet been installed. So we ascended in the time-honored way, climbing on foot from the forest at the bottom, at about 1,300 feet elevation, to the peak at 10,167 feet. We set out hiking through some of the most beautiful forest I have seen, a surviving representative of the rich forests that covered most of the warm-temperate Northern Hemisphere tens of millions of years ago. Having written and thought about these forests, it was a fabulous experience to be walking through them, surrounded by ancient, massive trees on all sides.

We carefully studied the *Epilobium* populations that began to appear in grassy clearings as we reached the middle elevations of the mountain. As we sampled them, we were able to gain valuable understanding of their patterns of variation, information that we would put to good use later when we wrote about the Chinese species of the genus. It was most informative to see living populations of plants I'd known earlier only from dried herbarium specimens.

The trail to the summit of Mount Emei includes stretches of stairs in the steep places—cumulatively about 3,000 individual stairs within a distance of less than three miles. To break the climb, we spent the night in a monastery halfway up, dining by candlelight and continuing to the summit the following day. The mountain is crowned by a large monastery—inactive in 1980 but reopened since—surrounded by low alpine bamboo and open grassland that is rich in high-elevation plants like fireweed, *Chamaenerion angustifolium*. After viewing the spectacular surrounding mountains with their characteristic clouds, we headed back down the trail, completely satisfied with our visit and what we had found and experienced on the mountain. At a souvenir shop at the foot of the mountain, Wu bought me a small picture of a bird perched on a rock as a memento of our trip. I asked what it was, and he said with a smile, "Maybe a raven!" The mutual trust we came to feel during this trip would prove to be crucial later, when we became the co-chairs of the editorial committee for the *Flora of China*.

Leaving Chengdu, we flew south to Kunming, capital of Yunnan Province, the so-called "Spring City," at 1,600 feet elevation. There were no hotels for foreigners in Kunming in 1980, and so we stayed comfortably in the Botany Institute's guesthouse outside of the city. Wu Zhengyi had been a major figure in building up the scientific and education institutes of the area after he fled there in 1943 during the period of Japanese occupation, and had made the Botany Institute in particular a leader among the Academy of Science's research branches. It would become a center for the *Flora of China*, along with Beijing's Institute of Botany. Because of that I would visit Kunming frequently during the coming years.

Tamra and I flew home after nearly a month in China, our heads spinning with the rich and beautiful scenes and interesting, friendly people that we had met. More than ever, I was determined to do whatever I could to enhance friendly collaboration between the scientific communities in our two countries. For several years I continued to try to organize additional major plant-collecting expeditions to China, but the Foreign Office of the Chinese Academy of Sciences had cooled to such projects given what they saw as the excessive cost of the 1980 Shennongjia expedition (Hubei Province officials had apparently used it to "harvest" funds from those the academy had provided for the field trip). In time, I would realize there was no obvious way to overcome their reluctance to organize another venture on that scale. I then saw that there was a simpler way of making fieldwork possible. A few Western botanists were getting into China individually and being allowed, under variable circumstances, to make collections. Seeing that this was working, I decided that the best strategy was for individuals to simply make the arrangements on their own with individual Chinese institutions, and I stopped intervening with the Academy.

ↄ

After the trip to China, and through the early 1980s, I continued to travel extensively as part of my personal mission to further develop the international botanical community and strengthen the Garden's role within it. On the occasion of the International Botanical Congress held in Sydney in 1981, for example, I was able to return to Australia with Tamra and reaffirm old friendships. At the Congress, we hosted a banquet for the Chinese delegates with some of our Australian hosts; this was part of my effort to help our Chinese colleagues make international friends who could support them in their further studies. After the Congress, and thanks to the Australian Academy of Sciences Rudi Lemberg Travelling Fellowship that I had been awarded, we were able to visit Brisbane, Heron Island on the Great Barrier Reef, Melbourne, Adelaide, and Perth, staying with Australian botanists at their homes everywhere we went.

Also in 1981, I made my first visit to Argentina, attending the annual meeting of the Sociedad Botánica de Argentina at the invitation of Professor Armando Hunziker of the Universidad de Córdoba. I delivered a paper at the meeting and was received as an honorary member of the group. Friendships made on this visit, it would turn out, helped to reinvigorate work on the publication of a flora of Argentina and, later, on a checklist of all the plants of southern South America. At that time, Argentina was prosperous economically but ruled by a military dictatorship and in the throes of its "Dirty War." Although as a visitor I was insulated from the war's consequences, I learned that three of Armando's four children had been murdered by the authorities in the years preceding my visit.

In early 1982, I returned to South Africa to attend the meeting of the Association pour l'Étude Taxonomique de la Flore d'Afrique Tropicale (AETFAT) in Pretoria. With the Garden's connections in South Africa and the collecting we had already done in Madagascar, we hosted enough Africa-centered activity to want to be represented at and recognized by this group. After the AETFAT meeting, Tamra and I traveled to the Cape Region to visit the fabulous Kirstenbosch National Botanical Garden and view the area's remarkable flora and vegetation. Here, a mediterranean (summer-dry) climate like that of California has stimulated the evolution of a remarkable assemblage of plants, including many stunning ones like proteas, bird-of-paradise, and agapanthus. South African plants have amazed botanists and delighted horticulturists since the time of their discovery by Europeans five centuries ago. Returning home, I dared to invite AETFAT to hold its next meeting, in June 1985, in St. Louis.

In 1983, my optimism about the Garden's programs in Africa proved to be well founded: we made a breakthrough in our relations with Madagascar, where the Socialist government had previously refused, in most circumstances, to admit visitors from any countries but China and the Soviet Union. Gerald and Lee Durrell, with their Jersey Wildlife Preservation Trust (now Durrell Wildlife Conservation Trust), had managed to maintain connections in Madagascar through its isolationist period. To investigate possible ways forward for the rest of us, a group convened at the offices of the Trust on Jersey, one of the Channel Islands off the coast of Normandy, from January 31 to February 2, 1983. In addition to the Durrells, our group included Alison Jolly, Allison Richard, Bob Martin, Elwyn Simons, Yves Rumpler, and Bob Sussman of the Department of Anthropology at Washington University, with whom I had collaborated previously. Significantly, the meeting was also attended by Madame Berthe Rakotosamimanana and Voara Randrianasolo from Madagascar.

Madame Berthe was the head of the Department of Paleontology at the University of Antananarivo; she also occupied one of the leading positions in Madagascar's Ministry of Higher Education and Research. Courageous in her decision to attend the meeting, given the political orientation of her government at that time, she clearly wanted scientific research in the country to be collaborative and open to all. Decisions made during our discussions in Jersey provided a way for us all to continue. Very soon a number of joint projects would be underway, and students from Madagascar once again allowed to work in the West.

An unexpected opportunity to deal with the plants of a new part of South America presented itself in 1984, when Julian Steyermark accepted my long-standing invitation to move to St. Louis. Julian had been born in St. Louis (in 1909) and grown up here, attending Washington University and earning his doctorate from the Henry Shaw School of Botany in 1933, a little later than my UCLA mentors Carl Epling and Mildred Mathias. During a twenty-one-year stint at the Field Museum in Chicago, he completed the *Flora of Guatemala*, underway there at the time of his arrival, and also authored, in his "spare time," the outstanding *Flora of Missouri* (published in 1963). In 1958, he had moved south to the Instituto Botánico in Caracas, Venezuela. There he had become a world expert on Venezuelan plants, collecting so many that he eventually distinguished himself as the most prolific collector of plant specimens *ever*. Steyermark had been working on an account of the plants in the ancient mountains of southern Venezuela south of the Orinoco River, but with the passing of years, his progress had slowed. Since I had a very high opinion of Julian and his pioneering work, I had invited him to come home to St. Louis, where we could help him complete his work in a more encouraging setting. Steyermark did move to St. Louis in 1984, after his wife Cora passed away. Here he worked on his *Flora of the Venezuelan Guayana* with renewed vigor, strongly supported by friends old and new. It was a joy for me to be able to encourage this fine botanist and be able to assure him that his life's work would be completed, an important contribution to our knowledge of the plants of Latin America.[5]

Given his well-known and very useful earlier work on the flora of Missouri, Steyermark's presence in St. Louis reminded me that while we were making significant contributions to knowledge of plants in many regions around the world, we might well be doing more in our own region. I added to my growing list of priority projects an updating of *Steyermark's Flora of Missouri*. It would take a few years for this project to get off the ground, but it would eventually come to fruition.[6]

Another project began following a reflection I had while being driven into a parking spot at the OTS headquarters in San José, Costa Rica, in 1984. The valuable flora of Costa Rica that Bill Burger had been pursuing at the Field Museum ended up seriously incomplete as Bill approached retirement. I continued to think that it would be useful, in view of the large amount of biological research being conducted in the country through the OTS and independently, to prepare a contemporary manual treating all the country's plants. Over the next couple of years we found two willing authors in the persons of Mike Grayum, recently graduated from the University of Massachusetts, and Barry Hammel from Duke. It was planned from the start that the manual would be written in Spanish and prepared jointly with the national museum of Costa Rica and INBio, the National Biodiversity Institute in Costa Rica. The project would get underway in 1987, aiming for completion by approximately 2020. Even though the *Flora Mesoamericana* effort was in progress, it was worth maintaining special attention on the plants of Costa Rica, and we were delighted to have the opportunity to do so. Both in Costa Rica and in the Central American area generally, far less was known about their plants than we thought when we initiated the projects, so there were numerous exciting surprises for those doing the work.

In 1985, Tamra and I were invited to travel to Japan to participate in the ceremony marking the first presentation of the International Prize for Biology. This award had been established by the Japanese government in honor of the sixtieth year of the reign of Emperor Hirohito (posthumously known now as the Emperor Shōwa). It was no coincidence that the International Prize was in the field of biology; the Emperor had studied and done research in biology for his entire life, publishing papers on the classification and biology of jellyfish, as well as on the plants that grew native at his country estate.

I was delighted that I would finally be able to visit Japan. The Japanese Garden had become one of the most popular features at the Missouri Botanical Garden. Everything I had heard about Japan from my parents, who had visited Nikko on their honeymoon, and from Dave

Boufford, who had seriously enjoyed his fieldwork there during his graduate career at Washington University, suggested that the country and its people would be to my liking.

The Prize was to be awarded to E. J. H. Corner. As the director of the Singapore Botanic Gardens at the start of World War II, Corner had decided on a course of cooperation, rather than confrontation, with the Japanese who occupied the city early in 1942. He wanted above all to do what he could to protect and preserve the gardens during a time of general looting and lawlessness. Happily, the Japanese occupiers shared this goal, and the decision to retain him as director was approved personally by the Emperor. Although Corner was accused by some of collaborating with the enemy, he was actually simply trying to keep the irreplaceable living and preserved plant collections safe in an institution that he had served since 1929. Well-regarded in Japan, Corner was a popular choice to be the first recipient of the International Prize.

Tamra and I traveled by train to Kyoto, where the prize ceremony was held, and had a chance to view the beautiful scenery along the route from Tokyo. Once there, it was a pleasure to get to know Corner personally, especially since I had studied his innovative works on plant evolution. I had even considered spending my postdoctoral year with him in Singapore to emphasize studies of tropical botany. The trip also began my personal acquaintance and lifelong friendship with Professor Kunio Iwatsuki, a student of ferns at the University of Tokyo and a leading internationalist with whom I would have a great deal of interchange in subsequent years. I also met again Dr. Shoichi Kawano, who I had first encountered at the International Botanical Congress in Montreal in 1959, and built a lasting friendship with him.

At this point, in the middle of the 1980s, I felt that I had accomplished much of what I had set out to do in the greater botanical world beyond St. Louis. The Missouri Botanical Garden had gained international recognition for its research and collections, and I had become thoroughly connected with an amazing, collaborative, global community of botanists in a network that I had helped to create and nurture. We were sharing information, supporting each other's work, and steadily building up our collective knowledge of the flora of our planet. At the same time, I was becoming increasingly aware, along with my colleagues, that this global botanical network would be a necessary foundation from which to confront the staggering losses of natural habitat, particularly in the tropics, that were threatening to destroy the biodiversity we were only beginning to understand. How could we continue to build up a sound knowledge of the plants that were disappearing so rapidly, and at the same time try to find ways to protect as many of them as possible from extinction?

11

Conservation Around the World

IN THE LATE 1970s, my concern about the increasing rate of biodiversity loss, which had been growing since the late 1960s, reached the point where it began to infuse nearly all my professional activities. By that time, my colleagues and I could see that explosive human population growth and ever-higher levels of consumption were driving biological extinction toward a rate unprecedented since the demise of the last remaining dinosaurs sixty-six million years earlier. In the decade since Norman Myers had circulated his prescient letter about biological extinction, the human population had increased by over 40%, to 4.9 billion. Signs of the destruction of natural habitats accompanied by the extinction of populations and species were everywhere.

The erosion of our planet's biodiversity, I knew, was not merely an abstract issue for species-counting biologists. We humans depend on the proper functioning of living ecosystems for our survival. Plants, specifically, are the source, directly or indirectly, for all our food, for most medicines used worldwide, and for building and clothing materials. Healthy, diverse ecosystems are necessary for making and maintaining topsoil, retarding floods, absorbing pollutants, regulating climate, and providing the beauty that raises life above the level of mere existence. We were in grave danger, I realized, of passing along to our descendants a world bearing little resemblance to the abundant one that we knew, a world on which simple survival might become a struggle. I was determined to do what I could to conserve plants and other living organisms and to help build a sustainable planet. Working, like others with the same concerns, against what sometimes seemed to be insurmountable odds, I was motivated by the knowledge that the Missouri Botanical Garden, with its now global reach, cooperative international research programs, and increased visibility, had the potential to become a real leader in the effort to conserve plants and ecosystems around the world.

In pursuing the cause of conservation, the Garden had a fundamental practical and scientific role to play. Most non-governmental organizations dedicated to conservation, then as now, endeavored to protect the more iconic animal species ("charismatic megafauna") from extinction and to preserve and protect wildlands so that their habitats and species would be saved from the inevitable losses that occur with almost any kind of devel-

A tropical forest in Sumatra, cleared for planting oil palms. Vast areas of forest in Indonesia and elsewhere in the tropics have been cut for this purpose.

opment. Among other activities, they worked to persuade governments to set aside lands as parks and preserves and to manage land to minimize the effects of human activities like farming, hunting, logging, and mining on wildlife and habitats. The Garden encouraged, supported, and often participated in such efforts. But conservation can, and must, be much more, especially for plants—and it was in pursuing these less glamorous activities that the Missouri Botanical Garden and other botanical gardens contributed best to conservation. Many plants endangered in the wild can be cultivated in botanical gardens, with enough individuals of a species conserved to maintain a population that is viable genetically. (Similarly, we can preserve seeds or living tissues of endangered plants for long periods, conserving their unique genotypes for cultivation or restoration in the future.) Botanical gardens are also equipped to undertake research aimed at improving our understanding of the biology and ecology of the plants being conserved, generating knowledge necessary to manage wildlands and dwindling populations. Perhaps most importantly, botanical gardens do fieldwork to determine how plants are faring in the wild. Only with this kind of basic inventory-level knowledge of the planet's plant life can it be decided which species need protection the most and what that protection should consist of.

Recognizing the critical importance of this basic cataloging activity for conserving biodiversity, I took every opportunity to step up and strengthen the Garden's plant inventory efforts around the world. Ever searching for the more poorly known parts of the world, particularly in North and South America, we combined our botanical efforts in these areas with partnerships with and encouragement to others. Our work in Mesoamerica, Costa Rica, Nicaragua, Ecuador, Peru, and Bolivia went forward steadily, as resident botanists accumulated new knowledge avidly and we helped to find funds to support their pursuit of these strategic goals.

We faced many challenges in our international work. Relatively few countries supported plant conservation programs, and many had strict regulations against exporting any living biological material. While we looked for ways to overcome these practical difficulties, activities like clear-cutting went on apace. With our sister institutions, we could only try our best to keep ahead of it. It had become tragically evident by the late 1970s that, with such rapid deforestation and loss of every kind of natural habitat, we were going to be the last people who would ever be able to collect plants at many of the tropical localities where we worked. Consequently, we had to make the best of it; the challenge is much more severe today, a half century later, than it was then.

I was glad to be director of the Garden, an institution that could contribute substantially to the cause of worldwide conservation through research, collecting, and direct conservation efforts. At the same time, I could a see a role for myself in promoting scientific collaboration, public awareness of the problem of biodiversity loss, stronger roles for nongovernmental organizations in advocating local policy changes, and in building the broader understanding that would be necessary if conservation was to be successful. Particularly important in all of this, I recognized, was supporting and encouraging talented botanists wherever we could, enabling them in every way possible to carry out their work.

❦

My interest in the role that botanical gardens might play in conservation was spurred on by a meeting in which I participated in 1974 at the University of Manchester. Professor David Valentine had called the meeting to discuss plant conservation, and I presented a paper on "ethics and attitudes." In it, I enumerated wide-ranging ideas on enhancing the role of botanical gardens in plant conservation and the reintroduction of plants back into nature. The immensity of the problem of global biodiversity loss was becoming shockingly apparent to me and indeed to anyone who was watching the situation develop.

I was thinking a great deal about biological conservation when in the same year, 1974, I was invited by the National Science Foundation to convene a workshop about future directions in systematic and evolutionary biology. The workshop was designed to help the NSF make its programs in those areas more targeted than they had been and hence as effective as they possibly could be. Held at the Missouri Botanical Garden from May 30 to June 1, 1974, it brought together many biological luminaries of the day. Its results, published both in *Systematic Zoology* and in *Brittonia*, proved to be both useful and influential. The rapid destruction of habitats throughout the world was a matter of great disquiet for the participants. They were especially concerned about the tropics, where most of the people in the world lived, along with two thirds or more of the Earth's species of plants and other organisms—the most numerous and poorly known anywhere. Accordingly, they recommended, among other resolutions, that targeted studies in the tropics be increased substantially so as to gain a solid base of knowledge while there was still time to do so.

As a follow-on to the original workshop report, I was asked to organize a second study ranking particular topics in tropical biology for priority funding in the future. Wanting to understand just what degree of threat tropical forests were facing, we commissioned a study by Norman Myers, since he had time to take on the task and was the one who had so usefully called attention to the problem of massive biological extinction a decade earlier. For the current study, Norman traveled throughout the tropics, trying to make sense of the written reports prepared by governments and other official agencies and to find out what the ground truth might really be. His 1980 report, *Conversion of Tropical Moist Forests*, was a genuine shocker. Not only were the forests being logged and otherwise destroyed much more rapidly than anyone had previously thought, these actions clearly were endangering tens to hundreds of thousands of species throughout the world. When our team of distinguished scientists published its report, *Research Priorities in Tropical Biology*, later the same year, there were more surprises.

Up to that time, biologists had extrapolated from the figure of 1.4 million known species of all organisms to estimate that the actual total might be about 2 million. At the same time, people were estimating that in well-known groups like plants, vertebrates, and butterflies, there were roughly twice as many species in the tropics as in temperate regions. But of the 1.4 million described species overall, only about 400,000 were tropical. Even a dull student of mathematics like me could see that we were seriously underestimating both the number of tropical species and the number of species in general. Based on known numerical relationships, the total number of organisms ought to be *at least* 3 million, and the tropics clearly harbored many hundreds of thousands of undiscovered species. We called for special concentration on groups of organisms that could help us understand the overall patterns of species distribution in the tropics; for emphasis on economically and environmentally important groups; for the establishment of comprehensive field stations in the tropics where all groups could be studied in detail locally and in relation to one another; and for conservation action where it could be effective.

Naturally, these findings affirmed the validity of the Garden's concentration on the tropics, where there were so many species of plants yet to be discovered, with most of them needing conservation attention if they were to be saved from extinction. Many of these plants are important for food, medicine, and other human uses. We were one of a handful of institutions that had a worldwide scope in its collections and a staff large enough to allow us to work in many tropical areas effectively. It was clearly necessary for us to try to build our research programs in the tropics as quickly as possible and with as many resources as we could find. This had been the thinking behind my successfully urging of Washington University, the University of Missouri–Columbia, and the University of Missouri– St. Louis to join the Organization for Tropical Studies and participate actively in its educational and research programs. It was also part of what motivated me to work so hard to launch the *Flora Mesoamericana* project in 1979–80. Completed, the flora would present a comprehensive guide to all the known plants of southern Mexico and Central America, which would be extremely useful for promoting both their sustainable use and their long-term conservation.

In the meantime, general interest in conservation had been growing rapidly. In 1980, the scientist Thomas Lovejoy introduced the phrase *biological diversity* as a reference to the whole array of species on earth and published the first charts showing rapid and widespread biological extinction. In 1984, the Center for Plant Conservation (CPC) was established at Harvard's Arnold Arboretum to coordinate the efforts of botanical gardens in conserving the plants of the U.S. and Canada. Each of its member institutions agreed to work actively to conserve threatened and endangered species in its own local region. Another key organization with similar goals but a worldwide scope, Botanic Gardens Conservation International (BGCI) was established three years later at the Royal Botanic Gardens, Kew. In 1986, the National Research Council convened the National Forum on Biological Diversity. E. O. Wilson effectively guided the organization of this meeting, and when its proceedings were published in 1988 the term *biodiversity*, used for the first time in the title of the work, was introduced, soon entering common usage.

With all this activity and growing awareness of the rapid destruction of nature, I was inspired to pursue an even broader and more active agenda in conservation. I began to speak out with ever more force about tropical deforestation, extinction, and the need for sustainable management of our resources. Leveraging my botany credentials, my prestige as director of a major botanical institution, and my solid network of connections, I advocated strongly for the cause of conservation, attempting to convince others of the need for forceful, coordinated action.

A turning point for me, as well as for some of those who attended, was my keynote address to the American Association for the Advancement of Science (AAAS), delivered at its annual meeting in Chicago on February 14, 1987. My talk, entitled "We're Killing Our World," attracted a great deal of attention and was subsequently published by the MacArthur Foundation. By that time, it had become clear that the implications of extinction and habitat loss for our food, our medicine, our economy, and our future were dire. In the address, I emphasized that nations would need to cooperate and offer mutual assistance if the world was ever to attain sustainability (although that word itself was not yet widely used). I carried the theme of cooperation and coordinated action forward and continued to develop it in a series of talks at universities, museums, and society meetings.

At the Garden, we continued to grow our commitment to conservation where it was possible in the places where we worked, an especially good example being our efforts in Madagascar. With Madagascar's political re-opening and adoption of a free-market economy in 1986, we were able to develop a resident program on the island. By that time, Pete Lowry had completed his graduate studies and was ready to move to Paris with his wife Hélène. Anxious to keep this talented man as a part of the

Garden's research program, I suggested that he establish a satellite office for the Missouri Botanical Garden at the Muséum national d'Histoire naturelle in Paris, where the connections he had made as a graduate student were very helpful. From that base, Pete would oversee our African programs in general, especially our growing program in Madagascar.

On the Grande Île, as the French call Madagascar, we were able to re-open our program and soon establish a per-

Garden scientist Armand Randrianasolo pressing a specimen of a new species of *Spondias*, a kind of tree, in the sacred forest at Analavelona, Madagascar, part of the Madagascar protected area network. The supervisor of this site, Tefy Harison Andriamihajarivo, is assisting, while two members of the community look on. The species was later named *Spondias tefyi* in the supervisor's honor.

manent office, hiring many Malagasy citizens to collect plants and launch conservation programs. In due course, we began to establish sites near villages, where we worked with local residents to restore habitats and reforest some of the adjacent land. We set up nurseries at most of the sites where the people could propagate native species for restoration of the local ecosystems. Groups of students from the Business School at Washington University, and specifically from the Skandalaris Center, visited and helped the villagers devise sustainable ways of supporting their economies. Although it remained difficult to pursue conservation in Madagascar, we were ultimately able to designate thirteen villages as centers for sustainability. At about the same time, the Royal Botanic Gardens, Kew, mounted a strong conservation program involving seed banking as well as re-growing rare native species. As we learned more about the remarkable plants of Madagascar, we began to realize that an incomplete firsthand understanding of the nuances and features of the local geography had led to a major underestimation of the number of species on the island.[1]

The development of my role as a global evangelist for sustainability and conservation coincided with deepening concern around the world about extinction, deforestation, unsustainable resource use, pollution, and climate change. The international movement to address these issues and find ways to reduce human impact on the natural world certainly encouraged my continued activism, and I like to think that I, in turn, played a role in fueling this movement as it developed.

An important indicator that people were finally beginning to understand the seriousness of the problems that we were facing and looking for ways to act on them was the so-called Brundtland Report, which appeared in 1987. This document, formally titled *Our Common Future*, was the report of the World Commission on Environment and Development, chaired by the Norwegian prime minister Gro Harlem Brundtland. Emphasizing the interconnectedness of nations and asserting that development and environmental issues could not be separated, it discussed how we could end our destructive activities for our common benefit. The report brought the word *sustainability* into common usage and defined "sustainable development" as "development that meets the needs of the present without compromising the ability of future generations to meet their own needs." In many ways, the Brundtland Report changed the conversation about the relationship between human activities and the environment, convincing many that we ignore our impacts at our own peril. (For many of us, however, it presented too rosy a view, greatly underestimating the degree to which we were already overusing the world's resources.)

In the late summer of 1988, I received word that the UN was attempting to organize an international agreement on biodiversity so that it could be protected as well as possible throughout the world. This work was initiated within the framework of the World Conservation Union (IUCN), but it was felt that the IUCN had neither sufficient diplomatic clout nor the jurisdiction to sponsor negotiations for an internationally binding treaty of the envisaged scope. It was also felt that a treaty negotiated under IUCN would focus only on conservation, failing to cover important related matters such as sustainable use or benefit-sharing for biological resources.

With Mike Soulé, on whose Ph.D. committee I had served at Stanford two decades earlier, and my good friend Tom Lovejoy, I was invited to participate in a preliminary meeting on the subject at UN Environment Programme (UNEP) headquarters in Nairobi. About two dozen people attended, including some who were particularly concerned with the situation in Kenya together with a few international specialists. The main logistical problem faced by global conservation was finding adequate funding for conservation efforts in the tropics, whose developing countries could not afford major efforts in this area among their many other pressing priorities. The world's industrialized countries seemed primarily to be interested in extracting natural resources from tropical countries at the lowest price possible, thus further adding to the extent of the problem through the related environmental devastation.

In Kenya, we learned that the income from admission to the national parks and game reserves was simply added to the nation's general fund, not applied to the management of those parks. Dependent on international funding sources, these parks had in effect become game reserves maintained for the benefit of wealthy people living in industrialized countries. We talked about taxing tourism and finding other sources of funds, but left feeling that we had not found a realistic solution. The money was in the north, the biodiversity in the south.

Following our meeting, an Ad Hoc Working Group of Experts on Biological Diversity was convened in November of the same year. Soon after, in May 1989, UNEP established the Ad Hoc Working Group of Technical and Legal Experts to prepare an international legal instrument for the conservation and sustainable use of biological diversity. The experts were to take into account "the need to share costs and benefits between developed and developing countries" and to find "ways and means to support innovation by local people." Masses of lawyers then entered the picture, developing a text that was agreed upon at a conference in Nairobi in May 1992. The "Convention on Biological Diversity" was opened for signature at the Earth Summit in Rio de Janeiro the following month. By the end of December 1993, the necessary thirty countries had ratified the Convention, and it entered into legal force. The United States was *not* one of those countries, and remains one of two globally, with the Vatican City, that have still declined to ratify it. At the time, those of us with a role in drafting the Convention had high hopes for its ability to at least diminish the rate of decline of biodiversity around the world.

☙

The first step in any conservation plan is knowing what it is you are trying to save. Crucially important for this endeavor is simply to catalog the known species and make this information available to other researchers and resource managers. Before the advent of computers and the databases that could be managed with their help, the cataloging effort was tedious and time-consuming and the results difficult to synthesize and to share widely. It was based on records in books or written out in paper files, with no ready way to compile them or to annotate the earlier ones. As time went by, however, increasing numbers of computerized databases were constructed and put online and thus made generally available—a great boon for systematic biology and conservation. One important plant database, Tropicos, was developed by the Missouri Botanical Garden and eventually made available globally in 1996.

Our database had its beginnings in the 1970s, when Marshall Crosby was developing a project intended to keep the recently published names of mosses available to all those studying the group. He worked with the computer science department at Washington University to produce a data collection card, of the sort the library used for accessioning books, that could be used to transfer the moss data easily to punch cards. That first system did not work efficiently, but it was a start.

In 1980, Bob Magill, another student of mosses who was then working as a scientist in South Africa, encountered a personal computer for the first time. Bob also found that the institution where he was employed, the Botanical Research Institute (BRI), had developed a computer system, PRÉCIS, to support its work with specimens for the national flora and for conservation as well. Bob was encouraged to enter his moss data into PRÉCIS. The institute's microcomputers were connected with the mainframe, and it all worked efficiently, adding up to a very useful advance in computerized record management.

Accepting a position at the Missouri Botanical Garden, Bob joined us in 1982. On arrival, he found that Marshall Crosby, who had just purchased his first PC, wanted to use it to produce manuscripts for his index of moss names. Working together, Marshall and Bob found a way to produce manuscripts that could be printed with the Word Star word processor. They built and expanded a database using PRÉCIS as a guide and began to enter data for all kinds of plants, not only mosses. The new system, which they called Tropicos, grew in usefulness over the years as computers became larger and more sophisticated. Soon, the Garden, which was amassing huge numbers of herbarium specimens, was able to produce herbarium labels directly from the computer, a major time-saving advance. In 1996, Tropicos went online as w3Tropicos, and the system's data became available freely and without cost to anyone who wished to use them. Tropicos quickly proved its value in advancing the management of data about plants worldwide.[2]

❧

Through the 1990s and beyond, I continued my work on behalf of plant inventories and conservation, emphasizing the need for us to work collectively to attain sustainability. I used every speaking engagement as an opportunity to remind my audiences of the growing crisis. As we steadily expanded the Garden's research programs in tropical regions, I increasingly used my membership on various boards, commissions, and other bodies to shape policies favorable to biodiversity conservation and reduction of the human impact on nature. A few highlights of this ongoing work stand out in my memory.

At the XV International Botanical Congress in Yokohama, Japan, 1993, Missouri Botanical Garden botanists were approached by their counterparts from Vietnam. They invited our institution to send botanists there and to carry out joint activities. Although formal relationships between our two countries would not be re-established until two years later, a substantial amount of interchange was already taking place, and the proposal was of interest to us. America's twenty years of fighting in Vietnam, with its high costs in human lives and devastated lands, had come to an end in 1975, and we were ready to engage in friendly and constructive relationships.

In pursuing these relationships, we signed a formal cooperative agreement with the Institute of Ecology and Biological Resources (IEBR) in Vietnam in 1994 and sent several of our botanists to begin collecting in the country. In 1998, along with the American Museum of Natural History, we received an NSF grant to inventory threatened and protected areas in Vietnam. We hired Dan Harder to establish an office in the country a year later and began to collaborate with several of its scientific institutions and conservation organizations. Our efforts were greeted with enthusiasm. Subsequently, Jack Regalado would take up our post in Hanoi and contribute a great deal to the advancement of botanical knowledge in Vietnam. Our work in the country represented an exciting opportunity to expand greatly our knowledge of a tropical region with extraordinary biodiversity. Although Vietnam is slightly smaller than the state of California (the most biodiverse state in the U.S.), it is home to about 2.5 times as many plant species (13,000 to 15,000) as California, many of them endemics that are found nowhere else.

In 1994, ten years after the Center for Plant Conservation was founded, the Arnold Arboretum determined that it needed the office space where the organization had been housed for other purposes. Being interested in stimulating the overall conservation program at the Garden, I persuaded the CPC leadership to relocate their headquarters from Boston to the Missouri Botanical Garden.[3] The Garden was already one of the Center's participating institutions, but moving its headquarters here allowed us to encourage conservation more widely and deeply that we had earlier.

In the late 1990s, I had the chance become personally engaged in environmental activities in India. Long fascinated with the country, I found it difficult to think about it without Paul and Anne Ehrlich's description of the large crowds on the streets of New Delhi on the opening pages of

Members of the Missouri Botanical Garden's Center for Plant Conservation team re-establishing the endangered *Astragalus bibullatus*, Pyne's ground plum, along the edges of a cedar glade near Murfreesboro, Tennessee.

The Population Bomb coming to mind. I understood well that India was an ancient center of culture and religion. I had glimpsed its beauty and complexity when Tamra and I visited Calcutta and New Delhi briefly on our way home from New Zealand in 1970. When a small boy, I had been fascinated by India's wonderful and varied animals while listening to my father read from Rudyard Kipling's *The Jungle Book* at bedtime and even then dreamed of seeing these amazing creatures myself someday.

My involvement in conservation and other environmental work in India began with an invitation from my long-time friend Kamal Bawa, an Indian botanist and ecologist who had joined the faculty at the University of Massachusetts Boston. Kamal, who had come to America in 1967, had worked in the American tropics for a number of years and then turned to see what he could do in helping promote biological understanding and sustainability in his country of origin. Kamal had conceived the idea of a private educational and environmental organization that would conduct research and training, promote sustainability, and provide advice on the proper management of the varied environments of the subcontinent. He had launched the organization in 1996, calling it the Ashoka Trust for Research in Ecology and the Environment (ATREE, now its legal name). Soon afterward, in February 1998, he invited me to India to see for myself what was going on and find suitable ways to become involved there. He arranged my trip and accompanied me.

We first visited Bengaluru (formerly Bangalore), where the headquarters of ATREE were located. There, Kamal had arranged for me to speak at a meeting of the Foundation for the Revitalisation of Local Health Traditions (FRLHT), which a close friend of his, Darshan Shankar, had founded. I was deeply impressed by what we learned about the paramount importance of plants as medicine in India, and also had good opportunities to meet with the staff members and board of ATREE. Leaving Bengaluru, we traveled to the Western Ghats, where the native people of the area, the Soligas, were adapting to the demands of the modern world with help and advice from ATREE.[4] I knew that India was complex, of course, but had not thought much about the large numbers of different groups of indigenous people that lived there, each with its own special way of life and needs.

Kamal and I then traveled to Chennai (Madras) to visit the research center that M. S. Swaminathan had organized there. Swaminathan, who is considered one of the fathers of the "Green Revolution," was as active as ever at seventy-three years old and still concerned with finding ways to feed a hungry nation. I was much taken by this charismatic man and the research of the institute that he had endowed with the proceeds from his various prizes. The researchers were focusing, among their other goals, on developing salt-tolerant rice by transferring genes from mangroves and other plants that grew well in salty habitats.

After our visit to Chennai, Kamal and I crossed over to the west coast of India, visiting the southwestern state of Kerala and the large and diverse botanical garden in Thiruvananthapuram. There, I learned how difficult it was to maintain a botanical garden in a warm and humid climate where everything seemed to grow over everything else as rapidly as possible.

Conservation, particularly in the tropics, remained an important theme of my work and career through the 1990s and into the new millennium. As time went on, however, my personal focus shifted subtly from limiting human impacts on the natural world to addressing the underlying causes of our destructive tendencies. Sustainability and a concern for people living in harmony with the biological resources of the world became my passions.

28. Compared with *Epilobium* species from other regions, many of those that occur in New Zealand are unique, including, for example, a group of species with creeping stems and erect flowers, like *Epilobium brunnescens*, shown here.

29. We were fortunate to be able to observe the tuatara (*Sphenodon punctatus*), in its native habitat on Stephens Island, New Zealand.

30. *Epilobium pycnostachyum* has sprawling stems that grow out under moving scree and pop up at the end of horizontal stems. This is a common habit for plants that grow on scree, but unusual in the genus *Epilobium*.

31. Redbud (*Cercis canadensis*) forms gorgeous displays in the early spring Missouri woods.

32. Flowering dogwood (*Cornus florida*) is a feature of woods all over midwestern and eastern temperate North America in early spring.

33. The luna moth (*Actias luna*), a lovely denizen of the central and eastern United States.

34. Flowering in Missouri summers, the butterfly weed (*Asclepias tuberosa*) is here being visited by spangled fritillaries (*Speyeria cybele*).

35. The Missouri evening primrose, *Oenothera macrocarpa*, forms showy colonies in some of the limestone glades around St. Louis.

36. Hybridizing strains of *Epilobium* early in my tenure as director of the Missouri Botanical Garden, in continuation of studies I had begun at Stanford.

37. Koichi Kawana, the architect of our Japanese Garden, painted this vision of what he thought the completed garden would look like and then worked hard for more than two years to make it a reality.

38. Japanese water iris adorning the Yatsuhashi bridge in the Japanese Garden, an island of peace in the center of the city of St. Louis.

39. Meeting Princess, the panda, at the Wolong Panda Center in western Sichuan Province, China, was a special treat.

40. Wu Zhengyi and I signing the agreement to produce the *Flora of China* near the Chinese dawn redwoods, *Metasequoia*, in the presence of our national delegations.

41. Here I was walking through a colony of the dwarf bamboo *Yushania niitakayamensis* on Hohuanshan, Taiwan, at about 10,000 feet elevation.

42. The Climatron, as shown in the 1990 *National Geographic* magazine article about the Missouri Botanical Garden.

43. The popular William T. Kemper Center for Home Gardening opened in 1991.

44. Japanese Emperor Akihito and the Empress planting a maple tree in the Japanese Garden during their 1994 visit to St. Louis.

45. The pavilion in the Margaret Grigg Nanjing Friendship Garden, completed in 1996, was constructed in Nanjing, sister city to St. Louis, broken down for transportation, and reassembled by workers from Nanjing in our garden in 1996.

46. The Ruth Palmer Blahke Boxwood Garden, opened in 1996, demonstrates a formal garden style, popular in colonial times, that was adapted from the traditional French parteres.

47. Our joy in travel was amply satisfied over the years to come. Here, Pat and I are standing above the shore at False Bay, along the coast of South Africa, east of Cape Town. Above us are the mountains of the Kogelberg Biosphere Reserve, home to the highest diversity of plants in any temperate part of the world.

48. The Sophia M. Sachs Butterfly House, Chesterfield, Missouri.

49. Bryan Haynes, noted Missouri artist, painted this symbolic commemoration of my life's work for my retirement from the Garden in 2010. The Garden's iconic Climatron (1960) dominates the painting's upper center, with a profusion of life illustrated along the equator on which the Climatron rests. Beyond the greenhouse's dome rises the Earth, held up by a pair of hands, symbolizing efforts to help create global sustainability. The painting is divided by an axis of pollinators with the flowers they visit, illustrating both my work on pollinators and my development with Paul Ehrlich of the theory of coevolution. To the left is symbolism suggesting the Japanese Garden (1977) and through it the work that took place on the Garden grounds during my years at the helm of the institution, and to the right are scenes of China, suggesting both my birth in Shanghai and the decades I spent

working in the country (1980 to the present), promoting botanical research and conservation while helping to develop global linkages for China's botanists. Overlaying the painting are swirls of *Clarkia franciscana* (left) and *Arctostaphylos montana* subsp. *ravenii*, plants that I rediscovered in the Presidio in the early 1950s, and also swirls of chemistry and biology suggesting the fragility of life on earth and our responsibility to protect it. In the lower corners are a Chinese dragon (right) and a raven with a gold ring in its beak (left) and two ravens (Pat and me) swirling upward into the sky. Novus International, Inc. generously commissioned this painting to mark the occasion of my retirement.

50. Raven family at a happy Thanksgiving gathering in 2019. Front row, Kate Raven with daughter Louisa; Pat, me, Noah Raven; back row, Daryl McQuinn, Liz Raven McQuinn, Alice Raven, Francis Raven, his wife Carolyn Kousky with Nate Raven.

Joe and Nesta Ewan, botanical historians, added a great deal to the research program at the Garden during the post-retirement years they spent with us.

JOE AND NESTA EWAN AND THEIR VALUABLE LIBRARY COME TO THE GARDEN

Particularly important for the expansion of the Garden's resources during the 1980s was our acquisition of Joe and Nesta Ewan's highly significant botanical library in 1986. Comprising some 11,000 books, along with a rich collection of peripheral materials such as letters and reprints, the collection was *the* archive for anyone who wanted to study the history of botanical exploration and discovery in the Western Hemisphere. The Ewans had built the collection with their own funds while Joe established himself as a leading historian of natural history and especially botanical exploration.

Joe and Nesta had always hoped that Tulane University, where Joe was a professor from 1947 to 1977, would formally establish their library as a research tool for scholars. This did not materialize, however—a real disappointment for the Ewans. I wanted to be sure that the library was conserved and housed where it would be easily accessible to scholars from throughout the world, so asked whether they'd be agreeable to moving it to the Missouri Botanical Garden if I could raise their purchase price. I did, and we acquired the collection in 1986, when Joe was seventy-seven years old. We were delighted that with the books came the Ewans, who contributed a great deal to botanical scholarship at the Garden during the eleven years they resided in St. Louis.

At the Garden, we installed the Ewans' collection in Henry Shaw's 1859 Museum Building. There, the Ewans held court until they finally moved to a retirement home near New Orleans in 1997. The years they spent at the Garden were a delightful period for them and their many visitors. Since the Ewans' collection was largely focused on the New World, it was an invaluable complement to the work we were doing at the Garden to expand the picture of our hemisphere's plants.

12

Champion of Science, Education, and International Cooperation

A<small>FTER A FEW YEARS</small> at the helm of the Missouri Botanical Garden, I was starting to understand what the job required. Leadership seemed to be a reasonable fit with my outgoing personality and native skills. Working with people and coordinating their activities provided a satisfying avenue for accomplishing goals that aligned with my values and vision. I recognized, too, that the more collaborative aspects of being a leader were what energized me most. I truly enjoyed connecting people, building bridges, generating enthusiasm for accomplishing common goals, facilitating other people's success, and organizing people to get things done. These were not activities one necessarily engaged in while collecting plants or studying herbarium specimens.

During the 1970s, I had happily exercised a leadership role in directing the Garden, building its new facilities, gardens, and exhibits; giving it a secure financial footing; and expanding its research programs around the world. But during that decade, I had also been increasingly drawn toward working in leadership capacities outside my home institution and even outside the field of botany. Opportunities had arisen, and for the most part I had seized them.

In 1973, for example, I was appointed to the National Science Foundation's Systematic Biology Panel, the principal government agency making grants in the area of biological systematics and evolution. At our frequent meetings, we considered grant proposals and rated them comparatively, with the help of reviews by selected specialists in each field. That was a good way to get to know people and started a substantial association for me with the foundation. Then in 1977, President Carter appointed me to the National Museum Services Board, the governing body of the newly created Institute of Museum Services (IMS, now IMLS). In this role, I represented botanical gardens as a special kind of living museum and helped define what criteria we would use for making grants.[1] In that same year I was also appointed chair of the National Research Council's Committee on Research Priorities in Tropical Biology, the body appointed at the request of the NSF to produce an analysis of that area.

During the 1970s, my scientific reputation was enhanced by my election as a member of the U.S. National Academy of Sciences, followed by terms on the Council and as Home Secretary. During this decade, I also served as president of the American Society of Plant Taxonomists, the Botanical Society of America (BSA), and the Society for the Study of Evolution.

So as the calendar flipped over into the 1980s, it seemed natural to continue on a trajectory that took me away from basic scientific research and increasingly deep into the worlds of politics, policy-making, institutional governance, and international collaboration. While continuing to direct the Garden and advocate vigorously for conservation and sustainability on a global level, I was able to play an increasingly meaningful role in helping to establish science policy, determine funding priorities, promote educational opportunities, bring crucial issues to public attention, and incorporate scientific knowledge into major national and international policy decisions.

My background in science, as well as the honors I had received, gave me credibility in other's eyes, as did my record of success at the Missouri Botanical Garden. The expansive network of colleagues, associates, and friends that I had built over the years also served me well, assuring many recommendations and nominations. And of course, the greater the number of committees on which I sat and the greater the number of offices I held in scientific societies, the greater the likelihood that I would be asked to do more along the same lines. All my many responsibilities kept me extraordinarily busy—perhaps too busy, in hindsight—but they fed my extrovert's hunger for interaction, keeping me motivated and energized.

❧

The election of Ronald Reagan as president in November 1980 did much to convince me of the importance of advocating for science at the very highest levels, even if it meant swimming in the often murky waters of politics. President Reagan was not particularly hostile to science in general, but his administration had the overall goal of shrinking the federal government and reducing discretionary spending, and many of his supporters were deeply suspicious of the scientific establishment and academia.

I first felt the effects of the new administration in my role as a member of the National Museum Services Board (NMSB). After President Reagan took office in 1981, he appointed Lilla Burt Cummings Tower as director of the Institute of Museum Services. Lilla was the wife of Senator John Tower, chairman of the Senate Armed Services Committee, and the sister of Samuel Cummings, the world's largest small-arms dealer. She was extremely conservative, and so working with her presented a considerable challenge on a day-to-day basis. From the outset, however, I was keenly aware that a path of confrontation would be counter-productive for the welfare—and even the future—of the newly established museum agency. So I worked hard to keep Lilla engaged and help her to feel comfortable in her new position. I received a great deal of criticism from the museum community for supporting her, but I considered the long-term survival of the IMS to be far more important than my own political views or hers. As a matter of record, the IMS went on operating much as it had earlier, despite some unusual statements that Lilla made from time to time. I got to know John and Lilla Tower reasonably well, and when she resigned her post in 1983 to work on John's re-election campaign, we parted as friends.

The Reagan administration had a budget policy that called for the elimination of every agency that had been formed by the previous administration, and that of course included the IMS. There are museums in every congressional district, however, and influential museum advocates who call on their elected representatives to support them, and so ultimately there was no change. When my initial term on the NMSB ended, Lilla Tower recommended that I, unlike most of the other members, be re-appointed. When he learned of the president's intention to reappoint me, Senator Claiborne Pell of Rhode Island, one of the sponsors of the legislation that had created the IMS, called me to his office. He reasoned that if President Reagan intended to re-appoint me as a member of the NMSB, he probably also intended to appoint me chair of the NMSB. Since Senator Pell had been so influential in the formation of the Institute, he wanted to see for himself what kind of a person I was. Apparently, he was satisfied, and his committee ultimately approved my appointment as chair in 1984.

By that time, the president had appointed Susan Phillips, sister of conservative activist Howard Phillips, as IMS director. I continued on, encouraging and helping Susan to the extent that I could. When I testified in Congress on our budget, I was instructed each year by the Office of Management of the Budget to testify that America's museums would do just fine with no appropriation (in line with the administration's policy on all newly formed government agencies). This would then lead to them asking me the question, "Well, that's all well and good, but what would you do if you did actually have a budget?" That got me off the hook and allowed a window for offering positive testimony, and the beat went on.

My next political challenge came in 1984, when I was called to serve as chair of the NSF Advisory Committee for Biological, Behavioral, and Social Sciences. This being deep into the Reagan era, the social sciences had become suspect; some considered them socialist, mainly because of the similarity in the words, although it was also true that many social scientists leaned to the left politically. Because of this antipathy toward the social sciences, almost all the committee's efforts had to be devoted to defending them. As a result, we had little time or energy left for biology. We heard testimony, wrote letters, and in general struggled to keep the social sciences alive at the NSF. What we learned was that for fields like economics and psychology, a very large proportion of the overall investment in basic research was actually coming from the NSF, even though comparatively small sums of funding were involved. We also came to realize that polling—a core data collection tool of the social sciences—was of importance to all politicians, and that became another line of defense for those fundamentally important fields.

The beginning of Reagan's presidency also marked the very height of the Cold War between the Soviet bloc and the West. His confrontational approach to the Soviet Union led to a military ramp-up for both countries with military spending so high that it was ultimately unsupportable. There was grave concern all over the world that a nuclear war was imminent. After Reagan assumed office in 1981, the Bulletin of the Atomic Scientists moved the minute hand of its Doomsday Clock—a gauge of the risk of nuclear annihilation—from seven minutes to midnight to only three.[2] This was the closest to midnight the Doomsday Clock had been since 1953. It was in this context that Carl Sagan and others organized a high-profile public discussion of what might happen to the earth and humanity in the event of a large-scale nuclear war. I was invited to participate. We gathered more than one hundred scientists in Cambridge, Massachusetts, in the spring of 1983 to discuss what we called "nuclear winter." Carl Sagan chaired the section on physical sciences and I chaired the one considering the life sciences. We then presented our findings at the Conference on the Long-Term Worldwide Biological Consequences of Nuclear War in Washington, D.C., on October 31 and November 1 that same year. The consequences, as we saw them, were dire: an enormous opaque radioactive cloud would likely cloak the earth in darkness, putting at risk many species, including our own. We hoped that our findings would bring some additional urgency into the conversation; at the very least, our efforts made Americans more aware of the issue, owing in part to the fame and notoriety Sagan had accrued from his starring role in *Cosmos*, the thirteen-part TV program he had produced and shown in 1980. In an effort to keep the issue in the public spotlight, Sagan collaborated with Paul Ehrlich, Walter Orr Roberts, and Donald Kennedy to publish a summary of the conference's proceedings entitled *The Cold and the Dark: The World After Nuclear War* in 1984.

ϛ∕Ͻ

In 1985, sitting in the living room of our log cabin on property Tamra and I had purchased along the Big River south of St. Louis the previous year, I received a call from Tom Smith. In addition to his role on the Garden's board of trustees, Tom had served for several years as a member of the governing body of the four-campus University of Missouri system, the Board of Curators. Tom was calling to ask if I'd be willing to stand for appointment to the group. Considering that serving on this board would be a good way for me to learn more about my adopted state and to ensure that the University of Missouri system was doing its best to be accessible to all Missourians, I agreed to be put up for consideration. I had always believed higher education to be essential to the advancement of society, and I was very appreciative of the education I had received, partially at public expense, in the University of California system. It was my turn to give back, even if was in an indirect way.

A couple of months later, I found myself sitting in the office of then Governor John Ashcroft in Jefferson City. He interviewed me carefully, and we shared stories of our respective backgrounds. He subsequently decided to appoint me for the Third District vacancy, which included part of the city of St. Louis. A hearing in the Missouri Senate followed, during which I was questioned for about an hour and subsequently confirmed.

The board met in different places around the state, most often in Columbia, where the system's flagship campus is located, and then once each year at the other campuses in Rolla, Kansas City, and St. Louis. We held vigorous discussions of many issues, most of them revolving around our individual visions of what the universities in the system should be. Conservative members of the board wanted to limit the number of departments and major subjects offered to those in which we could excel, as if we were a private university like Princeton or Yale. The rest of us believed that it was appropriate for a land-grant university to offer a wide range

The University of Missouri Board of Curators at a meeting in 1979. Standing, from left to right, are Jeanne V. Epple, Sam B. Cook, James C. Sterling, Eva Louise Frazer, Edward S. Turner, and Kevin B. Edwards, student representative; and seated are me, C. Peter McGrath, W. H. "Bert" Bates, Fred S. Kummer, and John P. Lichtenegger.

of subjects so that it could better fulfill the needs of all the state's citizens. For many years, our public universities and colleges had been important stepping-stones for people from the state's shrinking rural regions who wanted to move into the mainstream of society, and we wanted to keep it that way. Sometimes, the way was clear; for example, when the veterinary school was threatened with loss of accreditation, largely because of the need to upgrade its equipment to modern standards, the realization that we were serving several states in addition to Missouri ultimately made our decision to upgrade and maintain the school—the only logical choice. A more difficult case would be when the graduates of other schools were consistently better qualified on graduation than ours. For these majors, I would generally argue for diversity, with additional training after graduation on the job if necessary, so that Missourians could usually find what they wanted in state.

One topic we discussed passionately was divesting the university's holdings in South Africa because of that nation's awful policy of apartheid. Here I thought that we might do more good by staying involved, a characteristic position for me. Others pointed out that not to divest would be received so badly by our own African-American students that we simply had to do it. In time, I came to believe that my instincts on this issue had been wrong. Disinvestment by the world community turned out in fact to be a very strong influence in ending apartheid. During this same period, the University was able to initiate a cooperative program with the University of the Western Cape that helped build the capacity of black citizens of South Africa so that they would be better equipped to fill the many roles that opened to them with the end of apartheid.

We spent a good deal of time in closed session approving purchases on what was to be the future site of an industrial park in Kansas City, an effort that subsequently failed, and many other closed sessions discussing NCAA accusations that our basketball team had violated its rules. Notwithstanding these distractions, the time I spent as a member of the board of curators was deeply satisfying to me. In that role, I learned a great deal about the state, about people, and about higher education. I made lasting friends and deepened my appreciation for what society at its best could do.[3]

∾

A major milestone occurred in 1987–88 when the Flora of China project leaped over its last hurdles and finally was ready to become operational. This day had been a long time coming. The initial conversations with our Chinese colleagues had taken place eight years earlier, in 1979; since then I had held discussions on the matter in China on probably a dozen individual trips. During this time, fruitful botanical exchanges had continued between the two countries. In the early 1980s, for example, two botanists from the Institute of Botany, Chinese Academy of Sciences, Beijing, had each spent more than a year at the Garden. They were Chen Jia-rui, who had accompanied us during our excursion to Mount Emei and studied Chinese *Epilobium* with me, and Wu Pan-chen, a student of mosses who began a lifelong relationship with the Garden's moss specialists during his study visit.

At the XIV International Botanical Congress in Berlin, in the last days of July 1987, Wu Zhengyi, who recently had become the head of the *Flora Reipublicae Popularis Sinicae* (FRPS) Editorial Committee, drew me aside and delivered exciting news. The committee, necessarily with prior permission from the Chinese Academy of Sciences, had approved the *Flora of China* as a joint project of Chinese and foreign botanists. The editorial center for the flora would remain in Beijing, but the Kunming Institute of Botany (KIB) would also be important for the project, with Professor Wu taking over as its leader.

My plan always had been to have Harvard University lead the non-Chinese side of the project, since Harvard had had such a long and illustrious history with Chinese botany. When push came to shove, however, Harvard declined to take on the leadership role. Unwilling to let the project fail for lack of a non-Chinese editorial center, I decided to take it up at the Missouri Botanical Garden, even though we as an institution had not had one bit of earlier involvement with Asian floras or Asian plants in general. Once we had crossed this Rubicon and committed to playing a leadership role, I had to find help. That need was met through the appointment of Bill Tai, whom I persuaded to join the Garden's staff for this purpose. With help from the California Academy of Sciences, the Smithsonian, and Harvard, we were underway.

Once we got started, the Royal Botanic Garden Edinburgh asked to join us, and of course we were delighted. And the Royal Botanic Gardens, Kew, had so many people involved in the project, each needing individual permission to contribute as an amendment to his or her work plan, that I decided to approach Iain Prance, a long-time friend who was by then Kew director, to see if the institution might be willing to join as a partner as well. I was elated when he accepted, knowing by that sign that the Missouri Botanical Garden had truly become an institution with recognized global reach. The Jardin des Plantes in Paris joined later, and it was clear that many other institutions, such as the University of Tokyo and the Komarov Botanical Institute in Leningrad, were functioning as "virtual" members of our group. The Flora of China project clearly represented a global cooperative effort of the very best kind.

Coming back down to earth, we knew that we had to find a publisher and to develop an editorial policy that would work both for the Chinese and for their foreign partners. We began to discuss these matters and came together to sign the formal agreement for the project on October 7, 1988. The signing took place at the Missouri Botanical Garden, fittingly under the shade of the row of Chinese dawn redwoods that had been planted so carefully four decades earlier.

Those who gathered for the event included Gu Hongya (whose Ph.D. at Washington University I had supervised), Professor Wu, Chen Jia-rui, and two officials from the Chinese Academy's Bureau of International Cooperation, Su Fenglin and Li Mingde. In addition, there were botanists representing the different Chinese institutions involved, as well as Bruce Bartholomew and Dave Boufford, who had contributed much to help get the project to that point. Eager to get started on the real work, we coupled the agreement ceremony with our first official working session.

As we began to outline the principles by which we would develop and write the flora, it became apparent that we had a more challenging problem than I recognized initially. The FRPS, the Chinese-language flora that we were, it might be said, shadowing was eventually to be completed in 80 volumes and 40,000 pages. Considering how we would integrate our efforts with it, we learned to our dismay that Chinese text translated into English expands by roughly 50%, so that a literal translation would take up something like 60,000 pages. The prospect of dealing with a work of that size drove potential publishers virtually to tears. After all, a multivolume flora of China was hardly likely to become a best seller. And so we had to think harder about the scope of what we were doing.

We could, first of all, turn the long descriptions presented in FRPS into concise ones like those in the *Flora Europaea*, which I had long considered an excellent model. And since the illustrations had already been published in FRPS, we decided to leave them out of our revision. I calculated that these moves would allow the English-language work to be published in twenty-five volumes, which made the publication of our work seem much more feasible. We decided that its publication would be a joint venture of the Missouri Botanical Garden Press and Science Press, the publishing arm of the Chinese Academy of Sciences. This would prove to be a good arrangement. Important editorial matters awaited discussion and resolution, but we were off to the races!

Although I had been collaborating with botanists in other countries for years, the Flora of China project stood out as particularly significant. The scope of the project was massive, and because the U.S. and China had had a somewhat antagonistic relationship, we felt that the project exemplified the kind of cooperation, trust, and goodwill that should exist between the two governments.

~

Early on a sunny afternoon in 1987 I was sitting in a garden in Amsterdam chatting with friends when I received a call from Bob Adams, the archaeologist who had been selected, instead of me, to run the Smithsonian Institution a few years previously. Bob, I knew, was serving on the committee charged with nominating people to stand for election as officers of the National Academy of Sciences (NAS). He asked me, straight

When he visited St. Louis for the signing of the Flora of China agreement, the great Chinese botanist Wu Zhengyi enjoyed a picnic in the countryside. Although he had not been in North America before, he recognized the genus of almost every plant that he saw, because of their similarities with the Chinese ones he knew so well.

away, if I would be willing to run for the position of Home Secretary. Gasping a bit, I asked how long I had to think about the matter, since if I were elected it would represent a considerable investment of time. "Please get back to me within three hours," came the answer. Yikes!

I had known that Home Secretary was one of the positions to be filled in the upcoming NAS election, but I had not thought of the possibility of standing for election personally: my plate was already full. Indeed, I had had discussions with several people who *did* want to run for the position, which was why Bob's question took me completely by surprise. By that time, I had had a decade-long association with the NAS. I had been elected to the academy—itself a great honor—in 1977 at the relatively young age of 40. In 1983, I had been elected to the academy's Council as one of its twelve councilors, and in that capacity helped govern the affairs of the academy. I had found much satisfaction in being active with this venerable institution, which had been established by President Lincoln in 1863; I felt my work to be of importance because one of the academy's core functions was to provide "independent, objective advice to the nation on matters related to science and technology." I spent the next few hours thinking about the pros and cons of being an NAS officer, consulted with Tamra, and called Bob back. "I'll do it," I told him without hesitation.

In the months that followed, I had little idea if I would win the post or not, but when the votes came in, I had garnered a majority. As Home Secretary, I was to take care of everything for which one would normally expect an executive secretary to be responsible, except international matters, which are handled by the Foreign Secretary. In general, I was to tend to the interests of the members, an unruly but incredibly interesting lot if ever there was one. Two of my biggest tasks were to coordinate the annual meetings and oversee the process of electing new members. I was assigned an office in the impressive Academy Building on the Mall in Washington and began to travel there frequently for meetings and other activities.

The job of Home Secretary turned out to be interesting and rewarding— so much so that I would decide to run for another four-year term beginning in 1991 and yet another beginning in 1995. For the first six years of my tenure, I served under NAS President Frank Press, a dedicated physicist with style and intelligence. He worked hard for the academy, helped it to maintain its independence, and organized a variety of studies, many of them analyzing issues that were important in determining government

Welcoming President George H. W. Bush and his science advisor D. Allan Bromley to the National Academy of Sciences annual meeting in 1989.

policy. For the second six years, I would serve under President Bruce M. Alberts, a famous cell biologist who had a position at the University of California Medical School in San Francisco and did a great deal to promote science education and international collaboration, two of my own passions.

Serving as Home Secretary was not my only involvement with the academy. Prior to being elected Home Secretary, I had served for a few years on the Governing Board of the National Research Council (NRC), and I continued in this position for the first year of my term as an NAS officer. The NRC is the research arm of the National Academy of Sciences, National Academy of Engineering, and the National Academy of Medicine.

While I served on the NRC's governing board, from 1985 to 1988, I chaired an NRC analysis of where the life sciences (including agriculture and medicine) were headed in the 1990s—and what a job that was! The study was inspired in part by the coherent studies of the field of astronomy that appeared periodically, requesting ever-larger and more sophisticated pieces of equipment to fuel development in the field. It had also been stimulated by a very well-received 1985 study of opportunities in chemistry; the chair of the study, George Pimentel of the University of California, Berkeley, had relentlessly "sold" its conclusions to everyone who would listen and could make a difference to the future of chemistry.

As we looked into the matter, it turned out that biology was too diffuse, too multidimensional, and far too lively a field to reduce easily to a few major research objectives. We had a talented committee, the members of which collectively knew really well every corner of the areas that we were to consider, but we could find no stupendous conclusions that we could promote as such. The most active and clearly important fields of basic biology, molecular biology, neuroscience, and development, for example, were already heavily funded, partly because of their medical importance. While it was possible to foresee some future developments, it didn't make much sense to pour more money on them than they were already getting. A couple of findings stood out in our report, however.

First, we found that truly interdisciplinary majors (for undergraduates) and graduate programs, even more than sequential specialization in different fields, held great promise for advancing the life sciences. This silo-breaking idea met with widespread approval; unfortunately, specialization is so deeply embedded in science and academia that cross-fertilization among disciplines has not yet gained as much ground as we hoped it might. Second, our report called attention to the need for additional support for descriptive biology—as opposed to experimental, hypothesis-driven biology—because of its fundamental importance for so many fields. That recommendation did prove to have some impact. Chairing this committee taught me a lot, and I greatly appreciated the diverse and impressive talents of its members.

Two years after my election as Home Secretary, in 1989, NAS President Frank Press asked me if I would be willing to take up the additional post of chair of the Report Review Committee (RRC). My predecessor as Home Secretary, Bryce Crawford, a distinguished professor of chemistry from the University of Minnesota, had occupied that post for years and was about to step down. What would that job entail? A great deal, as it turned out.

I had a general idea what I would be expected to do if I accepted the appointment and worried about the additional time commitment that it would require. In the course of my deliberations about whether to do it or not, I consulted Phil Smith, President Frank Press's active and encouraging right-hand man. "I am sure you can do the job," said Phil, "but I am not sure you will have enough time to think about the meaning of individual decisions sufficiently." Considering my compulsion to keep so many balls in the air, it might have been natural for me to rush through the duties of this additional and very important responsibility without taking the proper care to consider the various aspects of each situation. I decided, however, that I did understand the implications of Phil's very important caveat and moved forward to embrace and accept the opportunity. Once Frank Press had assured me that if I chaired the RRC, every single article in the news that touched on science and technology would make more sense to me than it possibly could have otherwise, I became deeply interested.

The NRC publishes 250–300 or more reports a year. It is the job of the RRC to monitor them all, formally reviewing the ones that the committee decides would benefit from such review. The Report Review Committee consists of about thirty distinguished individuals drawn from the different disciplinary fields in which reports are being prepared. The committee's job, as described by my predecessor Bryce Crawford, is to "call the balls and strikes." In other words, we were never to argue for changing the conclusions of the expert advisory committee selected for the study under review; instead, we were to select the best possible reviewers—ones with varied opinions—to provide feedback to the committee and then monitor the committee's reactions to their comments. A typical report had anywhere from six or seven to perhaps as many as twenty reviewers, who summarized their conclusions and suggestions and submitted them in writing. Every single comment they made was required to be addressed specifically by the committee.

The reports we reviewed dealt with many important and sometimes controversial subjects. These included the use of laboratory animals in research, the ethical effects of modern biochemical procedures, preparations for disasters of various kinds, the monitoring of climate change, whether cell phones cause cancer, nutrition standards, the extent to which seat belts on school busses contribute to the safety of the passengers, the causes of terrorism, the health hazards of smoke for airplane passengers, AIDS, the safety of the nation's blood supplies, and how best to teach science. We also reviewed reviews of whole fields (such as the biology study I mentioned earlier). It was an exciting activity for me. I found, just has Frank Press had predicted, that being in the role of RRC chair was highly educational as well as intellectually stimulating. This experience informed my thinking about the broad value of science to the global public and the need to keep reports and reviews as apolitical and accurate as possible for the benefit of all.

As an officer of the National Academy of Sciences, I had a great many opportunities to interact with my counterparts in foreign academies. For example, we had a long-standing relationship with the Royal Society, Britain's national academy, which was established in 1662 under the patronage of King Charles II. We would meet with our British colleagues every other year, gathering in alternating countries, and the meetings were always interesting and enjoyable. One year, we were hosted at Trinity College, Cambridge, by Sir Michael F. Atiyah, the very distinguished mathematician who was at the time the president of the Royal Society. One evening, Sir Michael announced that we would convene on the second level, in the room where Isaac Newton had been knighted by Queen Elizabeth I. We gasped. Sir Michael told us that he had been impressed, too, until he learned that Newton was running for Parliament at the time, and the Queen had knighted him because she needed his vote.

❧

As an officer of the National Academy of Sciences, I also began to attend joint meetings with the equivalent officers at the Soviet Academy of Sciences. Our biennial meetings were fascinating events that helped to set the stage for renewed cooperation. They also brought me the opportunity to make connections with Russian scientists and institutions that would lead to a great deal of collaborative work in the years to come.

The 1986 Chernobyl disaster, which came not from conflict but as an accident, brought home the awful potential of nuclear arms for disaster. The Kremlin lost control of media, and the stage was set for the opening and then dissolution of the Soviet Union. By 1987, the year that I was elected the NAS's Home Secretary, our relationships with the Soviet Union were warming. President Reagan's strategy of re-armament and confrontation coupled with a continuing push for the relaxation of tensions were proving successful. At the same time, the NAS had been playing a role in ratcheting down the nuclear arms race. The NAS and the Soviet Academy had together founded CISAC, the Committee on International Security and Arms Control, in 1980. CISAC meetings had brought scientists and technical people from our two countries together to discuss the nuclear arms race and try to find ways to end it. The meetings were closed and discussion was correspondingly free. Contributing an essential common vocabulary helped greatly in making possible the eventual treaties, and the meetings had clearly led to major advances of general importance. Government officials from both countries had avidly studied CISAC's proceedings and suggestions.

My first joint officers' meeting with the Soviets was in mid-January 1988, in Moscow. Held in the age of General Secretary Gorbachev's *perestroika* and *glasnost*, the visit was filled with surprises. At one of our restaurant lunches, I found myself seated next to the famous Soviet scientist Andrei Sakharov, whose imprisonment had been a source of argument between the U.S. and Soviet governments for so many years. Although he and his wife had been released by Gorbachev a year earlier, I had never expected to see him personally. One of our Soviet counterparts remarked that we shouldn't be worried about the Soviet Union; even if they knew all our military secrets, they would not be able to use them because they were essentially bankrupt. Later that same year I was elected a Foreign Member of the Soviet Academy of Sciences.

More was to follow. My friends at the Tsytsin Main Botanical Garden in Moscow arranged a visit to Soviet Middle Asia in the summer of 1989, sponsored and funded by the Soviet Academy. I was invited to bring my family along. Francis at the time was eleven years old and Kate six—just old enough to be able to remember and appreciate the experience. With the help of Glenn Schweitzer, a tireless proponent of forming sound relationships with Russian and other Soviet scientists, we laid out an itinerary in Kazakhstan and Uzbekistan and set out for a great adventure.

Arriving at the Moscow airport on June 25, 1989, in the early evening, we were met by Igor Smirnov, with whom I had built a good friendship on earlier visits to his home institution, the Main Botanical Garden. He was to be our guide for the whole trip. When Tamra expressed her delight that we would have a guide who knew the areas we would be visiting, Igor quickly disabused her of her assumption. "Oh, I've never been there before," he said. "And I hear they kill Russians." It was hardly reassuring, but there we were.

The next day, we flew to the large city of Almaty, then the capital of Kazakhstan. The city lies in the foothills of the Tien Shan, a beautiful and mighty range that divides the country from China to the south. Professor Isa Baitulin, the leading Kazakh botanist and head of Almaty's Institute of Botany, greeted us genially, showed some of the city's sights, and brought us to the art museum, which opened after hours especially for our visit. The museum gave us a sense of the fascinating ancient history of the area and taught us something about Kazakh culture.

A special treat was the scenic drive up into the middle-elevation conifer forest of the Tien Shan Range above Almaty the next day. We enjoyed the rich and colorful mountain flora, identified for us by Professor Baitulin, and shared a picnic in a lovely meadow among the trees. Isa Baitulin's grandson Nicky, about thirteen years old, enjoyed carrying Kate around on his shoulders. Presently Isa asked, in jest, if the grandson could marry her. How many camels would I require for a bride price?

We enjoyed a sumptuous banquet in Professor Baitulin's home with his family, with its centerpiece a whole roasted sheep. Later we flew west to Uzbekistan and began to drive around the republic by car. In Tashkent, the capital, we visited the large and interesting botanical garden with its extensive array of interesting native plants, many of them endangered species, stretching out in orderly rows. Having occupied an important position along the Silk Road, Tashkent had a fascinating, 1,500-year-long history as a significant trading and commercial center.

We then traveled to Samarkand, one of the most interesting cities I had ever visited, with its 2,500-year history evident at every turn. One of the largest cities in Central Asia, it has been occupied by a diverse series of cultures, and was another important stop along the Silk Road. We celebrated Francis's twelfth birthday, a truly memorable occasion, in a rooftop restaurant looking out over the tomb of the legendary fourteenth-century conqueror Tamerlane. The final stop on our overall trip was Khiva. Glenn had urged us to visit this wonderful city in the desert of western Uzbekistan because, although fully inhabited, it retained most of its classical architecture; it was unique among the cities of the region.

On returning to Moscow, we took the children to enjoy the Old Moscow Circus, a real treat for us all. On our way back to the hotel in a cab, the woman driver turned to the children and remarked, "Forget about the Soviet Union; think about Russia. Russia is a great country filled with loving people who care about you, and you should always remember that." Those words had quite a bit of impact. On the way home, we felt extremely fortunate to have been able to take the trip.

It was a momentous year for the Soviet Union and its relationships with the countries it controlled. In Poland, the trade union Solidarity had won a partially free election shortly before we had left for our trip to Soviet Middle Asia, and later that summer Poland peacefully ended communist rule. This created momentum for increased civil disobedience in other Soviet Republics in Eastern Europe, and in late 1989 the Berlin Wall fell. In 1991, the Soviet Union failed. When it had been dissolved and ceased to exist that December, its constituent republics became independent nations.

In the aftermath of the collapse of the USSR, in late 1991, I received a telephone call from Alex Goldfarb, a politically active Russian emigrant. He was helping philanthropist George Soros to realize as much as possible of his dream—that of bringing economic independence, democracy, and capitalism to Russia, the other countries of the former Soviet Union, and the communist bloc generally. Recognizing the economic importance of the scientific community in the former Soviet Union and especially in

Russia itself, Mr. Soros had decided to help these scientists and their institutions survive during the incredibly difficult times they were experiencing. Later, I understood the importance of George's purpose more clearly when I spoke with the chief astronomer of Armenia, a fellow committee member. He was a very distinguished scientist whose circumstances exemplified the problem. He told me that he was having to sell his household furniture piece by piece to buy food for his family. He was not complaining, but pointed out that half of the furniture was already gone; he wondered what they would do when they had nothing left to sell.

Alex invited me to a meeting of experts to discuss how best to assist the region's scientists. After the first meeting, I was invited to serve on the Executive Board of the International Science Foundation for the Former Soviet Union (ISF). Representatives of each of the former republics and a cadre of scientists from the U.S. made up the group; our first organizational meeting was in January 1993.

Meetings and trips together to Moscow, St. Petersburg, and Kiev gave me a good chance to get to know Mr. Soros, a truly remarkable man, and to understand the situation regionally. To help as many people as possible in a meaningful way, our executive board adopted a strategy of making many $500 grants to individual scientists within the original borders of the Soviet Union. We set up an elaborate review system with many readers and panels to deal with the numerous applications we received. In total, we ended up making merit-based grants and awards amounting to approximately $130 million over a two-year period, most of it going to individual scientists. The average monthly salary of a scientist in the region at that time was approximately $18, so our seemingly small grants often equaled more than two years' salary and were very helpful to those who received them. Through these grants and in other ways, the International Science Foundation subsidized the work of approximately 80,000 scientists in the territory of the former Soviet Union, most of them in Russia.

Philanthropist George Soros generously supported the scientific enterprise in the former Soviet Union after its dissolution, in an effort to keep the regional economies running and the people fed.

Gerson Sher, who was assisting the ISF while on leave

from his regular post in the International Office of the National Science Foundation, became a close friend as we worked together on this important project. Of pleasant demeanor and fluent in Russian, Gerson was a good companion on many adventures during those years, and he made a major contribution to holding our activities together despite their inherent complexity.

While working with the ISF, I began to think about the decaying herbarium-library building at the Komarov Botanical Institute in St. Petersburg, wondering whether Mr. Soros might be interested in helping with its repair. I considered it prudent to begin by getting an accurate estimate of the scope and cost of the project using a non-Russian source. To do this, I enlisted the assistance of Sverdrup & Parcel, a St. Louis-based engineering firm. They took on the job, hiring a Russian-speaking engineer to help them in the task. I accompanied them to St. Petersburg (Leningrad had been returned to its original name with the collapse of the Soviet Union in 1991) for their initial visit, and they did a fine job, recommending a Finnish engineering firm that had been involved in similar projects in Russia to oversee the actual work. They came up with an estimate of $1.1 million to repair or replace the walls, ceiling, and heating and cooling systems.

With the encouragement of George Soros's friend Valery Soyfer, who consulted Iain Prance about the matter, I was able to obtain a grant of $500,000 from the Soros Foundation, which got the ball rolling. I then turned to USAID (the U.S. Agency for International Development), which was active in Russia at the time. Gaining the support of environmentally sensitive Colorado Senator Tim Wirth, we obtained an award of an additional $500,000 from U.S. AID. Rounding off the remaining $100,000 with gifts from several friends who liked me, even though they were puzzled by the project, I was able to give the go-ahead for construction. It was gratifying to have successfully raised $1.1 million in the U.S. to repair an old building in St. Petersburg, nearly two decades after first learning of its need for repair and restoration. The irreplaceable historical collection of herbarium specimens and library would now be safe.

As I soon found out, I was not quite through with Russia yet. As the ISF wound down its activities in 1996, Neal Lane, then director of the National Science Foundation, called me into his office to invite me to further service. He asked me to consider serving as chairman of the board of directors of the newly created U.S. Civilian Research and Development

Being able to raise the funds to repair and stabilize the Herbarium Building at the Komarov Institute in St. Petersburg, Russia, was one of my greatest satisfactions.

Foundation for the Independent States of the Former Soviet Union (soon abbreviated to CRDF). Congress had set up the CRDF along the lines of the model established by the U.S.-Israel Binational Science Foundation, which had been consistently successful since its formation in 1972. I told Neal I would serve as chair of the CRDF if he would allow Gerson Sher to take another leave of absence from the NAS to act as the principal operational officer for the foundation. Neal agreed, and we were on our way.

What we wanted to do was to create opportunities for scientific institutions and corporations in the U.S. to collaborate with their counterparts in Russia and other former Soviet republics and, when possible, to find ways to commercialize their ideas. An underlying motivation was to find peaceful and profitable occupations for the thousands of weapons specialists who had been displaced because of the disarmament agreements our two countries had signed. At CRDF, we worked hard to obtain funds for various programs but were not particularly successful, and Gerson and I moved on after a couple of years. Since that time, CRDF has greatly broadened its mission, which now is to "promote peace and prosperity through international science collaboration."

⌇

Closely allied with the geographically extensive flora projects that the Missouri Botanical Garden was pursuing in China, North America, Latin America, and elsewhere was the idea of establishing national biological surveys. The concept was to set up in each nation an institution that would strive to identify, collect, catalog, map, and assess the status of every species of plant, animal, and fungus, and microorganism that occurred within its boundaries. Whether the importance of these biological resources was realized or not, they were collectively just as important as the nation's mineral deposits and agricultural lands; it was impossible to properly protect and manage them without knowing what they were and where they existed.

Among the first national biological surveys was the one initiated in Costa Rica by the National Biodiversity Institute (INBio). The brainchild of Dan Janzen, INBio was founded in 1989 to carry out the survey and funded, on Dan's insistence, entirely through non-governmental sources. I personally persuaded Rodrigo Gámez, a professor of plant virology at the National University of Costa Rica, to take up the post of founding director. By the early 1990s, INBio had, under Gámez's direction, made excellent progress in surveying Costa Rica's rich plant and animal life, and it was proving to be an outstanding partner in the Garden's work on the *Manual de Plantas de Costa Rica*.

Impressed by the model of INBio, I started discussions with José Sarukhán about the possibility of creating a national biological survey in Mexico. José had been thinking along the same lines. In early 1992, he was able to persuade then President Carlos Salinas de Gortari to establish the Comisión Nacional para el Conocimiento y Uso de la Biodiversidad (CONABIO). The president announced the formation of CONABIO as a part of his presentation at the U.N. Earth Summit, held in Rio de Janeiro that June. In September, José Sarukhán convened a group of us—including David Packard, Dan Martin of the MacArthur Foundation, Tom Lovejoy, Rodrigo Gámez, and others—to present a public symposium in Mexico in celebration of the formation of the new organization. On the Saturday following the ceremony, September 15, 1992, we were all whisked off by helicopter to the magnificent pyramid of Yaxchilan on the Usamacinta River in Chiapas, where there was a further news conference and celebration. I was chosen to provide the response to the president. I asked José, Jorge Soberón (who was to become the first director of CONABIO), and others to help polish my Spanish text, a process that took us into the wee hours of the night. Somehow, I got through its delivery the next day, and President Salinas even kindly complimented my Spanish—a comment I appreciated even if I was wise enough not to take it too seriously.

Under the guidance of Sarukhán, Soberón, and others, CONABIO would flourish over the years, proving popular with successive federal administrations in Mexico. Rather than amass its own collections, it took the course of recording species in a database from which the information could be retrieved for any purpose. With its satellites circling the globe continuously, CONABIO was able to locate forest fires in distant regions and determine whether they were burning in environmentally sensitive areas, then demanding priority attention. It also began to develop formal and informal education programs that allowed it to provide important services to the people of Mexico. In this way, CONABIO gradually grew to be an indispensable part of Mexican society.

In 1993, at the start of the administration of President Bill Clinton, Tom Lovejoy and I held discussions with the new Secretary of the Interior, Bruce E. Babbitt, about the possibility of creating a national biological survey for the United States. Bruce did not need much persuading, as he was inclined to think that such a survey would be valuable. He reallocated funds and provided support for a National Research Council study of the matter, which I chaired in 1993. Despite our strong recommendation that biological research in several federal agencies would be much stronger if it were better coordinated, Congress refused to go along with the plan. The Department of the Interior was ultimately left with even less funding for biological survey work than it had had initially. Apparently, some in Congress were worried that if endangered species were documented on the lands of some of their constituents, the provisions of the Endangered Species Act would force the owners to take remedial steps and thus limit their use of the land they owned. By that time, Secretary Babbitt had already, partly on the basis of my recommendation, hired Ron Pulliam, a talented and experienced ecologist from the University of Georgia, to become the first head of the survey. Ron did a fine job, but the survey was not to be; he had to abandon his post when the whole structure came tumbling down around him after little more than a year. Despite our failure in the U.S., other countries were attracted to the idea of establishing national biological surveys. Such surveys would gradually become more widespread because of their ability to contribute a great deal of useful information about the best management of a country's natural assets.

— 13 —

Professional Recognition and Personal Tumult

THE MID-1980s and the 1990s, from my fifties into my sixties, were vibrant times for me professionally. Projects that had been in process for years began to produce tangible results. I received recognition for my work and that recognition helped give me the platform for greater impact. In my personal life, things were not as smooth. Several painful transitions would occur before the new millennium arrived.

Prior to 1985, I had received numerous honors and awards. The most significant of these, in my view, had been my election to membership in the National Academy of Sciences in 1977 at the age of 40. In the spring of 1985, I received word that I had been selected for an honor that rivaled the NAS membership for prestige: I was to be a recipient of a fellowship from the MacArthur Foundation, a significant five-year grant that I could use as I wished. Among others to be honored by the MacArthur that year were the scholar and author Harold Bloom, the environmental historian William Cronon, the choreographer and dancer Merce Cunningham, and the scholar and author Jared Diamond. These fellowships were not yet being referred to as the "genius awards," thank goodness, but the foundation awarded them to people who they felt had the creativity and vision to bring about progress in overcoming some of the major challenges faced by humanity. It was very satisfying to be selected as one of the recipients of the foundation's largesse and receive such a vote of confidence in my work.

John Biggs, a member of the Garden's board of trustees, decided that my award should not go unnoticed, and with the board set up a recognition dinner at the Log Cabin Club. It was a warm and lively event, with Buck Bush and Nora Stern serenading us with songs such as "I Talk to the Trees," one of my favorites. Dan Schlafly showed up with a bag of bones that he had obtained from a butcher and attempted to evoke memories of my Donner Party ancestors, rattling his bag while alluding to the rumors about what had transpired during that long-ago trek west. The event provided memories that I would treasure thereafter, but more importantly it forced me to take at least a moment to enjoy my own success. I didn't have the knack for savoring awards when they came; it was always my nature to hurry on to the next project, to the next connection, with a certain sense of urgency.

The following year, 1986, I was chosen as the recipient of the International Prize for Biology, commemorating the sixtieth year of the reign of Japanese Emperor Hirohito. The honor seemed especially sweet given my presence the previous year at the ceremony honoring the prize's first recipient, E. J. H. Corner; I understood its significance. Tamra and I traveled to Kyoto to receive the prize, which was awarded in a formal ceremony that involved the Japanese Prime Minister and other dignitaries. I was presented with a certificate, a handsome gold medal on which the sculptor had depicted one of the marine animals that the Emperor studied, and a silver vase displaying a golden chrysanthemum, the Japanese imperial sign.

Returning to Tokyo, Tamra and I traveled with our friend, Professor Kunio Iwatsuki, to the Imperial Palace for the very special honor of meeting the Emperor. The palace is located on a forested hill, like a park, and surrounded by a moat. Sitting in its central court facing a traditional Japanese garden, we were deeply impressed when the Emperor emerged from a door on the other side of the garden with his chamberlain, whom we were told was the descendant of a family that had served the Emperor in that capacity for more than three centuries. The pair of men walked gracefully around the garden and sat in chairs facing us. The Emperor was a dignified, kind-looking man with thick glasses. With Kunio Iwatsuki as interpreter, we began to discuss botany and learned about some of the Emperor's favorite plants. I knew that Professor Hiroshi Hara had tutored the Emperor in botany, but still was surprised at the level of his knowledge about plants.

As we spoke, my thoughts jumped back half a century to the posters I had seen in the grocery stores during World War II that featured this man's image with the caption "Enemy ears are listening!" The memory underscored how much the world had changed. The war had ended long ago, yet it was recent enough that I remembered it vividly, as the Emperor certainly did. It was amazing to be actually conversing with Hirohito, the universal language of science bridging our very different lives. The hour passed quickly, and we then emerged onto the bustling streets of Tokyo. A few days after leaving the Palace, we were told that the day we saw him must have been a very happy one for the Emperor, with the start of the year's Sumo tournaments and a good chance to talk about botany with an American scientist in place of politics. I left Japan feeling that my connections with the country and its botanists had been strengthened immeasurably, with memories of wonderful days in the company of friends old and new.

☙

Professionally, I sailed along through the remainder of the 1980s and into the 1990s, enjoying further honors and progress toward realizing goals. I was selected to join the prestigious Accademia dei XL (Academy of the Forty), one of Italy's two national academies, in 1987, and was invited to participate in a study week on the environment at the Vatican in early November of that year. I began my term as Home Secretary of the National Academy of Sciences, deepening my involvement in that great institution. The Flora of China project had been officially approved in 1987, and the formal agreements for its execution signed a year later, in 1988.

At home, however, life was not so harmonious. My relationship with Tamra was becoming more and more strained and uncomfortable. I had noticed the first signs of our marriage's deterioration a few years previously, though we had done nothing to address the underlying problems. Now the conflicts were harder to ignore, but still my response was not optimal. Rather than confronting the issues head on, I tended to avoid conflict by keeping even busier than my usually crammed schedule demanded and by booking a lot of travel. For her part, Tamra was juggling too many things, which caused her to be overly stressed, and she was reluctant to admit any kind of weakness. I didn't understand very clearly what was going on, but I could sense that changes were in the air.

While Bill Tai and I were visiting the Institute of Botany, Beijing, in the summer of 1989 to work on the Flora of China project, I was startled to receive word that my mother, in California, had been diagnosed with breast cancer and was soon to have an operation. I offered to pay for my telephone call, but Wang Siyü, who had guided us around China in 1980, placed it and assured me that they would not think of letting me pay considering all that I had done for them. This was gratifying and certainly made me feel less alone at that worrisome time. I hurried back to Walnut Creek so that I could be with my mother for her surgery. I was relieved to learn after it had been completed that her prognosis was very good.

At about the same time, my daughter Alice, then almost thirty, moved back to Columbia, Missouri, where she had lived with her husband Clark Powers in the early 1980s. After their marriage fell apart, Alice had taken off to live in Portland, Oregon, for a few years. Now she was returning to more familiar territory. Her move afforded opportunities for us to spend more time together. I was distressed to learn, however, that she had entered treatment for what turned out to be severe anxiety attacks. I recognized for the first time how difficult her relationship with Tamra had been. During all the years that Alice had spent growing up with Tamra as her step-mother, I had assumed that Tamra was offering the needed emotional support; now I realized how seriously I had misjudged the situation. I was the parent that Alice had needed, but for the most part I had not been there for her, either literally or figuratively. I was deeply upset that I had not expressed my love and support to both of my elder daughters more regularly and strongly during those trying years.

It was ironic—I realize now, in hindsight—that further recognition for my professional work intervened to help me not to concentrate on the guilt I was feeling about my neglectful parenting. It was, first of all, a great boost to the ego to be named the St. Louis "Man of the Year" in 1990 for my contributions to the city's economic and cultural life. At about the same time I learned that *National Geographic* planned to feature the Missouri Botanical Garden in its August 1990 issue. When the writer came to interview me, I remembered the inspiration I'd found in all those *National Geographic* articles I'd read when I was young. I hoped that some youngster might feel similar inspiration reading the article—"The Plant Hunters: A Portrait of the Missouri Botanical Garden"—when it came out.

The following year, 1990, I was surprised and delighted to be selected as a member of the Pontifical Academy of Sciences at the Vatican. My associations with scientists in Italy had begun a few years previously when my friend Bob Krukoff had introduced me to his professional associate Professor Giovanni Battista Marini Bettòlo Marconi. I had become friends with Giovanni and his family while participating in the study week on the environment at the Vatican in 1987. I was deeply impressed by his generous spirit and hospitality. Membership in the Pontifical Academy of Sciences is a lifetime appointment. Having been raised a Catholic, I was pleased to be doing something useful for the Catholic Church, even though the group was non-denominational. I also knew that being involved with the academy would open up meaningful new opportunities. Its primary role was to provide scientific advice to the Pope and thereby help the Church deal effectively with science and technology in this diverse and challenging modern world.

My first General Assembly of the Pontifical Academy occurred in 1992, beginning what was to be a long series of fruitful interactions with its members and others who attended the meetings and study weeks. Among the eighty lifetime members were many Nobel Prize winners and other outstanding scientists. Of these, I especially enjoyed getting to know the great physicist Nicola Cabibbo, the brilliant theoretical physicist who later became the academy's president; C. N. Rao, one of the most influential figures of Indian science; and Vladimir Keilis-Borok, an accomplished Russian geophysicist who by then was living in the United States.

At about the same time I received word about my selection for the Pontifical Academy of Sciences, I was invited by the Franciscan order to serve on a jury that would choose the recipients of a new prize that commemorated St. Francis's love for the environment. The prize was to be given at Assisi, with preliminary meetings in Rome. I believe that there can be no better model than that of St. Francis to inspire us to action on the environment and was pleased to help.

In 1991, the Missouri Botanical Garden celebrated the grand opening of the 9.5-acre William T. Kemper Center for Home Gardening. With its twenty-seven individual demonstration gardens, it has become a fine place for people to learn about how to grow and enjoy plants in their own gardens. The Kemper Center was one of the last major features of the master plan to be completed; in the planning for many years, its opening was a major milestone. In the course of realizing this project, it was a delight to work with David Kemper, thoughtful and always positive member of a banking family centered in Kansas City, who joined the Garden's board of trustees soon after moving to St. Louis as head of Commerce Bank. He provided the funding for the Center and has enhanced the Garden's program in more ways than I can easily count, as well as becoming a friend of many years standing.

My marriage with Tamra had continued in its decline through this period. In September 1991, Tamra established a friendship that would precipitate our marriage's eventual collapse. The person with whom she became friends was not, to my knowledge, a romantic interest, but rather a charismatic woman whose life, work, and interests captured Tamra's imagination.

Tamra and I were both attending a meeting of the Grupo de los Cien in Morelia, Mexico. The organization, formed six years previously by the writer and scholar Homero Aridjis, had by that time been recognized as one of the most influential environmental groups in Mexico. Among the dazzling array of writers, environmentalists, indigenous leaders, and scientists at the meeting, Tamra met Countess Alicia Spaulding Paolozzi, then seventy-three years old. Tamra offered Countess Paolozzi advice about her various gardens, and the two of them became fast friends. After the meeting, Tamra started volunteering at the United Nations and spending long periods with Countess Paolozzi in New York, at her summer home on Birch Island in Upper St. Regis Lake in the Adirondacks, and in Italy. With her late husband, the Countess had founded the Spoleto music festival in Italy and a counterpart festival in Charleston, South Carolina, and so Tamra visited these events with the Countess as well. Tamra often took Kate and sometimes Francis along on these trips. Her travel schedule became as busy as mine, if not busier. Soon we were barely spending any time together at all. With our interests and passions focused elsewhere, our lives became largely separate. Within a couple of years, our marriage had fallen apart completely.

Fortunately, not everything was difficult in my personal life in the early 1990s. After six years in Houston, my daughter Liz and her husband Daryl moved back to St. Louis at the beginning of 1992, buying a lovely older home on Juniata Street, just south of Tower Grove Park. Liz had met Daryl around the time she was finishing her degree at Washington University in 1984, and they had moved to Houston after getting married in 1986 so that Daryl could advance his career as radio engineer for Clear Channel. Back in St. Louis, Daryl continued working for Clear Channel, and Liz worked at C. V. Mosby, a publishing company. I very much enjoyed being able to so easily spend time with them.

☙

The years 1994 and 1995 were packed with events with great significance for my career, my personal life, and the Garden. This period began on a somber note with the death of my mother in April 1994. She had suffered from high blood pressure for some time and finally succumbed to a stroke at the age of eighty-five. Alice, Liz, Kate, Francis, and I were deeply saddened by her death. Alice and Liz had been particularly close to her; she and my dad had taken care of Liz and Alice for weeks after their own mother's death. Fortunately, Alice had been able to spend a good deal of time with my mother before she left us, collecting many stories about her life and the family's past.

Mom's funeral, a sad and rather formal ceremony, was in St. Anne's Catholic Church in Walnut Creek, California. Her interment, though, was in the beautiful old civic cemetery in San Juan Bautista, where Peggy and Patrick Breen of the Donner Party and many other Breen descendants had been buried over many decades. It was a comfort for me to know that she was at rest with her people.

In the wake of my mother's death, some of my parents' most special possessions came to the children and me. Alice and Liz particularly, who had seen the memorabilia of their childhood swept away after Sally's death, were happy to be able to have some items by which to remember their grandparents. Alice received a large piece of Chinese ancestor art, for example, and Liz a golden bird-shaped box of my father's and a lovely carved Chinese sideboard.

A couple of months later, in June, Emperor Akihito, Hirohito's son, along with Empress Michiko, honored the Missouri Botanical Garden by visiting our Japanese Garden. They planted a Japanese maple tree to mark the occasion, touring the Garden with a huge crowd of admirers. The maple was planted near a stone boat that had been Koichi Kawana's last contribution to the Garden's design, directly across the lawn from my father's memorial benches. While we were touring around the Garden, the Empress admired some lovely weeping birch trees planted near the Lehmann Building, noting how much she would like to have some like them for herself. After she left, we began to conspire to see how that could be accomplished. First, we determined that such weeping birch trees were indeed unknown in Japan. Then we took cuttings from our trees and got them to the Botanical Garden at Tokyo University, where Professor Kunio Iwatsuki could grow them up. In October the following year, we would be able to present them to the Empress during a formal ceremony at the Imperial Palace.

Also in 1994, I received notice that I had been accepted as a Foreign Member of the Chinese Academy of Sciences, a member of the first group of foreigners to be recognized in this way. The election was a high honor of which I was deeply proud, and one that I could not have imagined during all the years that China was essentially closed to outsiders. I understood it as my Chinese colleagues' way of expressing their appreciation for my role in all the cooperative work that we were carrying out to produce the *Flora of China*, particularly my efforts to open up avenues for collaboration between Chinese and foreign botanists.

In a way, they were also thanking me for other forms of assistance I had offered in the past, such as participating in the evaluation of the then newly formed National Natural Science Foundation of China at a meeting in 1987. I had been the only non-Chinese person present in that group of perhaps seventy-five people charged with developing the evaluation; Bill Tai sat next to me the whole time, whispering into my ear reports about what was going on. At the end of our meeting, we reported its results directly to Jiang Zemin, then president of the People's Republic of China. He listened with interest to what we had to say and then engaged in a discussion with the group to get their thoughts on the development of science in China. I was impressed with Jiang's understanding of the need to attract some outstanding Chinese scientists back from abroad. He insisted that it would be wrong to offer outsize salaries to these scientists to attract them back, stating clearly that if they wanted to return they should do so on the same terms as those of the scientists who had been working in China all along. He wanted, however, to offer these scientists the best possible laboratory facilities for their studies, indicating that he really understood the scope and nature of their work.

In late 1994, I was awarded the Tyler Prize for Environmental Achieve-
ment, given by Alice Tyler in memory of her husband. I sat at Mrs. Tyler's
table not far from Gregory Peck, with Mildred Mathias as my "date."
Mildred was in fine fettle that evening, her usual bright, energetic self.
Looking back, I treasure that last evening with her; she died a few months
later. Consistent with the pattern that I had developed in childhood, I
didn't spend a long time in a state of satisfaction, dwelling on my achieve-
ment. Instead, the winning of this major award was followed by feelings
of restlessness and insecurity. Outwardly my star was rising, but inwardly
my personal life was crumbling slowly to bits.

By the beginning of 1995, Tamra and I recognized that our marriage
was beyond repair. We had both had had enough of a relationship that
was bringing only frustration and sadness. A marriage doesn't work unless
both people are devoted to it—and neither of us had been so for some
time. Francis and Kate were shocked when we told them we were getting
divorced. I purchased a house for Tamra in Clayton, a St. Louis suburb,
and we arranged for Francis and Kate to move between our residences
regularly. We shared driving the children to school, an activity that I enjoyed greatly because I got to know them better during the drives. Francis
was finishing up at Clayton High School and Kate was attending The
College School, an outstanding private school in Webster Groves that
both Alice and Liz had attended earlier.

Divorce proceedings were completed late in 1995. The divorce was
painful for everyone and on every level. In the settlement, I gave up the
cabin and farm on the Big River we had bought in 1985. This was very
difficult because of the strong attachment I had to the place. Even more
difficult for me was learning more in the course of this process about how
much I had let Liz and Alice down. I should have acted more as their ally
and advocate through all those years. The recognition of my failure to
them made me feel guilty, angry, and sad. It was one of the most difficult
discoveries I have ever had to face.

After our separation and before the divorce was final, I had met a divorced woman with children of Kate's age who had a passionate interest in
the environment. I was immediately taken by her attractiveness and intelligence. She had been instrumental in reviving Earth Day in St. Louis in
1990 and had co-founded the EarthWays Center. Her name was Kate Fish.

Early in my acquaintance with Kate, she became a member of an external advisory committee at Monsanto of which I was also a member. We

realized that we had many common interests and began dating occasionally. Our attraction to each other grew, and I was emotionally very needy at the time. We decided to get married, tying the knot in the spring of 1996. It didn't take very long, however, for us both to realize that the marriage had been a mistake.

Kate was an introvert who valued her privacy and peace. I, on the other hand, was a needy extrovert who wanted lots of attention and kept too busy to provide what Kate needed. The more she wanted to enjoy quiet down-time, the needier and more talkative I became—a clear recipe for disaster. There were really no lasting elements that could satisfy either of us over the long term, even though we shared a common interest in the environment that gave us something to talk about from time to time. We spent the latter half of June on a memorable safari in Kenya, with Kate, Francis, and Kate's two girls, Clara and Alexandra. But it had become clear well before we returned to St. Louis that my wife Kate had moved on psychologically.

From then on it was a matter of largely separate travel schedules, driving the kids to school, and seeing separate marriage counselors. The problems we were theoretically trying to solve proved intractable: our marriage was never really there. Kate didn't want me along on her business trips and didn't accompany me on my trips, so we soon became two people living together but going our separate ways. I hoped that it still would somehow work, but it simply didn't. It would take only a few years before we were ready to call it quits.

എ

With another personal relationship failure to contend with emotionally, it was fortunate that 1996 was an auspicious year for the Garden and the beginning of a several-year period during which many goals and plans were realized. This notable period began when the Garden hosted the annual meeting of the American Association of Botanical Gardens and Arboreta (now the American Public Gardens Association) in the last week of April. Participants in the meeting were thrilled with the beautiful horticultural displays and our new landscape features. We had achieved a horticultural program of national quality, and I was finally confident that we could be compared favorably with any other botanical garden in the country.

That summer we completed construction of the pavilion in the new
Margaret Grigg Nanjing Friendship
Chinese Garden and then opened the
garden to the public. It was judged
authentic and lovely by experts and
visitors from China, standing as a true
symbol of our lasting relationships with
Chinese colleagues at all levels. For the
Chinese Garden's design, we were fortunate to have obtained the services of
Yong Pan, a Chinese landscape architect living in Atlanta; he came to us
highly recommended and amply lived
up to his reputation. He designed our
garden on the model of the scholar's
gardens of China's southern provinces.
The garden holds a lovely Tai Hu stone,
a serene memorial to my mother, recognizing her five years of adventure in
China.

Tai Hu stones, weathered limestone
from Lake Tai, are a standard feature
of Chinese gardens. This one was
dedicated to the memory of my mother,
Isabelle (1908–1994), who so loved her
five years in China. The Chinese
inscription reads "A son should love his
mother as much as the inch-high grass
loves the spring sun."

In conjunction with the opening
of the Chinese Garden, we collabo-
rated with the Chinese Culture and Education Services Foundation to
produce Chinese Culture Days. The festival was intended to broaden
visitors' understanding of Chinese culture and history. Its success insured its continuation as an annual event. Although Henry Shaw had
modeled the garden on ones in England, it had always seemed appropriate to me to present exhibits and events that broadened the horizons of
our citizens and opened the gates for international understanding. The
strategy worked, with the new annual event allowing us to reach out in
yet another direction.

The year 1996 also saw the completion of the Ruth Palmer Blanke
Boxwood Garden, a beautifully enclosed parterre situated between the
Japanese Garden and the Anne L. Lehmann Rose Garden. Its designer,
Chuck Freeman, modeled it on the boxwood gardens of colonial Amer-
ica. Those gardens, in turn, were fashioned after classical French gardens,
with geometry, symmetry, formality, and restraint predominating. In ours,

we included an alcove dedicated to the great Garden scientist Edgar Anderson, who had brought hardy boxwoods back from the Balkans in the 1930s. The Garden's central boxwood hedges spell out the letters H and S, for our founder Henry Shaw.

On February 5, 1997, at a meeting of the National Research Council's Report Review Committee, for which I was serving as chair, I received a major surprise quite unrelated to the business of the committee. One of the committee members—Washington University Chancellor Bill Danforth, whom I had recruited to the committee—drew me aside, along with Virginia Weldon, a committee member and Monsanto officer, to discuss a proposal. While on a safari in East Africa the preceding summer, Bill told us he had developed some important plans. Having noticed that it was difficult to retain plant scientists at the various academic institutions in St. Louis, all of which had relatively small groups of people with such interests, he had conceived of a way to strengthen the area's role in plant science research. He proposed the formation of a new center for plant research in St. Louis that would be dedicated to improving agriculture and finding ways to reduce world hunger. It would be an independent organization, governed by its own board, which would select its own goals. Would Monsanto and the Missouri Botanical Garden, he asked, be willing to join Washington University as co-founders for this exciting new project?

Both Ginny Weldon and I liked Bill's idea very much. Thus began a multiyear project that would keep me quite busy, and happily so. Ginny and I had first to convince the recently appointed CEO of Monsanto, Hendrik Verfaillie, that it would be in the company's interests to participate in the formation of an independent plant sciences center. With the help of Sam Fiorello, Hendrik's chief of staff at the time, we started the process. Eventually we succeeded in persuading Hendrik that the formation of the center would benefit both the company and the region. In due course, Monsanto agreed to donate an excellent piece of vacant land that it owned, immediately northeast of the company's headquarters, to provide a site for the eventual building. We received some development funds from the state government and a grant from the Danforth Foundation. When we invited the University of Missouri–Columbia to join in the effort, they readily agreed. Then, in consultation with Roger Mitchell, the very capable dean of the University's School of Agriculture, we recruited the University of Illinois and Purdue to join the Center and help us with its development.

My best personal contribution was the idea that the new institute be named the Danforth Plant Science Center. Like everyone who had thought about the matter, I was deeply impressed by the outstanding civic contributions that the Danforth family had made, especially since Ralston Purina founder William H. Danforth had established the Danforth Foundation with his wife in 1927. Persuading the family to accept the name was not simple, however. Senator John Danforth agreed with the proposal, but his brothers Bill and Donald Jr. were too modest to feel comfortable with the center carrying the family name. But ultimately they came up with the excellent idea that the Center be named for their father, Donald Danforth Sr. This was a fine result, and one much appreciated by the St. Louis community. Less than five years after Bill Danforth proposed his idea to Ginny and me, the Donald Danforth Plant Science Center building was completed, ready to be occupied by a growing staff of first-rate plant scientists.

The Missouri Botanical Garden had seen many infrastructural improvements through the 1990s, but none matched the importance of the new herbarium and research center, which was completed in 1998. Named the Monsanto Center to recognize a substantial grant by Monsanto Company, the building was renamed the Bayer Center after the acquisition of Monsanto by Bayer AG twenty years after its completion. It is a state-of-the-art facility featuring rolling compactors to store specimens and an environmental control system that can keep the herbarium section at a temperature low enough (62°–65°F) to prevent the pests that would otherwise eat the dried specimens and keep them from reproducing. A remarkable earthquake-protection design called "base isolation" cushions the rigid box of the building on top of some thirty-eight concrete pillars attached to the bedrock far below. Each pillar is topped with a multilayered rubber-and-steel "sandwich" that allows enough movement to keep the building from cracking. The building was also designed in such a way that it could accommodate modules of expansion to be added to the west, expansion that was becoming necessary two decades after its completion. The Center is located just a couple of blocks west of the northwest corner of the Garden proper, an easy walk to and from the institution's other facilities.

The need for a new structure to house part of the Garden's herbarium had become evident by the early 1990s, when our rapidly growing collections began straining the capacity of the John S. Lehmann Building. A suitable new structure would be too large and visually intrusive to place on the Garden grounds, and so we had sought a new site while raising the funds needed to make the facility a reality. We had decided early on that the new building would house only part of the herbarium collection, with the remainder kept in the still very useful Lehmann Building. The entire library, however, would be moved to the new building. The Bayer Center opened as the principal home for the Garden's Research Division, becoming an outstanding example for other herbaria and a concrete sign of the Garden's strength and growth. More than a third of the $22 million needed to complete the project came from the Department of Agriculture, in recognition of the Garden's major role in advancing the plant sciences; Senator Christopher ("Kit") Bond was a strong advocate for these appropriations.

On Earth Day that same year, 1998, I had the honor of being selected one of the first four of *Time* magazine's "Heroes for the Planet." I was very proud of this honor, as it recognized the work I had carried out on behalf of Earth's biodiversity for almost three decades. Roger Rosenblatt, an outstanding writer, visited St. Louis and interviewed me for the article in the magazine. He quoted me as saying one of my very favorite utterances: "We have relatively short lives, and yet by preserving the world in a condition that is worthy of us, we win a kind of immortality. We become stewards of what the world is."

In February 1999, I was visiting Jane Davenport Jansen, founder of the Quarryhill Botanical Garden in the Sonoma Valley of California, seeking funds for the *Flora of China*. I received a telephone call from John Fahey, the National Geographic Society's recently appointed president, who told me that I had been selected as chair of the Society's Committee for Research and Exploration (CRE). Would I accept? I had been a member of the CRE since 1982, having been drafted by my friend Don Duckworth, and my work with the committee had become a gratifying and important part of my life. I readily accepted John's offer.

Earlier, I had not participated in the annual field excursions of the CRE, but now that I was chair, I felt that it necessary to do so. What a burden! The first trip, to Belize later that spring, was a delight, featuring a valedictory dinner for the previous committee chair, archaeologist and scholar of the ancient Maya George Stuart. Many other very special and memorable trips would follow, among them visits to grantees and their research sites in Cuba, Peru, China, Vietnam, Madagascar, Turkey, Egypt, South Africa, and India.

❧

I knew the Missouri Botanical Garden had arrived as a major force in world botany when our bid to host the 1999 XVI International Botanical Congress at the St. Louis Convention Center was accepted. It would be the first such Congress to be held in North America in thirty years, following the Seattle Congress in 1969. When the early-August opening date finally arrived, about 4,700 delegates from 85 countries and every state and territory of the U.S. flooded into St. Louis—the largest convention to meet in our city that year.

If someone had suggested upon my arrival in St. Louis in 1971 that twenty-eight years hence the institution I was joining would be hosting an International Botanical Congress, I would have stared at them in disbelief. But the Garden's global botanical program, herbarium, and research staff had made great strides in those twenty-eight years, and it all added up to recognition at the highest levels. Of particular importance in landing the Congress had been our growing visibility in Europe, since it was the people and institutions there that had primary control over these meetings and what took place at them.

Two individuals made contributions of special importance in planning the Congress and then executing those plans effectively. One was the secretary general, Peter Hoch, whom I had first met when he entered Stanford as my graduate student three decades earlier, and who had been a member of the staff at the Missouri Botanical Garden since then. The other was Sir Peter Crane, former director of the Royal Botanic Gardens, Kew, who served as program chairman, assembling the many dozens of symposia and hundreds of individual papers presented at the meeting. They have both been lifelong friends, and we made a strong team, despite the confusion that arose from the fact that we were all named "Peter"!

To signal our pleasure in welcoming our international visitors to the Americas, we presented the St. Louis Symphony performing Antonin Dvořák's *From the New World* and a medley of Latin American musical pieces. At the opening session, the governor and mayor spoke, our twenty-six vice presidents were honored, and I welcomed the huge crowd to St.

Louis. Then I received the top award from the International Association of Plant Taxonomists (IAPT), the Engler Medal in Gold, given once every six years. I thought, of course, of the challenge Eric Godley had set out for me in his Christchurch garden so many years previously. The great German botanist whom the medal honored had just happened to be the model that Eric had had in mind for my career.

A major highlight of the Congress for me was the chance to recognize Ledyard Stebbins, the great evolutionary biologist who had been such a good friend and supporter for a half century. Following a series of tributes by colleagues and former students, Ledyard presented a marvelous speech and even sang a little ditty on evolution. In his remarks, he said memorably that he would like nothing better than to be a young professor again, with a group of graduate students, and to live forever in the study of plants that he had pursued with such enjoyment for all of his long and productive professional life. It would be Ledyard's last public appearance: he died at his home in Davis, California, five months later.

I was very pleased that my keynote speech at the Congress and its Resolution I, which was adopted unanimously, soon became key influences in plant conservation. I spoke out on the need for international

Peter Hoch, long-time curator at the Garden (1972 to present), was secretary general of the XVI International Botanical Congress in St. Louis (1999).

Sir Peter Crane, former director of the Royal Botanic Gardens, Kew, served as program chair for the 1999 Congress.

agreement for plant conservation in order to provide a context within which individual countries could effectively act. These sparked the eventual development of the Global Strategy for Plant Conservation, which has helped substantially to advance the cause ever since.

Hosting the Congress that I had first attended forty years earlier gave me a feeling of satisfaction and accomplishment. It was the culmination of many decades of hard work and a fitting end to the millennium.

A half century after I first met Ledyard Stebbins, I had the pleasure of introducing him at the International Botanical Congress in St. Louis, in August 1999.

— 14 —

New Adventures at Career's End

AGING BRINGS MANY UNWELCOME CHANGES, but one of its distinct rewards can be gaining greater insight into many facets of life. By the year 2000, when I turned 64, it seemed I had accumulated enough wisdom to approach mine a little differently. I recognized that, always eager to finish some current project, I had often not paid enough attention to my own emotional needs nor those of the people around me. The excellent advice contained in the eden ahbez song I had listened to as a boy, that everyone needs to learn to love and be loved, had finally sunken in. With this revelation, I felt smarter about romance and intimacy, ready for a relationship that really worked. I was also more confident about my own success—at least enough to be able to relax a bit and not be so constantly driven toward new achievements. So, as the new millennium loomed close, the idea that it represented a new, hopeful beginning was to me especially relevant.

After my birthday in the middle of June, I eagerly anticipated the upcoming World Botanic Gardens Congress. It was held in Asheville, North Carolina, during the last week of the month, in conjunction with the annual meetings of the American Association of Botanical Gardens and Arboreta (now the APGA), Botanic Gardens Conservation International, and the Center for Plant Conservation. The overall theme of the meeting was to be plant conservation, which was of course very important to me; the program emphasized the idea of working together to elaborate a worldwide strategy for saving threatened plants and their habitats, a goal toward which I had been working much of my life. Since this was a time when the botanical garden conservation movement was finally reaching its stride, I was expecting a new sense of optimism and opportunity to pervade the gathering. Personally, I was also looking forward to the meeting's social aspects because I was very much ready to pull things together in a romantic direction.

The historic Grove Park Inn, an impressive National Historic Landmark building, served as headquarters for the meeting. Following an excellent opening address by North Carolina governor Jim Hunt, we enjoyed a memorable talk by the famous author Wilma Dykeman, who had saved the French Broad River from dams and other development and memorialized it in her fine book, *The French Broad*. The rest of the day was busy with eight concurrent sessions featuring papers about different aspects of conservation and botanical garden management. They were interesting, and I attended many of them.

The following day was reserved for field trips. In the morning, we all visited the North Carolina Arboretum, where a proud George Briggs showed us what he, his staff, and volunteers had accomplished over the first few years of its existence. In the afternoon, the group moved to spectacular Chimney Rock Park, then a private park owned by Lu Morse, a friend from St. Louis and Garden trustee, whose grandfather had built the park. There, the delegates swarmed over the trails, thoroughly enjoying the local plant communities, massive granite cliffs, and views across the valley. It started to rain, and one of the meeting delegates, Laura Wyatt, had the misfortune of slipping off a new trail, breaking her leg in three places and dislocating her hip. Those nearby came to her rescue immediately, getting her ready to be transported to the hospital. Waiting, Laura lay shivering in the back of an open truck, sopping wet, covered with mud from head to toe, in excruciating pain and going into shock. Patricia Duncan, executive director of the Mercer Botanical Gardens in Houston, went through the crowd asking people for a few of the souvenir lap robes they had been given that morning so that Laura could be covered against the cold. Bian Tan, a staff member at the San Francisco Botanical Garden, was one of those who willingly contributed his. Pat, who had been injured herself on a field trip two years earlier, accompanied Laura to the hospital, where she was soon joined by George Briggs.

Meanwhile, I had made my way back to the Grove Park Inn with the rest of the group and eaten dinner. I was sitting with Bian Tan in the lounge near the great stone fireplace sipping Scotch while discussing Bian's adventures hunting plants in Indonesia. At the time, I was wondering how to move the cause of plant conservation in Indonesia to a higher level so that the country's wonderful and diverse flora could be saved from logging, mining, and clearing of land for oil palm plantations and other forms of agriculture. Bian and I talked about the botanical garden in Bali, and he told me how beautiful that island is: it made me think about the possibility of traveling there. Someone came up behind my chair, and Bian said to her, "Pat, we're having a conversation about plant collecting in Southeast Asia. Would you like to have a seat and join us?"

It was Patricia Duncan, who had demonstrated so much caring and presence of mind earlier that day. She moved around to a chair at the side of ours. We had met before, briefly, in earlier botanical garden meetings, but this was our real introduction. Somehow, I sensed it was important to get to know her better, and asked jokingly "Pat, have you ever been to Bali? Would you like to go?" She asked when, and I suggested the next morning. Rising to the occasion, she replied, "Sure, I packed black silks; I can go anywhere." We continued with light chatter, and eventually turned in. I went to sleep with Pat very much in my thoughts.

We met the next morning, and Pat pointed out that I was scheduled to give a keynote address Friday, so we couldn't leave for Bali. Kidding along, we flagged down Bob Bruenig, head of the Lady Bird Johnson Wildflower Center in Austin, Texas, and asked if he would be willing to give my talk because we needed to leave the conference early for our trip. Proving he was good for a joke, Bob gasped and then, recovering quickly, said, "O.K.; do you have slides? I'll need them if you do."

Pat and I returned to our respective homes after the meeting, but stayed in touch by email. I would learn later that Sarah Reichard, an attendee of the meeting and a close friend of Pat's, had given Pat some advice after the meeting. "Why don't you just give in gracefully?" she said. "Peter wants you, and what Peter wants Peter gets. Your goose is cooked. I will come to your wedding, and I will wear powder blue." I certainly wouldn't have believed that at the time if I had heard it. In any case, I returned to my busy life, distracted by travel and meetings, but with Pat often present in my thoughts.

In mid-July, I attended a meeting in Trondheim, Norway, where Walt Reid was busy hatching what became the Millennium Ecosystem Assessment. We had a good discussion, with me contributing the idea of drafting Hal Mooney to be co-coordinator of the effort, a role that he carried out with his usual distinction.

Back at home, I carried on with visits to Kimmy and Steve Brauer at their summer home in northern Wisconsin and with friends Art Ortenberg and Liz Claiborne in Montana, all the while carrying out an extended email conversation with Pat. Surprisingly, Pat and I discovered during this correspondence that we were ninth cousins, both of us descended from Stephen Hopkins, a passenger on the *Mayflower!* After two months of this virtual courtship, Pat finally agreed to my visiting her in Houston at the very end of August. It was a good visit, and we started seeing each other as often as we could.

The more time we spent together, the more we recognized the many dimensions of our compatibility. We were both extroverts who enjoyed interacting with other people a great deal. We had enough interests in common to occupy ourselves in endless conversation and could easily make common cause in many areas. A fine horticulturist, leader, and scholar in her own right—she holds a Ph.D. from Ohio State University—Pat did not feel in the least intimidated or overpowered by me. Grounded in her self-confidence, she could be appropriately challenging while at the same time being supportive. I had been vulnerable and very needy when we first met, but over time she was able to coax me gently back to a much better place. During the late summer, I threaded my work schedule around bringing my daughter Kate to visit potential colleges and visiting Pat.

Pat invited me to meet her family in late September. The occasion was a cousin's wedding in Charlotte, her home town, but it was an appropriate opportunity to meet the Duncan-Frick clan. By then, I was already making matrimonial noises. As a classic Southern lady, Pat suggested that my one and only chance to meet her estranged father, Duke Duncan, would likely be at that event. At the reception, I met him and asked for his daughter's hand in marriage. He assented. As luck would have it, the deal was sealed when Pat caught the bride's bouquet and I had the garter land in my outstretched hands!

Later that night Pat agreed wholeheartedly to marry me, but laid down one staggering condition—that we have an engagement of at least two years. Naturally, I immediately began to scheme about how to shorten the time. In the morning, we shared our exciting new plans with Pat's mother Todi and her stepdad, Tom Frick.

Back in the "real" world, the Garden announced on November 1 that we had completed the negotiations for the EarthWays Center to join the Garden. The Center had renovated a building in the Grand Square area of St. Louis where they demonstrated the principles of sustainable architecture to the public. I was excited that in joining forces with EarthWays, the Garden could amplify its efforts to promote sustainability in the region.

On November 20, Pat and I participated in the grand opening of the Millennium Seed Bank at Kew, where we were presented to His Royal Highness Prince Charles. We then went on to take in several shows on the Strand and have a wonderful time in London. We were both pleased and excited that our relationship was working very, very well. Pat gradually phased out of her job in Houston and began to have more time for us to be together.

On November 28, we enjoyed a romantic dinner at the Hay Adams Hotel in D.C., looking out across Lafayette Park to the White House. We had already agreed informally to get married, but I wanted something more formal—and I had a handsome diamond ring in my pocket. I got down on one knee and officially proposed. She accepted and I slipped the ring onto her finger. We both seemed to know it was right. Pat had already softened on her demand that we wait a couple of years, so we set the date for June 2001.

We were in Washington for another exciting event. On November 30, President Clinton awarded me the National Medal of Science, our nation's highest award for scientific accomplishment, during a ceremony in the White House Oval Office. Pat was there, along with my children, Pat's mother Todi, and Todi's husband Tom. Receiving the Medal was a very great honor, and I was thrilled to have been accorded such extraordinary recognition. It was followed by a gala dinner and public presentation ceremony.

A few weeks afterward, Pat and I traveled to Charlotte to spend Christmas with her parents. I had a fine time with them and other relatives I was meeting for the first time. It was family custom for the person newest to the clan to wear a felted antler headpiece, and I happily went along

with the tradition. I was soon able to reciprocate, introducing Pat to relatives in San Francisco when travels brought us west. Meanwhile, Pat
resigned from her botanical garden directorship and tied up her affairs in
Houston, beginning a slow transition to St. Louis.

We wasted no time in setting the tenor for our upcoming lives together. Soon after New Year's Day, 2001, we were flying to New Delhi for
what would be my fourth trip to the subcontinent and Pat's first. My
friend Kamal Bawa had invited me to India, graciously including Pat after
he became aware of our engagement. I was excited to have Pat along on
the trip.

We began our time in India by participating in that year's Indian National Science Congress. Afterward, Kamal guided us on an exciting tour
through the country, stressing the Western Ghats and Bengaluru, where
the ATREE headquarters were located. At first, Pat was incredulous about
many of India's contrasts—the beauty of the buildings in the midst of
poverty, and the extraordinary nature that flourishes everywhere. I was
surprised by the changes that had occurred since my first visit in 1970.
Large areas of pastoral countryside and scattered traditional villages were
now almost completely filled with houses, factories, gas stations, and stores,
and the highways were said to be dangerous places to drive at night. Considering that the population of India had grown during the preceding
three decades from about 550 million to over a billion people, one could
understand what lay behind these changes. Pat and I enjoyed both the
trip and our time together very much. We had no significant conflicts
and, for both of us, sharing the experiences enhanced them greatly.

Pat and I both thought that our happiness on the India trip boded
well for the future, and it made us want to travel even more. Fortunately,
opportunities to do so abounded. Before our wedding in June, we enjoyed no fewer than three wonderful trips overseas. In March, we visited
Peru on a National Geographic CRE trip organized by Ann Judge, director of the Society's travel program. It featured a tour of Machu Picchu,
mountain trekking, a scary walkway in the canopy of the rainforest, and
a fifty-mile boat trip on the Napo River, a tributary of the Amazon. In
April, it was off to Australia for visits with old friends and a chance to
show Pat some of the natural wonders of the country.[1] In May, we flew
to China, where, among travels to Beijing, the Great Wall, Nanchang (the
capital of Jiangxi Province), and Lushan, I helped work out the plans for

the part of the Flora of China project that included the ferns and club mosses.

We were married on June 23, 2001, in the Garden at a lovely spot overlooking the lake in the Japanese Garden, near the benches I'd given in memory of my father and the Japanese maple planted by the Emperor and Empress of Japan. Blending the new and the old, Pat wore silk from our trip to India and pearls from China, but carried lilies and flowers from a scarlet pitcher plant from North Carolina in her bouquet. When I had asked Don Duckworth to be my best man, he fixed me with a hard stare. He knew my ex-wives Kate Fish and Tamra and was familiar with my troubles with Sally as well, so he had every reason to be skeptical. "Peter," he said, "I will agree to be your best man—on one condition. This *must* be your last marriage. You've got to be sure. You've got to get this one right." He wasn't the only friend to have doubts about my romantic life, myself included. But in Pat I had found a real soul mate. I had indeed gotten this one right. Moreover, I had finally learned that a marriage needs to be nurtured, like a garden, and given constant maintenance and attention, if it is to succeed and flourish.

Right after our wedding we took off for the place we had met— Asheville, in the beautiful mountains of North Carolina—for a short honeymoon. We visited Pat's family in Charlotte and then explored the countryside, stopping at every overlook along the scenic Blue Ridge Parkway to examine some new kind of plant and thoroughly enjoying every minute. We stopped at the Pisgah Inn, where we had dinner with George Briggs and Tod Morse. Later, we visited Tod's parents, our friends Lu and Bonnie, and were welcomed to a delightful picnic on the rocky flats above the great waterfall at Chimney Rock. It was a memorable honeymoon in every way, not yet the promised trip to Bali—but that was to come!

Pat and I nestled into married life and slowed down the pace of our travels, and the Garden recorded several more milestones. In July of 2001, the Sophia M. Sachs Butterfly House, in Chesterfield's Faust Park, became a part of the Garden family, adding another type of display and attraction to our growing list.[2]

The Butterfly House had been the brainchild of Evelyn Newman, the patron of countless good things in St. Louis. She had been enchanted by a butterfly house she visited in Chiang Mai, Thailand, and, more or less on the spot, had decided that we should have one in St. Louis too! She had shared her exciting vision with me soon after she returned and asked if we would like to build such a structure in the Garden. I loved the idea but couldn't see where it might be placed within our walls. Never one to be easily deterred from her dreams, Evelyn had pushed on; after consulting the Zoo and considering other possibilities, she had made a deal with St. Louis County Parks to build the 8,000-square-foot conservatory in Faust Park, where it opened in September 1998. Later, when Evelyn fretted about the future of her "baby" and told me she wanted to put it under the wing of a larger institution that could see it through thick times and thin, I agreed that was a good idea. Some might think that a butterfly house is better suited for a zoo than a botanical garden, but for me, plants and butterflies have always been a natural combination. As I've said, their relationship had become clear to me at about age seven, watching butterflies collecting nectar from the flowers in San Francisco, and was cemented when Paul Ehrlich and I studied the relationship between the two groups of organisms in the early 1960s and came up with the concept of coevolution. The Butterfly House was a perfect place for children and adults to come to learn about and appreciate nature, and the Garden was delighted to have it become a part of our display facilities.

Our next advance came through people associated with Pioneer Hi-Bred International, a major corn-breeding company based in Des Moines, Iowa. The company had been founded in 1926 by farm journal editor and future U.S. Vice President Henry A. Wallace, who through his own work hybridizing corn had become convinced that hybrid seed corn would become important. Staffing his company, Wallace selected two students of the great Garden scientist Edgar Anderson, who was actively involved in corn breeding. Of these, William L. Brown became the company's CEO and Don Duvick its director of research. Subsequently, they were joined by Indian scientist Suri Sehgal, who became the lead in opening up Pioneer's global business. An endowment had been formed to support publications, mostly initiated by Henry Wallace and continued by Bill Brown, but they were failing by the summer of 2001. In this atmosphere, I was able to work with Suri to transfer the endowment to the Garden and use it to found a center for economic botany, the William L. Brown Center, which has flourished subsequently and allowed the Garden to open a whole new area of scholarship effectively. Carrying on in the tradition of Edgar Anderson, this center has conducted research throughout the world, but especially in the eastern Himalaya, Latin America, and North America.

That same summer, the Garden began to realize an opportunity to consolidate and extend its leadership in the area of plant conservation. We had strong research programs in many parts of the world where extraordinary plant diversity was being threatened by increasing human pressures, but we lacked coordinated means for making available the knowledge we had accumulated about the plants in these areas or helping to translate that knowledge into sound management, policy, and planning. In the developing nations in which we worked, the considerable biodiversity was all out of proportion to the relatively small number of trained conservation professionals and scientists. I saw a huge, unfulfilled need for training, information sharing, and capacity-building.

The primary barrier to the Garden becoming more active in conservation had been a lack of funds. We prioritized our basic floristic and taxonomic research, and the budget had little room for other endeavors. I suspected, however, that a solution to this problem might exist within the philanthropic leanings of Robert Brookings Smith. Bob, the energetic and innovative nephew of the legendary, pioneering St. Louisan Robert Brookings, had resigned from the Garden's board in 1962, after the completion of the Climatron, when the plans he was making for the Garden's future with then director Frits Went were thwarted by other more conservative board members. Fortunately retaining his fondness for the institution, he had been persuaded to rejoin our board as an emeritus trustee in 1981. We had become good friends and I trusted him as a keen ally.

Over the years, Bob and I had had many conversations about projects at the Garden that he might support. That summer, when he was ninety-eight years old, I persuaded him to donate $5 million to establish, within the Garden's Research Division, what we would call the Center for Conservation and Sustainable Development (CCSD). I pushed this objective hard with Bob both because I wanted to secure the future of conservation at the Garden, and because I knew the demands for conservation of plant diversity and sustainable use of resources would only grow over time with all the mounting pressures on the environment. The Center worked consistently toward meeting the goals of the International Strategy for Plant Conservation that we had gotten off the ground at the International Botan-

ical Congress in 1999. It put us in a good position to emphasize conservation and training programs in Latin America, North America, Madagascar,
and Vietnam, thus helping to promote sustainability in these countries
and throughout the world. With an endowment substantially expanded
in particular through the subsequent generosity of the Smith family, the
CCSD would flourish under the outstanding leadership of Olga Martha
Montiel. Over the years, it has provided an important link between our
scientific research and conservation efforts and the practical world of governance and management.

Following a short trip to the Kunming Institute of Botany in China to
celebrate Wu Zhengyi's eighty-fifth birthday, Pat and I thought we were
settling down to a quiet autumn and anticipating our long-awaited trip
to Bali, scheduled for a few weeks hence. On September 9, 2001, I gave a
keynote address on conservation at the meeting of the American Zoo and
Aquarium Association at a downtown St. Louis hotel and then made a
quick one-day trip to Washington, D.C., the next day for a regularly
scheduled meeting of the National Geographic's Committee for Research
and Exploration.

Back home in St. Louis on September 10th, I got up early the next
morning to meet with Bill Conway, George Rabb, and other notables of
the zoo world at the convention hotel. We were to have a panel discussion
later that morning. But when I walked out of the room where we were
having breakfast, Charles Hoessle, director of the St. Louis Zoo, stopped
me and asked "What are we going to do now?" I asked about what, and
he motioned me into a side room where a television set was focused on
the World Trade Center in New York. A plane had crashed into the north
tower of the Center at 7:45 a.m. and, while I was watching on television,
a second plane hit the south tower at 8:03 a.m. St. Louis time. The zoo
meeting was instantly over, though we had to figure out how the thousands of delegates gathered at the meeting could manage to return to their
homes, with all flights halted until the situation was better understood.

A couple of hours after the initial attack, American Airlines Flight 77
was crashed into the Pentagon in Arlington, Virginia. We soon learned
that our dear friend Ann Judge, who had organized the trip to Peru for
the CRE a few months earlier, was on board. She was accompanying three
sixth graders and three teachers who were on their way to participate in a
Society-funded research project on the Channel Islands off the California
coast. Then United Flight 93, apparently headed for the Capitol or for the

White House, was crashed in Pennsylvania by brave passengers who seized control of the plane back from the hijackers. All in all the September 11 attacks killed 2,996 people and injured more than 6,000 others.

Pat and I had been scheduled to head to China late that Tuesday evening. Some feared that we might have left for the West Coast on the plane that crashed into the Pentagon, but thankfully we were safe at home. All travel off, we were left to wonder what was going on and what might happen next. Along with other botanical gardens and many similar institutions, the Garden opened its grounds to the public free of charge; we were mobbed with people seeking a peaceful, beautiful place to be for a few hours to process their shock and grief.

After things began to get back to normal, we carried on with our lives, greatly shaken by the events of September 11th. Kate began her university career at Stanford, and we were happy to see her settled there. I continued to travel to attend the meetings of the various boards of which I was a member, but our greatly anticipated trip to Bali had necessarily been canceled. On November 2, there was a grand opening of the beautiful facility at the Danforth Plant Science Center, attended by former President Jimmy Carter and other officials. Considering that it had been less than five years since Bill Danforth had drawn Ginny Weldon and me aside to tell us about his dream, this represented remarkably rapid progress.

In early January of 2002, Pat and I traveled to a meeting of the Flora of China Editorial Committee in Paris. We were delighted to see how the Garden's satellite office there had flourished over nearly twenty years under the guidance of Pete Lowry, playing a key role in the development of our programs in Africa and also being a great help to the office's host institution, Muséum national d'Histoire naturelle. Then, in an upcoming trip to Australia to make some speeches, scheduled for the end of June, I saw an opportunity to make good on the offhand remark that had started my relationship with Pat: I would invite her to accompany me to Australia and then surprise her with a new plan for a side trip to Bali. I had to let her know just before the trip that she should bring some light clothing, not just the warmer attire needed for the Australian winter, so landing in Denpasar after our time in Australia wasn't a complete surprise. We stayed in a villa on the nearby Jimbaran beach for a couple of days, and then moved to Ubud nearer the center of the island, enjoying the beautiful countryside, the Gamelan musicians, and a complimentary dinner and couple's massage as a wedding gift from Liz and Daryl. With its warm, hospitable people and orderly beauty, Bali was much more of a paradise than I had realized, and the days we spent there were magical.

After our return from Bali, we traveled to Guemes Island, just off Anacortes on the coast of Washington State, for our first blended family vacation. Pat's folks Tom and Todi, her brothers Ted and Richard, nieces Sloan, Carson, and Clare, Francis and his wife Carolyn Kousky, Kate, Liz and Daryl, Alice, and Pat and I all converged at a lovely house we had rented with plenty of room for everyone. The "children" took turns going out with me to collect plants, and I assembled a very nice collection for the island—including a plastic rose stem that jokester Alice had placed among the plant specimens that I was drying. In the evenings we took our cocktails on the deck and watched mule deer grazing on the meadow just up the coast.

It was a lovely family vacation, but also the beginning of an unwelcome medical odyssey. My left foot, the one on the leg I had broken skiing with Tamra in 1968, had been giving me increasing trouble, and the pain was really prominent during the stay on Guemes Island; at times I could hardly walk. I didn't know what the trouble was, and pushed on, thinking that it might get better with exercise. When we returned to St. Louis, however, I consulted Dr. Vilray Blair III, who informed me that it was acute osteoarthritis. Looking at the X-rays of my foot, he called in a colleague and remarked, "Here's a picture of the foot of a man who is really in pain." Dr. Blair said he could stop the pain by fusing the bones to immobilize my heel, and so an operation was scheduled for later that summer.

In the meantime, I met with Jack Taylor, a warm and fascinating individual with a lively intellect who had become a good friend over the years. Jack had founded Enterprise Holdings, which was initially based on replacement car rentals and then expanded as the core company grew to become the world's largest vehicle rental agency. Consequently, Jack was very wealthy. I wanted to see if I could interest him in supporting our worldwide programs of exploration, research, and conservation of plants, particularly in the tropics, since I knew we could do a much better job with additional funds. Having reached the age of eighty, Jack had phased out of active management of his company, leaving that to his son Andy, also a good friend and a trustee of the Garden. Impressed with what we were doing, Jack responded to my plea with the largest single gift ever made to the Garden—$36 million, to be delivered over four years. What

Jack Taylor, the founder of Enterprise Rent-A-Car, made the largest gift ever granted to the research program at the Missouri Botanical Garden, $36 million, so that many plants could be found and understood before they disappeared.

we were able to do with part of his generous donation demonstrates the importance and power of philanthropy: we accumulated indispensable data that could not have been gathered later because of the rate at which forests were being destroyed worldwide and used the data to inform conservation choices in a number of countries in Latin America, Africa, and Asia. It was a transformative gift in every sense of the word.

On August 15, Dr. Blair performed the surgery on my foot. My recuperation began with what was pretty much a month at home, since I could put no weight on the heel until it had healed completely. Our living room became my new office and I continued with my work there. I tried to get around a bit with crutches, but it was miserably difficult, and I eventually fell on the front steps, dislocating my shoulder, leaving me unable to use the crutches at all. Not being able to get around after the fall, I started using a motorized wheelchair and had to be carried up and down the stairs in our house by a transport ambulance crew that stopped by twice a day. During one of their visits, in the early evening, I felt a pain in my side and told Pat about it. She asked the crew leader to check me out; doing so, they immediately called a full-service ambulance that rushed me to the nearby Barnes-Jewish Hospital emergency room.

Soon after I arrived at the ER, our close friend Tom Woolsey showed up in his white medical coat and badge. He simply stood there for the hours it took for the testing to be done, to make it evident that I had an advocate. Several of the ER staff had been his students, and so I'm sure that his presence was important in speeding the process and making it more efficient, which turned out to be a very good thing. Around 10:30 p.m., I was diagnosed as having a large pulmonary embolism. I began treatment with the blood thinner heparin immediately.

A few hours later, at about 2:00 a.m. that night, a physician making the rounds read my chart, looked at me, and remarked, "Oh, it's a good thing that you're alive; people usually die when they've had a pulmonary embolism like yours." Not a cheery thought, but it was the truth. If the clot had moved from my lungs up to my brain, I would almost certainly have died in short order. After eleven days in the hospital, tethered to an IV pole for regular heparin treatments but with Pat in faithful attendance as my patient advocate night and day, I was finally ready to go home and try again.

For a number of weeks, I could get around only in an electrically driven wheelchair, so we moved temporarily to one of the ADA-equipped Garden apartments across Alfred Avenue on the west side of the Garden so that I could avoid stairs. Garden staff members were as kind as they could possibly have been, helping me in every way they could every day. I will always be immensely grateful to them for their attention and assistance. We had clearly become a kind of family. A few weeks later, I was well enough to announce the opening of the annual Systematics Symposium, although I continued to need a wheelchair for almost six months. As a result of my condition, I was able to check the Garden's degree of accessibility personally and was generally well pleased with the results.

❧

Time passed, and a normal routine, if it can be called that, began to re-emerge. I returned to regular work duties and once again began attending various committee meetings scattered around the country. One great comfort was that Francis returned to St. Louis to begin a master's program in philosophy at the University of Missouri–St. Louis after earning his bachelor's at Evergreen State College, in Olympia, Washington—an unusual and successful public college that had turned out to be near perfect for him. It was good to have him close by and see him thriving.

In February 2003, Pat and I resumed our former pace of regular travel with a most enjoyable trip to Costa Rica with several of the Garden's trustees to take stock of our programs there and share something about why we were institutionally so active in the tropics. In March, we flew to San Francisco to attend the fiftieth reunion of my high school graduating class at St. Ignatius. Shortly thereafter, in April, we traveled to Scotland, where we participated in an institutional review of the Royal Botanic Garden Edinburgh, at the behest of RBGE's director Steve Blackmore. A month later, we returned to the British Isles for a Flora of China meeting at Kew Gardens. I was still secretly tickled by the fact that Kew was acting as a partner in a program about China that was centered at Missouri. As part of that trip, we visited the fascinating Eden Project in Cornwall with Iain Prance, the former director of Kew, and then scientific advisor for Eden.

Later the same year, I received word that I was to be awarded the International Cosmos Prize, which had been established in Osaka, Japan, with the proceeds from the successful Floral Expo held there in 1990. In October, Pat and I traveled to Osaka to receive the award at a beautifully organized ceremony that represented in a meaningful way our common desire for world peace and harmony. I had very much enjoyed being present and speaking when my friend Wu Zhengyi was awarded the same prize three years earlier, and it came as a surprise when I was subsequently chosen as the recipient myself. The motto of the Cosmos Prize, "The harmonious coexistence between nature and mankind," expressed a goal that was to me more important than any other. Traveling on to Tokyo, Pat and I had an opportunity to meet privately with Crown Prince Naruhito, representing the third generation of the Japanese royal family with whom I had been privileged to meet personally. He and Crown Princess Masako were genial and good hosts, and we enjoyed our time thoroughly.

The year 2003 was also the time we laid the groundwork for a new and important attraction at the Garden. During this period we had continued to think seriously about building up our attendance. We wanted to become an even more attractive destination for the people of St. Louis and those who visited from outside the area. We knew that the Garden was a great place for children, but we were also aware that some other places, including the Zoo, were attracting far more young people than we were.

A number of other botanical gardens had built various kinds of children's gardens—places that were especially attractive to young people. The board and I were intrigued with this idea; to develop it further we appointed a committee consisting of trustees, staff members, Pat, and volunteers. The committee traveled to some of the existing children's gardens to see what had been done in each of them and how effective they seemed to be. With their findings at hand, we hired Patrick Janikowski Architects that autumn to develop a plan. We decided that we would develop a relatively complex children's garden based on the theme of Missouri in the 1800s, the time of our founder Henry Shaw. Such a theme would allow us to celebrate Lewis and Clark and Sacajawea, Mark Twain (Samuel Clemens), Daniel Boone, and Shaw—each of them pioneers in their own way. The garden would weave around between the venerable

Osage orange trees that once lined one of Henry Shaw's roads, thus forming a powerful link with the past. We would finish the four-acre garden with structures and other attractions intended for children and plant it entirely with Missouri native plants.

When it came time to find the funds to carry out our plans, the civic-minded Schnuck family stepped forward, making a generous joint contribution to the project. Construction began in 2004. The Doris I. Schnuck Children's Garden would open to the public two years later, on April 1, 2006, living up to our expectations as a draw for large numbers of children and their parents.

Before 2003 came to a close, Pat had her own health scare. We were in Madagascar for the annual field trip of the National Geographic Society Committee for Research and Exploration (CRE), which I chaired. Naturally I was thrilled to be visiting that fascinating island, where I had initiated a program for the Garden thirty years earlier. After several days of meetings and discussions, however, Pat awakened in the night with a terrible pain in her abdomen that obviously needed attention. We rushed to the medical center at the U.S. Embassy, where Pat was treated by an extremely competent Ph.D.-RN. After running diagnostic tests, she concluded that there was no physician in Madagascar with the expertise to deal with Pat's condition, suspected to be a perforated ulcer. We flew out to Johannesburg, South Africa, at dawn, the soonest we could travel there. After four days of treatment at a hospital in Pretoria, Pat was well enough to fly home. Years later, her condition recurred and was diagnosed as pancreatitis, a very dangerous and potentially lethal problem.

The years 2004 through 2008 passed by with what seemed like unprecedented speed, marked mainly by memorable trips to some of our favorite places in the world. In 2004, we traveled to Glasgow and sailed on the Celtic Sea, returned to China on a National Geographic CRE field inspection trip, and made another visit to Australia. The highlight of 2005 was a tour of Vietnam during which we were updated on the progress of the Garden's program in that country; in 2006 it was a trip to Brazil for a symposium on biodiversity after which we also celebrated Pat's fiftieth birthday with a memorable stay at the Copacabana Palace in Rio de Janeiro that included a visit to H.Stern!

In April 2008, as the *Flora of China* was winding down toward the publication of its last few volumes, we held one of our last joint editorial meetings at Zhejiang University, Hangzhou. After the meeting, our colleague Hong Deyuan, who held an appointment in the university, wanted to show us the part of the country from which he had come. He brought us to the famous and beautiful West Lake in Hangzhou, which my parents had visited on holidays some seventy years earlier. Walking around the lake or boating on its surface easily reminds one of why artists, philosophers, and poets have loved it for so many centuries. With spring flowers in full bloom, it was simply wonderful.

Leaving Hangzhou behind, we traveled on westward through forests, fresh green rice paddies, fields golden with mustard, and villages with dark timbers accenting their fresh white walls in the characteristic architectural style of the region. Eventually we reached our first destination—Tien Mu Shan, a historically important mountain area with peaks nearly 5,000 feet tall, dotted with active and ruined temples and nunneries and clothed with a rich subtropical forest. The giant Japanese cedar trees that grow in this forest are extraordinary, but even more so are the native maidenhair trees of the genus *Ginkgo*. This is one of the few places where these ancient trees have survived outside of cultivation.

When we moved to the foot of Huangshan, a beautiful high mountain range in Anhui Province, we could immediately understand why the range,

On the summit ridge of Huangshan (Yellow Mountain) in Anhui Province, China (May 6, 2008) with Libing Zhang (Missouri Botanical Garden), Hong Deyuan (Beijing Institute of Botany), William McNamara (Quarryhill Botanical Garden), and Fu Chengxin (Zhejiang University), after the joint editorial meeting of the Flora of China project in Hangzhou.

celebrated for centuries in poetry and art, was chosen as a UNESCO World Heritage Site. Huangshan's jagged granite peaks, the highest of them rising 6,115 feet above sea level, had been extensively molded by glaciers. Clouds frequently form and touch the mountains from all sides and at all levels: they are particularly beautiful during the sunrises and sunsets for which the mountain is justifiably famous. A cable car climbs directly up to the summit ridge of the mountain, and then 900 stone steps lead the way to overnight accommodations. Pat, laden with camera gear and dark chocolate, strongly suggested that we accept the two-man sedan chairs being offered at the base of the stairs. When I declined, she whispered sharply that *she* needed the chair but couldn't take one unless I did. The view from the summit was extraordinary and is perhaps the most beautiful place I have ever seen in China, a fitting culmination of nearly three decades of frequent travel to the country of my birth.

☙

With thirty-seven years at the Garden's helm in the summer of 2008, it seemed logical for me to bring up the issue of finding my successor. I discussed the matter with the chair of the board of trustees, Nick Reding, a former Monsanto executive. I told Nick that I'd like to consider retiring in the summer of 2011, which would coincide with my seventy-fifth birthday and the completion of forty years at the Garden. Nick thought that identifying my successor was such a good and timely idea that he began working on it actively. At his suggestion, the board selected and hired the search firm Isaacson, Miller that October; it began work in November 2008. Nick then announced the formation of a search committee to the board at its February 2009 meeting. I knew the process of finding suitable candidates, selecting one, and then negotiating the terms of employment would take same time, but it began to seem as if it would all fall into place well before the summer of 2011.

The committee met on June 13, 2009, briefing the board later that month. Over the course of the summer, the firm identified three possible candidates and interviews began in the fall. On December 12, 2009, the committee informed the board of trustees that they had selected Dr. Peter Wyse Jackson as my successor. The board unanimously approved. Dr. Wyse Jackson, an Irish botanist and plant conservationist, was then the director of the National Botanical Gardens of Ireland; among other professional credentials, he had been instrumental in the development and administration of the organization that became Botanic Gardens Conservation International. He would start on the first of September 2010, and a series of gatherings were planned to welcome Peter and Diane Wyse Jackson to the Garden and the St. Louis community.

On April 17, 2010, the Garden held a gala celebration in honor of my retirement. At the gala, Rex and Jeanne Sinquefield sponsored a performance of a symphony that they had commissioned in my honor, by the New Music Ensemble of the University of Missouri–Columbia. It evoked parts of my life and places in the Garden, and we all enjoyed it greatly. Novus International, Inc., through its president Thad Simons, commissioned a painting by Bryan Haynes that illustrated stages in my life, from boyhood in California and study of Onagraceae to features of the Garden (see color image, #49, and the cover of this book). And it pleased me greatly that a scholarship fund for horticulturists, the Dr. Patricia D. Raven Horticulture Training Award, was being established in Pat's honor. I was delighted to see her receiving this well-deserved recognition of her own after all she had done for the Garden herself. Various other ceremonies of congratulation were held in connection with my tenure at the Garden over the next few months; for example, an appreciation day was held August 6, 2010, when the mayor and county executive presented me with certificates of appreciation and hundreds of people turned out.

Even though I was ready to retire, it was tough to relinquish the habit of responsibility to an organization I loved after nearly four decades of service. And leaving the director's house, which had been my home for just as long, was a wrench. But as my time at the helm wound down, I worked hard to wrap up or reassign everything on my task list. The day came when, miraculously, everything on the list was done—but of course a myriad of other projects, plans, and obligations on behalf of the planet awaited me.

When I formally stepped down as Garden president and director, I was honored with the title of President Emeritus, and that of George Engelmann Professor of Botany Emeritus at Washington University. The Garden's world-class library was renamed the Peter H. Raven Library, an honor for which I am deeply grateful. I very much appreciate the Garden's generosity in assigning me an office near the south end of the Lehmann Building, where I moved during August that year. In addition to providing a place to work, the office allows me to continue to enjoy the beloved

Garden grounds, facilities, and people that have been part of my life for
so many years. Finally, the Garden has kindly provided me with an assistant, which has enabled me to accomplish far more in retirement than
would have been possible otherwise.

℘

It is natural in retirement to occasion a look back at what has been accomplished. Proving the degree of leadership required for the success of
the institution is a pleasure, but is possible only when many people blend
their talents in a common effort. Likewise, the many communities—city,
state, national, and global—of which the Garden is a part, play an indispensable role in whatever is realized. In particular, the citizens of St. Louis,
who embraced what we did and supported it with so much enthusiasm
and generosity, were the real heroes in the story. Without their support,
we would not have been able to accomplish anything.

As I look back at my tenure as director, and in the context just discussed, I am, however, certainly very proud of what we achieved. On my
arrival in 1971, the Garden's budget was approximately $850,000 and the
employee roster included about 150 full- and part-time staff members.
There was no tax support and the endowment was valued at approximately $5 million. There were 2,500 members and about 150 volunteers.
Our scientific program was based on the efforts of only a handful of staff
members, some temporary. The herbarium that we moved to the Lehmann
Building in 1972 consisted of just over two million specimens. Only
about one fifth of the Garden grounds could be considered developed
gardens worth visiting. No organized fieldwork programs had been conducted by Garden staff outside of North America and Panama during
more than a century of Garden history.

When I retired in 2010, the Garden's budget was approximately $38
million and the employee roster included about 500 people. The Zoo-
Museum District contributed about $11 million to the budget (almost
25%), and the endowment, even after the tough years following the 2008
recession, was well on its way to $100 million. There were about 38,000
members and more than 2,000 regular volunteers. Our scientific staff
included about fifty Ph.D.-level research scientists. The herbarium had
tripled in size to become one of the largest in the world and certainly the

most active during the period of my directorship. We had amassed some 1.3 million specimens from Africa and Madagascar, making our collection from that region one of the largest anywhere, and certainly indispensable for anyone studying African plants. We had well over one hundred employees working in Madagascar, and were conducting conservation, reforestation, and sustainability programs at thirteen sites on the island. The Garden grounds had been fully transformed to gardens, lawns, and structures, developed carefully with the guidance of a master plan that we had followed from 1973 on, and they were consistently judged attractive and interesting by visitors from near and far. Among the features added during the period I served as the Garden's director were the Japanese Garden, the Cherbonnier English Woodland Garden, the Ridgway Visitor Center, the William T. Kemper Center for Home Gardening, the Ruth Palmer Blanke Boxwood Garden, the Margaret Grigg Nanjing Friendship Chinese Garden, the Temperate House, the Commerce Bank Center for Science Education, the Monsanto (now Bayer) Center, the Edward L. Bakewell Jr. Ottoman Garden, and the Doris I. Schnuck Children's Garden. We had acquired the Sophia M. Sachs Butterfly House, were operating the Litzinger Road Ecology Center, acquired the EarthWays Center and transformed it into a solid, Garden-wide program in sustainability, and had enlarged the Shaw Nature Reserve by a third to nearly 2,500 acres in size. In addition, we had restored the Climatron, the Shoenberg Administration Building, and Henry Shaw's Tower Grove House.

During the period from 1971 to 2010, the Garden's research productivity had come to rival or exceed that of the largest comparable institutions in the world. We completed the *Flora of Panama* and *Flora de Nicaragua*, guided the ongoing production of *Flora Mesoamericana* and housed the editorial center for the *Flora of North America*, published checklists of the plants of Peru, Ecuador, and Bolivia, and built a substantial database on the plants of Madagascar online. We were actively working on the *Manual de Plantas de Costa Rica* and had substantially completed the 49-volume *Flora of China*, which I was coediting. Demonstrating continued attention to our own locale, we had published two of the three volumes of an updated version of Julian Steyermark's *Flora of Missouri*, a project begun in 1987 and executed by George Yatskievych. Endowments for both the William L. Brown Center for Economic Botany and the Center for Conservation and Sustainable Development had been secured, so the future of these programs would never be in doubt. In our encouragement of and collaboration with individuals and institutions around the world, we had accomplished a great deal more.

℘

Since I'm not the kind of person who could actually retire, I shifted my focus from running the Garden to myriad other ongoing projects as well as a great deal of international travel with Pat. I still didn't have enough time to do all the things I imagined doing. Nevertheless, I was able to enjoy a little more time at home and devote more energy to projects that previously had to take a back seat to other priorities, such as working with the National Geographic Society and the Center for Plant Conservation. The Missouri Botanical Garden remains a central part of my life; I use my office in the Lehmann Building almost daily and delight in touring around the grounds in one of the resident golf carts, watching the seasonal changes and visiting familiar features with personal significance, such as the statue of Henry Shaw we had commissioned to celebrate our founder's 190th birthday in 1990. I felt a certain kinship with old Henry, as both of us had spent the greater part of our lives focused on the Garden.

As one might imagine, friends and acquaintances began to ask me after my retirement if I missed being director of the Garden. My ready answer was that I certainly did not miss dealing with personnel problems or the headache of preparing and presenting complex budgets when there never seemed to be enough money to do everything that we would have liked! Upon reflection, I realized that what I did miss was helping those who came up to discuss a worthy botanical project—studying a certain plant family in a faraway place, for example—and sometimes needed just a few thousand dollars to make that project become a reality. It was exhilarating to be able to help people, young people especially, who had high aspirations but little cash, and to see them be able to do work about which they cared so passionately. As director of the Garden, I had appreciated the ability to help their dreams come true. After retiring, that power diminished, but I was still able to encourage and connect people, activities that have always given me great satisfaction.

One of my greatest post-retirement joys is spending time with Pat. With her, I found welcome hours of peace at our home in Wildwood, listening to music, reading books, and watching crime shows or movies. Although I am still emotionally needy, Pat continued in her remarkable way to satisfy that neediness and allow me to feel much more comfortable and self-assured and for the most part to act that way. She continues to be tolerant and understanding of my restlessness and tendency to keep too many balls in the air at once. We both work hard to support one another and maintain a good life together. I had come to understand—and put into practice—the wisdom expressed in a fine song about love by Clint Black and Skip Ewing, called "Something That We Do." The song points out that love must be something we do actively; if we just sit and expect it to come to us automatically, we'll always be disappointed.

From my retirement onward, all my children have continued to do well, despite having experienced various difficulties along the way. I very much enjoy having more time to visit them and their respective partners and spouses and, perhaps, to make up a little for my less-than-ideal parenting. Kate worked for a few years at the non-profit Global Exchange in San Francisco and, after graduating from the UCLA law school, became a public defender in Alameda County, California. Her daughter Louisa Frances Raven was born June 2018. At first, Francis lived in Washington, D.C., with his wife Carolyn, when she was a fellow at Resources for the Future. She has since moved to lead a center on the economics of environmental disaster at the University of Pennsylvania Wharton School of Business. Francis homeschools their two boys Noah, born August 2010, and Nate, born January 2015. Lizzy and Daryl live in Glen Carbon, Illinois, near St. Louis, Lizzy working at the Federal Reserve Bank and Daryl as chief engineer at the public radio station in St. Louis, KWMU. I was happy for Alice when in 2011 she reconnected with a fine man she had known in college, Sam Stulhman, and began a serious relationship that was, and has continued to be, a great comfort to them both. Over the years, our family has continued to grow and prosper.

Retirement afforded Pat and me even more time to travel. Although some trips were undertaken on our own for the simple pleasure of the adventure, most were associated with ongoing projects or continuing work with various organizations and committees. As the invitations flowed in, we packed forty-eight international trips into the first forty-eight months after my official "retirement."

In 2011, Pat and I had the good fortune to visit India on a National Geographic CRE field inspection. By then, Kamal Bawa had become a member of our committee. After studying the monuments and urban history of Mumbai (Bombay), we flew to Guwahati, in Assam—in the subtropical "arm" of India that projects out to the northeast. Driving to nearby Kaziranga National Park, we viewed spectacular assemblages of elephants, rhinos, deer, and waterfowl on the lakes and got to go out wildlife viewing on the back of an elephant. The next day I was fortunate enough to see a tiger, but the jeep in which Pat was riding was on the other side of the park so she didn't get to share my experience. Opening her computer that evening, she requested views of tiger sightings in Kaziranga, and found a video clip that recorded a recent incident there in which a tiger had jumped up and attacked the people riding on an elephant. Yikes!

In the autumn of 2012, Pat and I began our love affair with the incredibly interesting and magical Himalayan kingdom of Bhutan, our first trip there resulting from an impulse purchase at a local charity auction; we returned on a field inspection with the CRE two years later. Bhutan, just a bit larger than the state of Maryland, borders China to the north and India to the south, east, and west. It lies along the southern slopes of the Himalaya, with most of its cities and towns at or just above 7,000 feet elevation.

Bhutan is a thoroughly Buddhist country that is dominated by India, which supplies a major proportion of its budget either as foreign assistance

Pat, Katie, and I, along with Dave Boufford, were hosted in August 2011 on a memorable trip to Tibet by Gu Hongya, like Dave a former student; the trip was in honor of my retirement. There we got to know *Saussurea*, a member of the sunflower family (Asteraceae), which has radiated into many species, such as *S. medusa*; the flowers in many of them are protected from the constant cold with beautiful webbing.

or in payment for hydropower. It has a population of about 700,000 people, scattered along valleys and ridges running out perpendicularly from the crest of the Himalaya range. One of only a few countries that attempt to assess Gross National Happiness as a measure of success, its people are very much concerned about how far to go in maintaining traditional ways versus modern culture, how many people from outside to admit as tourists or otherwise, and how to deal with its powerful neighbors to the north and south.

More than 70% of Bhutan is forested, with about a quarter protected in nature reserves. Agriculture is varied, but as the friendly and open Bhutanese people like to point out, they can grow almost any crop there but only a little of it: up and down the steep valleys can be found almost any climate in Asia except the extreme tropics. Paddies of red and black rice fill the flats and low slopes of every valley. Nature is largely unspoiled and very rich. Bhutan has been a constitutional monarchy since 2008, with King Jigme Khesar Namgyel Wangchuck and Queen Jetsun Pema both deeply appreciated by their subjects. On the occasion of our National Geographic expedition in 2014, we were able to meet with the King for a couple of hours; the Queen attended one of our evening events, dancing with some of us to local folk music under the beautiful pale light of a nearly full moon.

During the course of our two trips to Bhutan, during which we traveled widely in the western and central parts of the country, I had the idea of trying to promote a national biodiversity inventory like those with which I had been associated in Costa Rica, Mexico, and elsewhere. There was already a National Biodiversity Centre near Thimphu, the capital, and Jan Salick of our Missouri Botanical Garden staff had, over the course of several years, worked closely with that center. Getting a more comprehensive inventory program going with international support was difficult, however. There are very few trained scientists in Bhutan, and most of them have their hands full with their current duties. Eventually, the exploration of biodiversity there will open up to a wider circle of foreign collaborators in formal and informal relationships, who will be very fortunate indeed to work on those projects.

In between the two trips to Bhutan, Pat and I had a most enjoyable return visit to Yunnan Province in China. Starting with a symposium at the Kunming Institute of Botany (KIB), we traveled to the mountains in the northwestern part of the province where, north of Lijiang, we visited

the Jade Dragon Field Station, a joint project of KIB with the Royal Botanic Garden Edinburgh. The field station is located within the highest
botanical garden in the world, which extends up the mountains from
8,000 to about 14,100 feet elevation. In this lovely setting, I rushed about
collecting *Epilobium* and refreshing my knowledge of the Chinese species,
which are abundant in the region. After this trip, we continued on to
Beijing for the celebration of the completion of the *Flora of China*.

During this period, I also worked with my Russian colleagues for the
last time, co-chairing a joint U.S./Russian NRC committee organized by
Glenn Schweitzer. It resulted in the publication in 2013 of an NRC report
entitled "The Unique U.S.-Russian Relationship in Biological Science and
Biotechnology." We had an excellent team of leading scientists and ad-
ministrators representing the two countries. However, the situation in
Russia had become so confused through the ascendency of oligarchs and
eventually the belligerence of the government that it became difficult to
carry on with even the relatively few concrete goals that we identified. As
I have said, science by its very nature can be a leading force in bringing
together countries; so we had to be content with the hope that people of
goodwill would be able to find ways for serious and extended collaboration in the future.

The freedom Pat and I were enjoying in retirement received a temporary setback in January 2014 when I fell face down on a stair at home. I
was bleeding so badly that it was impossible to tell what the problem
might be, and so Pat rushed me to the hospital. There I was diagnosed as
having longitudinally fractured C1 and C2, the two vertebrae immediately below my skull, a very serious injury. Dr. Jacob M. Buchowski performed surgery, installing titanium plates and screws along the sides of the
two vertebrae, and supervised the outstanding care I received at Barnes-
Jewish Hospital. For some weeks after the operation on my neck I could
scarcely crawl around, much less get into a regular bed, but experts in
physical therapy allowed me to return to fairly normal activities within a
year. Pat was a rock during the whole recovery period, taking good care of
me every single day.

I recovered slowly, but in June 2016, we were back on the road. Pat
and I traveled back to the Institute of Botany in Kunming for a symposium commemorating the 100th anniversary of the late Wu Zhengyi's
birth. With more than three million people, Kunming had become a
much larger city than the one I had first visited in the summer of 1980,

essentially filling the valley where oxcarts and water buffalo had been frequent sights earlier. Skyscrapers had sprouted all over the area, some for offices and some for housing, making the overall appearance was very different from what it had been thirty-six years earlier.

At the airport in Kunming, our friends, led by Professor Sun Handong, the director of the Institute of Botany, met us at the plane, escorted us to a luxury VIP waiting room, collected our checked bags planeside, and whisked us away to our hotel. During the course of the next two days, we participated in the seminar, and then we were treated to a lovely birthday party—my eightieth and Pat's sixtieth—held in our hotel around the largest "lazy Susan" I have ever seen to that point, with more than thirty people sitting around its periphery. Hong Deyuan, my long-time friend, had come from Beijing to host the party, at which my former graduate students Gu Hongya and David Boufford were in attendance. It was a grand feast, with roast pig, many toasts, multiple renditions of "Happy Birthday to You," and a very large cake.

The next day, Kunming botanist Sun Hang and one of his students flew with us to Pu'er (Simao), named for the dark, pressed, caked tea that comes from that area, and we drove north to Jingdong, located at middle elevations (about 6,000 feet) in the mountains. From Jingdong, we first drove east to the Ailao Shan Nature Reserve and, under a slight drizzle, enjoyed a rich array of plants along the road. It was the kind of broad-leaved evergreen forest that Dan Axelrod and I had written about years earlier—a survivor of the much more widespread forests that had covered central and southern China for tens of millions of years. Dominated by evergreen oaks and laurels, it included lovely members of the camellia family and scattered deciduous trees including maples, alders, and birches, with many different kinds of ferns, including some tree ferns. Epiphytes grew densely on the tree branches. Strange white-flowered Commelinaceae sprawled over the shrubs and trees; gesneriads with lovely pale orange flowers, gingers, aroids, and other attractive herbs grew on the banks. As I walked along the muddy path, our Chinese friends supported me on either side and provided a cane so that I had no problem at all navigating the rough terrain.

That afternoon we drove west to another reserve, Wuliang Shan, and the next morning to one nearer Pu'er, Taiyang He National Forest Park. They all presented a marvelous feast for a botanist's eye, even though they were only fragments of the forests that had covered much of the province even half a century earlier. Around Pu'er, along with the corn and tobacco, there were extensive tea plantations, some of them established within the last decade or so, and only a little of the original forest left.

Overall, Yunnan Province is biologically the richest in China. In addition, the tropical trees and other plants found there are often ones at the very northern end of their ranges, which means that the local Chinese strains of these plants might have special importance in the changed climates of the future. So much of the natural vegetation has been destroyed, however, that our chances of finding them, learning about them, and preserving them are decreasing rapidly with every passing year. For that reason, I have urged for many years that a detailed botanical survey (or more general biological survey) be undertaken in Yunnan Province. Unfortunately, this has not yet proved possible. Even though the Chinese government is doing an increasingly effective job with the preservation of natural resources, its natural forests are continuing to disappear, and the opportunities to understand them well continue to decrease.

Our entire time in Kunming was marked by care and comfort. I had quite a lot of trouble getting around, with my fused left heel and titanium-reinforced neck, but everyone took such good care of me that I was able to see and do a great deal. That care seemed perfectly in keeping with the nurturance I had received in the loving arms of my amah some seventy-nine years earlier. It certainly made China feel like a home, in which I was very welcome, and I was filled with the most pleasant thoughts during our memorable visit and beyond.

We traveled again to China to attend the XIX International Botanical Congress, held in Shenzhen in July 2017. It was the first Congress to be held in a developing country. A tropical storm had just passed through, the sky was the brightest blue, and the whole city sparkled. When night fell, an enchanting floral-themed light show in honor of the Congress appeared on both sides of the downtown skyscrapers. The meeting was a major success, with the largest attendance ever at such a gathering—about 7,000 people. The papers and other presentations were superb; as a demonstration of the state of development of botany in China, it could not have been better. Clearly, China will continue to grow as a world leader in botany, an amazing development considering that botany, like all other sciences, was at such a low ebb in the country forty years ago, when Chinese scientists started to rebuild contacts with foreigners.

At the Congress, Hong Deyuan and I were the honorary co-presidents. I presented the first public lecture, on our common need for preserving our planetary home. Later, I was deeply honored to be chosen as the first recipient of the Shenzhen International Award in Plant Sciences, endowed by the Shenzhen government to be awarded at each subsequent International Botanical Congress in recognition of someone who had made deep contributions to the plant sciences.

᠙

Several ongoing projects and activities are keeping me busy and happy as I finish writing this book. One project into which I am investing quite a bit of energy is the development of a *Checklist of Indian Plants.* The work was conceived by Kamal Bawa and his Indian colleagues some years earlier, and its early stages were carried out by means of a literature survey conducted by personnel at ATREE and the University of Agricultural Sciences in Bengaluru. Subsequently, Kanchi Gandhi, a distinguished Indian botanist at Harvard, had added a number of missing species to the list and begun the tedious job of editing its technical aspects. Kamal asked me to coordinate the completion of the list, and, given that completion was in sight for the *Flora of China* at the time, I took it on. On the occasion of my retirement, the James S. McDonnell Foundation generously awarded me a special grant to continue with my work on the checklist, and thus it seemed that I had all my ducks in a row.

What I hadn't fully comprehended was the high degree of isolation of taxonomists in India from those in the rest of the world, and the resulting chaos in the names of their plants. The Botanical Survey of India is committed to advancing and publishing up-to-date knowledge of the country's rich flora, but the process has been extremely slow, at least in part because of the attitude that everything should be accomplished within India. Unfortunately, plants do not respect national boundaries, and those who classify them are continually striving to bring their names in line throughout their worldwide ranges. The rules that are in force, made even more complex by the implementation of the Convention on Biological Diversity (CBD) a quarter century ago, and combined with a severe bureaucratic approach, have in effect meant that it is very difficult for foreigners to study Indian plants—even if they have been actively studying the same group throughout the countries bordering India. It is possible to travel to India and examine the material in many of the herbaria, but extremely difficult to study and collect the plants in nature. When Indian taxonomists find a distinctive set of populations, they have often gone right ahead and described them as a new species. When it comes time to try to evaluate such entities on a worldwide basis, it is often difficult for foreigners to examine the original material from India. As a result, the taxonomy of many groups of Indian plants turned out to be somewhat garbled and difficult to rectify.

Having worked in China, where foreign collaboration is generally welcomed and where I had, for a long time, coordinated a major work that involved Chinese-foreign collaboration on every group of plants, I was genuinely surprised by these discoveries, finding it much more difficult to get to the end of the road when investigating the names of Indian plants. Moreover, many of the plant groups in India had not been revised in detail for many years, and the names and distributions that we had gathered from the literature tended to be dated and confusing. Under the circumstances, we did the best we could; certainly, for some groups of plants (including ferns, grasses, and orchids), we were able to advance the state of understanding a good deal. Fortunately, my McDonnell Foundation grant allowed me to hire J. Richard Abbott, a young man with extraordinary botanical knowledge, to work with me for two years helping to put the material we had in order. In the final phase of the work, we enjoyed additional support from Suri Sehgal, who gave generously to allow the work to be completed. We will begin putting the checklist online and publishing it in 2021, thus providing a base for working effectively with the Indian flora in the years to come.

Another piece of the changing kaleidoscope of my "retirement" work moving forward was the Center for Plant Conservation. As discussed earlier, I had brought this important organization to the Garden in 1994 to become part of our conservation program. I remain enthusiastic about its purpose, having worked hard in attempting to get similar organizations established in Australia, China, and elsewhere. The CPC's strategy is to assign specific conservation regions to different institutions and act as the central coordinating body, so that the overall effort can be as effective as possible and with the least duplication of effort. In this arrangement, the individual institutions can watch more closely over a particular set of plant species, bring them into seed banks, cultivate them, and do whatever proves to be necessary to ensure their survival. They may of course also work to conserve plants from regions other than the one to which they have been assigned if they wish to do so.

A close friend of long standing, Kathryn Kennedy, had moved to St. Louis to become the president of the Center for Plant Conservation in 2000, the year before Pat and I were married. We were happy to welcome her with her family, and I worked with her to try to keep the CPC successful. After my retirement in 2010, I had agreed to join the CPC board; three years later I had become its chair. Managing a small not-for-profit organization, particularly a widely dispersed one with a national headquarters to fund, is a very difficult task at best, and the affairs of the organization became increasingly complex. Eventually Kathryn decided to return to government employment, and we had to look for her successor. In the late summer of 2014, we appointed John Clark to the post. John, who had had wide experience in botany at several institutions, soon organized a move for the Center's headquarters physically to the San Diego Zoo Global in San Diego and a partnership with that great institution.

In January 2016, after extensive debate and thorough consideration of the various aspects of the move, we formally announced the partnership, on terms whereby the Zoo would provide substantial new funding for the CPC. The arrangement is working well and, although I miss the Center's presence in St. Louis after more than twenty-one fruitful years, I feel that it is very well suited to its new cooperative arrangement on the West Coast. Under strong leadership, the CPC continues to move forward and play a key role in the conservation of plants in the U.S. and Canada.

The other organization to which I dedicated a good deal of my time is the National Geographic Society. As noted earlier, I had been appointed to the Society's Committee for Research and Exploration in 1982, becoming chairman on the occasion of George Stuart's retirement in 1999. I served in that capacity until my eventual term as an NGS trustee (2008–2019) ended. In 2008, I was elected to the Society's board of trustees, subsequently serving as a member of several board committees. I was particularly honored to have been awarded the Hubbard Medal, the Society's most prestigious award, in 2018.

With the appointment of Gary Knell as president and CEO in January 2014, the Society began an extensive and fast-moving effort to adapt to the rapidly changing needs of the modern world. We selected Gary with an enthusiasm that soon proved to be well founded. Gary had served as president and CEO of National Public Radio for the preceding three years, doing an excellent job of smoothing some troubled waters there. Earlier, he had worked for twenty-three years at Sesame Workshop, a project that has over the years made a truly original and effective contribution to children's education.

The National Geographic Channel, formed as a joint venture with Fox Cable Networks in 2001, subsequently contributed the major part of the Society's income. By 2008, Gary devised and saw to the implementation of an expanded joint venture between the two entities that controlled all the outreach products of the NGS, including publications, TV, travel, and so forth, a creation that promised and delivered an enhanced diffusion of our message. Fox paid the Society $725 million for an additional 23% of the joint venture, thus ending up with a 73% financial share, the Society with 27%. Management has been shared equally, with alternating chairs for the governing committee. With the additional $725 million in its endowment, the Society was able to engage more widely and effectively in communicating about "the world and everything in it." The overall arrangement functioned well for a couple of years, but in March 2019, the Disney Company closed on its purchase of many Fox assets, including the Fox share of our joint venture. This partnership will continue to develop over the years to come, and its ability to help build a sustainable world is potentially enormous.

As has always been my bent, I face the future with plenty of goals. I want to complete the *Checklist of Indian Plants* and continue to fret about the completion of other projects that I have been involved with for a long time, such as the *Flora of North America* and the *Flora Mesoamericana*. Meanwhile, I gain profound satisfaction from seeing projects for which I helped plant the seeds come to fruition. One of the most significant of these moments occurred at the very end of 2017, when a team of twenty-four botanists (including me) published a checklist of all the known vascular plants of the Western Hemisphere, some 125,000 of them. The effort to produce this remarkably comprehensive work began, in a sense, more than 500 years ago with the first European botanical exploration in the Americas and continued with further collecting and the construction of regional floras over the subsequent centuries. But it received a major boost when we at the Missouri Botanical Garden began in the 1970s to encourage, fund, and coordinate further exploration and collecting in Latin America and to nurture with all of our strength the development there of educational and research programs that proved effective in advancing these overall goals. It is no accident that the checklist study's lead author, Dr. Carmen Ulloa Ulloa, originally from Ecuador, is a Missouri Botanical Garden scientist and that the searchable online database form of the checklist is housed within the Garden's Tropicos database.

I was brought closer to the Garden again on the occasion of my eightieth birthday by a gala evening—a fundraiser for the Center for Conservation and Sustainable Development. Longtime friend and garden supporter Buck Bush and newer trustee Dan Burkhardt chaired the event, which brought several hundred people to the Garden. Burkhardt, avid Missouri land conservationist, was elected to the board just as I retired and proved, along with his wife Connie, to be a great friend and companion in conservation adventures locally.

I know I'll never finish everything on my list of things to do. Sometimes in the wee hours of the night I dream of trying to find my way through difficult and unknown territory, sometimes accompanied by Alice and Liz as little girls, or Francis and Kate, never really being in danger but always having trouble finding my way, and having urgently to get somewhere. It's a dream that speaks to a core tension of my life. Day by day and year by year, I try to move forward on as many good and important purposes as possible, understanding that the joy and meaning are in the path rather than the destination. Yet inevitably as I grow older, I'm more and more aware that the destination also matters, that many goals are in fact urgently in need of realization, and that they will never all get done. Thus I feel compelled, yes, driven, to continue to write and speak out in an effort to draw attention to and spur action to protect the life on this planet that supports us all. I work to promote sustainable practices, protect biodiversity, mitigate climate change, and promote human diversity in the belief that we must act if our civilization is ultimately to be saved.

That's part of the reason for encouraging others, most of all the young people who will take up the cause long after I've departed this beautiful and fulfilling world with all its joys and sorrows. Bringing scientists and organizations together, encouraging others, helping people in different countries strengthen their involvement in the conservation of their native plants and animals, attempting to promote the empowerment of women and children throughout the world—all of those causes are important to me. And of course, encouraging and enabling others is simply the right thing to do. I so deeply take to heart the original motto of the United

Negro College fund—"a mind is a terrible thing to waste." I know that every person has something to give for our common benefit, and that helping them to share their talents will fulfill them and help save the rest of us. And so I live each day trying to help as many people as possible find the path that allows them to develop, use, and contribute their unique abilities for our common benefit.

We are all part of a grand enterprise, the limits of which we can scarcely imagine. John Muir was right when he wrote, "When we try to pick out anything by itself, we find it hitched to everything else in the Universe."[3] My job, my mission, my reason for being on this earth, is to raise the awareness that we're all part of the great web of nature, connected to each other and to every flower, bird, spider, stone, volcano, virus, and rainbow. I want to encourage people, to inspire them to love, to lead, and to work for the benefit of our planet. If I don't do that well, the rest doesn't matter very much. I believe we all must deeply value and love one another before we can get anywhere collectively. That our common good matters to each of us seems to be the hardest lesson of all for humanity, but I am convinced that we cannot survive without learning it.

— Epilogue —

TOWARD A GREEN AWAKENING

We live today in a world that is very different from the one in which I grew up. In the 1940s and 50s, large portions of the Earth did not seem to have been impacted by human activity. We didn't worry much about the capacity of our air and water to absorb all the pollutants we were cranking out, and we had yet to deploy some of our most destructive technologies. As I became fascinated by the natural world as a child and went out to collect all sorts of organisms in San Francisco and around the Bay Area, it never crossed my mind that the same forces that energized my city and turned on lights around the world might also threaten my beloved butterflies, beetles, and plants.

One of the key lessons of my career as a scientist was gradually recognizing the depth of my naiveté and that of my fellow humans. The size of our population, with the associated demands for consumption, having grown rapidly over thousands of years, had reached unsupportable levels in terms of the pressure they were putting on living ecosystems. Despite this, we were largely unaware of the collective global effect that our local actions were having. Individual species like the passenger pigeon and the dodo had been lost over the previous centuries, but these extinctions, I came to realize, were simply the early warning signs of a growing catastrophe that was affecting the entire planet—a wave of massive extinctions of plants and animals for which we are the sole cause.

And if that lesson isn't worrisome enough, it was followed by another: human beings' awareness of our impacts and their consequences inevitably lags well behind the reality. In 1936, when I was born, we had already felled vast tracts of forest, converted a third of the world's land surface to farms and pasture, radically altered ecosystems around the globe through the transport of invasive species, and transferred immense quantities of carbon from ancient geologic deposits into the atmosphere. What we hadn't done was to come to terms with what all that meant. Now, despite greatly intensifying our destructive activities, we human beings have become no better at recognizing the severity of the crisis that we have created. During my lifetime, *the global human population has more than tripled*, an extraordinary fact with huge consequences. With this massive increase in number of people worldwide, we have far more than tripled our demands on natural resources and exceeded their ability to satisfy our needs on a sustainable basis. While the wave of environmental awareness in the 1960s and 1970s has in some ways embedded itself in our culture, it has not translated into the necessary collective will to alter the ways we conduct our lives on this fragile world. The situation we face today is far more challenging than any that has confronted our species since we left our African ape relatives behind in the course of our evolutionary journey six to eight million years ago, and yet we go on from day to day as if very little is awry with the natural systems that sustain us.

It is from the perspective of a scientist who has lived during a span of decades that have proved very consequential for the planet that I assess where we stand today, discuss what the future is likely to hold, and propose some principles I think we would be wise to adopt as we face up to what we have already done and are continuing to do in damaging our only home.

ↄ

As described in the preceding chapters, I've spent a great deal of my life working, in one way or another, to help improve our knowledge of Earth's biodiversity—to understand and appreciate the myriad species with which we share our planet. I have experienced great joy in discovering more and more about our rich biological heritage, a legacy that is the product of billions of years of evolution. At the same time, however, there has been an inescapable element of tragedy and sadness in recognizing that the more we know about what we have biologically, the more acutely we become aware of what we are losing and how quickly it is disappearing.

The possibility of stemming these losses has certainly motivated me, along with many others, to attempt to conserve natural habitats and try to make our human footprint less impactful by adopting sustainable practices. It is with the firm conviction that we *must* do all we can to protect biodiversity, regardless of how limited the prospects for success might seem to be, that my colleagues and I, together with countless other environmentalists, have lobbied governments, raised funds, taught principles, advised authorities, undertaken surveys, and appealed for wiser approaches to land use. Thankfully, we have, in some ways, enjoyed a measure of success. We helped pass major environmental legislation in many countries, brought into existence numerous NGOs dedicated to conservation and sustainable development, protected large swaths of wildlands from development and exploitation, and spawned an international environmental movement reaching from the grassroots to government ministries. We helped to make many people aware that when we pollute the air and water, bulldoze natural habitats, cut down forests, and poach rare plants or animals, we are ultimately damaging our own future prospects because of the critical role that biodiversity plays in supporting our existence. There is certainly reason for us to be proud of our efforts, but as I look at the world around me, I can't help but think that we have fallen far short of what's needed and feel an extreme urgency to accomplish much more.

One of our leading achievements was the ratification of the Convention on Biological Diversity (CBD), the legally binding international treaty that went into force at the end of 1993. Originally hailed as salvation for the world's forests, wildlands, and species, the CBD has failed to realize its potential. Funding for it has been so inadequate that more than *a quarter* of the world's tropical forests have ended up being logged in the twenty-eight years since its adoption. Moreover, it is difficult to demonstrate that many species have been saved as a result of its existence. Confounding our intentions, the regulations created under the treaty in the name of national sovereignty have, very often, had a damaging effect on our scientific efforts to understand the biodiversity that still exists. By limiting or preventing access for researchers to many countries, the regulations often act as a barrier to the accumulation of knowledge about species, even as they disappear quietly and permanently. The CBD helped to call widespread attention to the importance of biological diversity and resulted in the establishment of numerous protected areas, but whether these will be maintained over the years to come remains to be seen. I would have to say that considering the CBD as a unified mechanism that could solve the problem of how to preserve biodiversity worldwide has turned out to be a costly delusion. Perhaps it can be improved to the point where it is highly useful, but that is not yet certain.

The disappointing impact of the CBD is just one example of how conservation efforts have been blunted in recent years and how other concerns—many of them legitimate and pressing—have become obstacles toward conserving biodiversity. This logjam comes at a crucial juncture for life on this planet—a time when we must act or lose the window of opportunity forever. We can ill afford a weakening of our efforts to protect biodiversity because it is fundamental to our existence and that of all other species that make up the biosphere. As serious as the world's other problems are, and despite the suffering, loss of life, and stunting of human potential caused by war, famine, oppression, poverty, and natural disasters, biodiversity loss is what should most trouble those who care about the future of the world and of our civilization in particular. Massive biodiversity loss will be catastrophic for human prospects. Ecosystems will collapse as their fragile networks are undermined by the loss of key species, and the future of our own species will ultimately be endangered.

In the 11,000 or so years since *Homo sapiens* developed agriculture and began to build permanent settlements, we have had a disastrous impact on an ever-growing number of the species around us. Extinction is a natural process, but according to conservative estimates, we are at present driving to extinction thousands of species of plants, animals, fungi, and other organisms—most of them not yet discovered—*every year*. What's more, the number going extinct each year is rising rapidly. This amounts to a terrifying rate of loss, a massive dying of organisms at a rate that the world has not seen since the end of the Cretaceous Period, sixty-six million years ago, when the last dinosaurs disappeared and the character of life on earth was altered permanently.

Many human activities push species toward extinction: clearing land for agriculture, draining wetlands, destroying natural ecosystems, deliberately or accidentally moving species to places where they have no natural controls and can overwhelm native ecosystems, generating massive amounts of pollution, and selectively overharvesting animals and plants in nature. At the same time, global climate change, caused primarily by our burning fossil fuels, is rapidly altering the conditions for life on earth, making it likely that countless species stressed directly by other human activities will disappear even more quickly than they would have done otherwise.

Scientists who have tried to project extinction rates have generally considered that about one fifth of existing species are at risk of extinction now, and that the changes accompanying the expected growth in human population and consumption levels could drive half of the total to extinction by the end of this century. There is no doubt that we will suffer from the consequences of this destruction of the living world, which provides the context that makes our life on this planet possible. We depend on other organisms for the composition of the air we breathe and, to a large extent, its quality; for the condition of the soil in which we grow crops and the pollination of many of those crops; for most of the fresh water that we drink; for all our food, directly or indirectly; for about half of the medicines that we use; for most of our building and clothing materials; and for the beauty that inspires us every day. Together, Earth's organisms form one great living web that provides the conditions of life for every species. A single member of this web—our one species—is multiplying so rapidly and consuming so much that we are endangering all the others and ourselves as well. I agree with Ed Wilson's eloquent assessment:

> The worst thing that will *probably* happen—in fact is already well under-way—is not energy depletion, economic collapse, conventional war, or the expansion of totalitarian governments. As terrible as these catastrophes would be for us, they can be repaired in a few generations. The one process now going on that will take millions of years to correct is loss of genetic and species diversity by the destruction of natural habitats. This is the folly our descendants are least likely to forgive us.[1]

The causes of biodiversity loss and the concomitant erosion of the Earth's ability to support the human species are so complex and multifaceted that I've found it very helpful to think in terms of a single metric—that of the human ecological footprint. The ecological footprint of a single individual, a family, a corporation, or of the whole human species is the biocapacity of the land it occupies—the inherent productivity of that land and its capacity to generate biomass and absorb wastes—compared with the demands it places on that biocapacity for resources and waste absorption. When biocapacity is greater than the demand placed on it, the unit of human society in question is sustainable. When demand exceeds biocapacity, that unit, or the world as a whole, runs an ecological deficit. When a place has an ecological deficit, it meets the shortfall by using up its stored biocapacity or by importing resources from elsewhere. The overuse of natural systems degrades their productivity and reduces overall biocapacity permanently. These ways of compensating for an ecological deficit are, by definition, not sustainable.

Mathis Wackernagel and William Rees, then at the University of British Columbia, conceived of the ecological footprint idea in 1990. Subsequently, with Wackernagel's leadership, the Global Footprint Network

(GFN) (www.footprintnetwork.org), a think-tank in Oakland, California, has done much of the conceptual and measurement work needed for us to make meaningful assessments and comparisons of our ecological footprints at every level from the individual to the entire human species. The results of their analyses are not encouraging. Since the1970s, humanity as a whole has been running an ecological deficit, one that the GFN estimates at more than 175% of the Earth's productive capacity. This means that every year we are consuming 75% more than what the Earth can produce on an ongoing basis. One way of looking at this is that humanity is currently using the equivalent of 1.75 Earths to support itself. To make up the "overshoot," we are taking more fish from the oceans than the oceans can produce, cutting down more trees than forests can regenerate, putting more carbon dioxide into the atmosphere than can be absorbed by natural sinks, and exhausting the fertility of the soils where we farm and graze animals. To cover our ecological deficit in these ways, we are essentially stealing from the future—running down a "bank account" that's taken millions of years to build up and will take just as long to restore. In addition, the burden of our ecological theft falls primarily on the poorer countries of the world, as the richer nations are supporting themselves by importing resources and raw materials from other countries to make up the difference between what they need and what their internal biocapacity can produce. Inequality on all sides, within and between nations, can only grow if selfishness continues to prevail. In fact, the charity Oxfam has calculated that eight individuals, six of them in the United States, possess as much wealth as the world's 3.6 billion poorest people—a symbol of global conditions that can only lead to instability and unrest.[2]

If we continue down a path of business-as-usual, our ecological footprint and deficit will only increase over time. The world's population is growing at a net rate of more than 200,000 people per day, and each new individual adds to humanity's total footprint. At the same time, most of the 7.8 billion people already living on this planet are striving to increase their standard of living—and thus their use of resources—regardless of how wealthy or poor they may be initially. Therefore, average per-capita resource use will increase in the future along with our total numbers—an explosive combination and a nasty recipe for disaster in a world already living on an acute ecological deficit.

As the ecological footprint of our species grows, the Earth's climate continues to warm and to become less predictable. Climate change, part of our ecological footprint, is a juggernaut that we have set in motion and have little power to alter over the short term. Much of the carbon dioxide and other greenhouse gases that we have pumped into the atmosphere will remain there for centuries, continuing to exert their warming effects. Meanwhile, dozens of positive feedback loops insure the continuation of warming for many centuries at least. Melting of sea ice and snow on land in the Arctic, for example, reduces the reflection of sunlight and increases the absorption of its associated heat, ensuring that more warming creates still more warming. At the same time, the melting of Arctic permafrost releases more carbon dioxide and methane, a particularly potent greenhouse gas—another ominous feedback loop. Cutting our emissions significantly now would reduce the magnitude of future warming, but we are collectively finding it very difficult to accomplish this essential measure. In fact, despite international efforts, the emissions of carbon dioxide, an important greenhouse gas, have risen in 2017, 2018, and 2019, demonstrating the inadequacy of global efforts to cut back on them, and it is generally conceded that the 25th Conference of the Parties meeting, held in Madrid in December 2019, failed to achieve its goals.

Climate change has devastating effects on plants and animals all over the earth. The increased severity and frequency of droughts that accompany warming will push species in many places beyond their limits of tolerance. In some places in the world, it will simply become too hot for many species to continue to exist; in other places, too dry. Pests and diseases will spread to new areas, the former controls imposed by cold winters gone. Rising ocean levels will flood the biologically rich transition zones between land and sea, eliminating mangrove swamps and estuary habitats that are the breeding grounds for many fish and other marine species. As it absorbs carbon dioxide from the atmosphere, the world ocean will become more acidic, eventually reaching the point at which corals, mollusks, and other animals can no longer precipitate calcium carbonate from seawater to make their protective shells. With many agricultural zones becoming too dry or too hot or too inundated with salt water to maintain their productivity, people will need to seek new land on which to grow food, destroying additional forests and other natural communities in the process. The effect of all these stressors compounding on species already pushed to the brink of extinction by habitat degradation, uncontrolled harvesting, conversion of rainforest to grazing land, and pollution is clear—many of them are going to disappear forever. And as more species blink out, the ecosystems of which they are constituent parts are harmed and their functions progressively weakened, becoming simpler, less stable, less capable of providing the ecosystem services on which all species, including human beings, depend for survival.

In some ecosystems, including the Amazon rainforest, the combination of land clearing and climate change may lead to the collapse of the entire community, with the great majority of the species living there unavoidably facing rapid extinction. On the basis of the intricate relationships between the millions of species that live in its forests, Amazonia is a very large sink for atmospheric carbon dioxide, which is absorbed to create new biomass. As the forest is burned or otherwise destroyed, this capacity is correspondingly reduced, and a vast quantity of carbon dioxide is liberated to the atmosphere. In addition, the water-regulating functions of the trees and other vegetation progressively decline. This in turn ultimately results in the replacement of rainforests by dry savannah, ultimately affecting both local climates and those farther afield. Overall, the effects of forest removal include increased climate change everywhere. Thus the preservation of the Amazon and other forests is urgently needed both to protect the rich biodiversity living there and to hold back the rate of global warming.

With the outbreak of the viral disease COVID-19, a disease that had already infected tens of millions of people and killed more than a million by the autumn of 2020, we fully grasped the dangers posed by diseases spreading from animals in tropical forests to people. In fact, about two viral diseases per year have spilled over from natural hosts to humans or our farm animals for the past century, with SARS and MERS being recent examples. Tropical deforestation greatly increases contacts between wildlife and humans, and when we keep wild animals in captivity or handle them for other reasons, such as traditional medicine, the pet trade, or as bushmeat, we escalate the chances of humans contracting their diseases. From the initial zone of infection they spread readily in our densely populated world and become difficult to control.[3]

How can we reduce the risks of future pandemics? In the short term, we could help to control these diseases cost-effectively by limiting deforestation, slowing the trade in wildlife, banning commercial production of wild animals for food, and effectively modeling the early stages of disease spread in human populations to find ways to reduce the speed with which they are transmitted to new areas. Overall, controlling the spread of diseases, as with virtually all of our contemporary problems, is a matter of attaining overall sustainability.

In the year 2000, Nobel laureate Paul Crutzen introduced the notion that we humans have unknowingly brought into existence a new geological era, which he aptly termed the *Anthropocene.* Crutzen's idea is that we have created a world in which human activity has become the dominant influence on climate, the environment generally, and on all natural processes, and that our effects will leave a radically distinctive mark on the geological record. Certainly the loss of biological species—so many of them that most of us consider that we have already entered the Sixth Great Extinction since the origin of life some four billion years ago—is one of the most telling features of the Anthropocene. Some date the beginning of the new era to our development of crop agriculture 11,000 years ago; others to the Industrial Revolution, which began some 250 years ago; and still others to the period following World War II, when consumerism completed its conquest of our culture and the effect of everything we do was so greatly intensified. Regardless of which starting point is chosen, we are clearly living in a new kind of world. In essence, we have assumed powers like those the ancient Greeks and Romans ascribed to the gods. "We are as gods," wrote Stewart Brand wryly, "and might as well get good at it."[4] Unfortunately, we seem to be failing rather spectacularly to meet that challenge.

Among most world leaders and a great many average citizens, these problems are in general not taken as seriously as their mitigation, in view of the enormous threats that accompany them, clearly deserves. It has become clear that the future of our civilization depends on our ability to cope with them. People may acknowledge that challenges exist, but they tend to think about how humanity has successfully met other challenges in the past and tend to assume that we can find ways to deal with the current ones, too. But we have reached the point where technology alone cannot fix the situation; reduction in our population levels and consumption rates will be necessary if we are ever to attain stability. In the face of the dangerous threats that confront us, counting on some form of "working it out" and getting on with business as usual is roughly the equivalent of drinking a large glass of Scotch whisky, adding some sleeping pills, and going to bed. If our very civilization is to survive, we must soon wake up and actively confront the self-destructiveness of our ingrained habits, and then swiftly and systematically begin to change them for the better.

ℰℬ

We humans like to celebrate our adaptability, intelligence, imagination, moral character, farsightedness, and determination. If we do indeed possess these qualities as a species, how have we managed to work ourselves into such an untenable situation? Part of the reason, I would say, is the tendency I've described for us to be insufficiently aware of the depth of the hole we are digging for ourselves, of not seeing beyond the boundaries of our own short lives. But this is just one reason that we are failing to respond appropriately to the demands of the times. Another reason is that we continue to let our baser instincts—selfishness, greed, fear, anger, and the quest for dominance—guide the activities and direction of our societies. As President Abraham Lincoln so appropriately pointed out at the conclusion of his first inaugural address, with the country on the brink of a devastating civil war, now is a time when we must find "the better angels of our nature" because they offer the only way out of our predicament.

The capacities for altruism, cooperation, self-sacrifice, empathy, moderation, and concern for future generations have been a part of being human for as long as our species has existed. These traits were necessary for the development and maintenance of the complex social groups that gave early humans a survival edge in a world filled with threat and danger. But they coexisted with the more primal emotions and behaviors our earliest ancestors inherited from their evolutionary forebears. Human beings have therefore always exhibited both cooperation and selfishness, trust and treachery, violence and peace, good and evil. The decisive factor in the human drama has always been to see which side of our nature is going to predominate.

Fear and perceived threat, working through the ancient parts of our brains, activate deep anxiety that is easily channeled into aggression, violence, hate, and distrust of those outside of our own social group. These emotions take control unless our higher brain functions—forbearance, critical thinking, patience, empathy for others, compassion, circumspection—rein them in. That is why it is easy for demagogues to come along and manipulate our fears and anxieties, for cunning, self-interested actors to persuade us to fixate on false enemies and misleading promises of wealth and power, for nationalist ideologies to rise. We therefore head into the future guided by beliefs, values, and priorities that are more likely to pull us apart than to bring us together, more likely to keep us focused on the same drives to consume, accumulate, and grow that got us into this mess in the first place. Selfishness, greed, and rapaciousness can only exacerbate the challenges facing us and cause their horrific consequences to become both more likely and to arrive sooner.

Another barrier we face is that we are not well equipped, cognitively, to deal with slow-moving disasters like global biodiversity loss and climate change. We evolved to respond as quickly as possible to the threat of a charging mammoth or saber-tooth cat, not abstract threats with consequences in what we consider some distant future. We are indeed good at anticipating what's to come and planning for different possible outcomes, but our timeframe for that skill is mainly focused on days or a few years at best. Too many distracting everyday demands are placed on our cognitive processing for most of us to think in the long term, much less beyond our own limited lifespans. That is a primary reason why our planning for the actual, very scary future that we face is exceptionally short-sighted.

This limitation was brought home for me when Pat and I visited New Orleans in 2006, less than a year after Hurricane Katrina left the city devastated. New Orleans and the area to the south are subsiding, in part because a great deal of oil has been pumped from beneath the region's coastal flats. Indeed, much of the road to the south of the city runs on top of the only remaining dry land, a narrow ridge, even though maps show the land on either side of the road as being much more extensive. Perhaps because of the legal relationship between coastline and state oil-drilling rights, the maps have not been updated to match current reality. Meanwhile, sea-level rise related to climate change is continuing, with the best estimate that can be provided now anticipating a rise of perhaps seven feet by the end of the century. When we visited, New Orleans was already beginning to rebuild, and it is now much as it was before Katrina, just with better levees and a lot of misplaced optimism. Between subsidence of the land and rising sea levels, this city is on the front lines of change and will face an increasingly difficult struggle for survival in the years to come.

All around the rivers and coasts of the U.S., people are pushing new developments out into the salt marshes, wetlands, and catchment basins that naturally serve to ameliorate the effects of flooding. Each county seems to want to do this because the new malls, stores, and housing add to local tax revenue. This is all very foolhardy, because it is only a matter of time before *all* these developments will be inundated by the oceans or rivers along which they have been built. But, blinded by greed and the false vision of prosperity fueled by indefinite growth, we consider only that the developments will bring in so many tax dollars this year and generate so many dollars of profit for the private owners. With every levee constructed and wetland filled, the problem waters are simply moved onto someone else's property and often end up costing millions of dollars to people who realized no financial gain whatever from the original project. Thus, the costs of the eventual failure of the enterprise will end up being borne almost entirely by the general public. Planning for a future in which whole cities will have to be moved is not the sort of thing that politicians and elected officials can usually bring themselves to even consider, but it is *exactly* the kind of issue that our society must grapple with to be able to respond to the changes in store. It is beyond astounding to me that denial of anthropogenic climate change has during the past few years actually been the official position of the U.S. government.

𝒞𝒶

What I have written up to this point may very well have inspired some fear, dismay, or even depression, and for this I offer no apology. Fear can motivate action, especially when it is combined with hope. And I do believe there are grounds for hopefulness, despite the severity of our situation. What we can't afford are despair and resignation. Inaction is certainly not an option that will bring a good outcome.

As a biologist I may be biased, but I am convinced that biology—and science in general—gives us much cause for hope. At its core, science is a collaborative way of developing the most accurate hypotheses possible about aspects of the world around us. It is motivated by a desire to find truth, not to gain advantage over others. In these ways, it operates apart from the selfishness and ignorance that underlie our greatest challenges and provides the best information that we can assemble for dealing with those challenges. The biological sciences, in focusing on the living things that make our life on earth possible but are so rapidly falling victim to our mindless human growth, are particularly important for our survival. If the challenges we face on the fronts of conservation, biodiversity, and sustainability are like Goliath in their apparently overwhelming strength, to my mind biologists are our Davids, quietly slinging stones, improving their aim, trying new techniques, practicing under all circumstances. The results of their efforts may not always be apparent immediately, but in the end they can help keep intact both the living systems of the world and the best features of our civilization.

Biology gives us hope not just because of what it can tell us about the world but because of what it demonstrates about the human spirit. The biological sciences are flourishing because human beings are curious animals; we share a passion to understand our place in the living world. By simply doing our work, we life scientists are providing the foundations for building a deep compassion and respect for all the species on this planet. Such a stance, which I often think should be labeled with the overused but potent word *love*, is perhaps the single most important force working in our favor.

I think of what Doug Stevens, curator of Central American botany at the Missouri Botanical Garden, wrote recently as part of the Garden's research communication efforts:

> [T]he most important part of my work has been plant collecting. Most of the collections I have made over the last half century cannot be repeated with any amount of effort or money because those plants no longer occur in those places. The same will be true about the collections I make this year and next; the rate of landscape deterioration will not slow and the time to complete the inventory is rapidly running out.

What else but love can explain Doug's dedication to his work in the face of the catastrophic losses he is witnessing? He goes on collecting and documenting plants that he knows are disappearing because to do anything else would be to deny life itself. All over the world, biologists are seeing many different forms of devastation—dying coral reefs, putrefying lakes green with algae multiplying on fertilizer runoff, animals killed solely for their horns or tusks or organs, the last individual of a species vainly seeking a now non-existent mate—and yet they go on doing their work because abandoning it is unthinkable and despair is not an option. This is a critically important message for the world at large, and clearly one that if understood should stir us to much greater levels of action. Fully comprehending the shocking condition of the world and the importance of our last chance to save it can appeal to people's emotions and moral sense— you must grab their hearts. Throughout history, nothing has proved better at this than religion and the value systems associated with it. Scientists and people of faith are often seen as enemies, but I believe that religious institutions can play a pivotal role in bringing about the needed changes in humanity's values and priorities and that science can function as their ally. All the world's great religions have ethics and morality at their core. They offer sound prescriptions for living together with others peacefully and productively. And every religion, through its moral teachings, preaches the importance of respect for all of what the creator (or creators) brought into being. With this common moral foundation, every religion is only a step away from promoting environmental ethics that place respect for all species at the center of everyday living and spiritual practice.

On June 18, 2015, the Pope issued an encyclical titled *Laudato Si'*, or "Praise Be to You," a quote from the "Canticle of the Creatures" by St. Francis. In a remarkable departure from the encyclical tradition, *Laudato Si'* is devoted entirely to ecology and the environment; further, it is addressed to all people rather than just Catholics and moves beyond the traditional focus on Catholic doctrine. In this text, Pope Francis embraces the concept of climate change, talks about the disproportionate effect of ecological despoilment on the poor, links "human ecology" to that of nature, and makes it a moral as well as practical imperative to care for our "common home." I am proud to say that the Pontifical Academy of Sciences, of which I have been a member since 1990, played a part in informing this important document.

In writing *Laudato Si'*, the Pope made it very clear that he was building on the Church's well-established tradition of concern for the environment —and with the related issues of war and poverty—that began with his predecessors. Paul VI, John Paul II, and Benedict XVI are all quoted extensively in these regards. Pope Francis reaches outside of Catholicism to acknowledge the very strong stand that the leader of the Eastern Orthodox Church, Ecumenical Patriarch Bartholomew, has taken on ecological destruction. He agrees with the Patriarch that destroying wetlands, cutting down forests, and contaminating the land and water are all sins. All this has the effect, I think, of broadening the encyclical's authority: the directive to care for the planet comes not from Pope Francis or the Roman Catholic Church, but in effect from the Almighty himself. That is powerful— even if you don't believe in the existence of God.

By issuing the encyclical *Laudato Si'* and pursuing related initiatives, the Pope greatly elevated his moral authority in the world. I know that many non-Catholics and atheists see him as the great leader on the environment that the world so desperately needs. To break the logjam of political and economic assumptions that allow for the unrestricted destruction of our environment, we will need more leaders like him. It is our responsibility to encourage them.

&

During my lifetime, humanity has made some progress in eliminating or lessening the impact of many of its ancient scourges. We have lowered the proportion of people living in poverty from more than half to about 10%. We have reduced the proportion of people who suffer from chronic hunger and famine. We have lowered the incidence of some chronic infectious diseases and increased access to health care for many. Women are now more highly empowered than ever before. Infant mortality has been reduced in a number of regions. We have avoided world wars for seventy years, and overall the planet might be a less violent place than it once was.[5] Although we still have far to go in all of these areas, these successes show that we *can* muster the will and the wherewithal to make progress in solving major and seemingly intractable problems when we want to do so badly enough.

Climate change and the crisis of biodiversity loss are unlike any of these problems in their scope and eventual global consequences. But we can use our record of progress over the last eighty years to generate the hope that these problems, too, can be addressed, along with the many others that humanity faces. In today's hyperconnected world, it's becoming increasingly clear that threats to the environment, rising violence, growing inequality, social unrest, racial and other discrimination, and political upheaval are all interrelated and can't be confronted piecemeal. We cannot afford to use our record of progress as an excuse for complacency and inaction. We can't just assume that "they" will work out solutions to all these problems. We—all of us collectively—can change the world for the better and we must begin now.

Coordinated efforts are needed on all sorts of fronts, but I see a few priorities. Nothing is more frightening to me than to witness the world's nations, ethnic groups, and religions coming into conflict, inciting violence, and pursuing their own self-interest at a time when just the opposite is required. We need a renewed effort at bringing the world together to focus on finding solutions to the problems—environmental and otherwise—that threaten *everyone's* prosperity and well-being. As Pope Francis notably said, what we need to build are bridges and not walls. Only by helping people to realize that our fates are intertwined and that differences in skin color, culture, religious belief, and lifestyle are great blessings can we conjure up Lincoln's better angels, build the foundations for peace, and secure the future of our civilization. The concrete representations of international cooperation that exist today—trade agreements, the United Nations, the World Health Organization, international scientific collaborations, the rule of law, and many more—must be strengthened and used as the foundation for new means of cooperating to promote the common good. Act as we may, we *are* one people, one species, and our common cause is a singular one.

The international cooperation I have in mind is not the economic globalism that has left behind many of the world's most disadvantaged people. In fact, reversing the growing inequality that globalism has fostered must be another of our highest priorities. We need a new kind of universal globalism that has as its reason for existence not the enrichment of a small global elite but the leveling-out of the vast inequalities that exist between nations, between the developed world and the global south, and between the rich and the poor within nations. One way to achieve this goal is to end the ecological and economic colonialism that has the richer nations appropriating from poorer nations the resources they need to maintain their own currently unsustainable ecological footprints. At the same time, we must recognize that much of the progress that has been made in reducing poverty in developing nations has come at the expense of the environment, and find robust ways to deal with that dilemma also. The environment is neither a bottomless cookie jar nor an inexhaustible bank account.

More than a half century ago, in 1965 in Geneva, Adlai Stevenson, U.S. Ambassador to the United Nations, in his last major speech to the Economic and Social Council of the U.N., spoke words that I consider to be one of the most succinct and eloquent pleas for addressing both inequality and ecological damage at the same time:

> We travel together, passengers on a little spaceship, dependent upon its vulnerable reserves of air and soil, all committed for our safety to its security and peace; preserved from annihilation only by the care, the work, and, I will say, the love we give our fragile craft. We cannot maintain it half fortunate, half miserable, half confident, half despairing, half slave to the ancient enemies of man, half free in a liberation of resources un-

dreamed of until this day. No craft, no crew can travel safely with such vast contradictions. On their resolution depends the survival of us all.[6]

To address the poverty, powerlessness, and attenuation of human potential that exists across Africa, in large parts of south Asia, in some of the countries of Latin America, and in much of China and Oceania, will necessarily involve raising standards of living, developing sustainable infrastructures, and increasing access both to education and to health care. These improvements mean higher levels of consumption and larger ecological footprints for the countries concerned. For this to happen and not further expand the footprint of all humans collectively, people in the advanced industrial countries will need to find ways to significantly shrink their ecological footprints. Achieving the goal of more balanced levels of consumption across the world requires that those in the wealthier nations reduce fossil fuel use dramatically, use land much more efficiently, develop new technologies that are less damaging to the Earth's ecosystems, and change their diets to, among many other things, greatly reduce the consumption of meat. Clearly, we cannot use the equivalent of 1.75 Earths mainly to support the resource-intensive lifestyles of the wealthy nations indefinitely.

Reducing and ultimately eliminating the use of fossil fuels and other non-renewable resources does not have to entail enormous sacrifice or reversal of progress in developed countries. A good life is not built on consumption. We must re-define our needs and re-evaluate the criteria that define a sustainable standard of living. It is time to bring back repair shops, "classic" fashions, and durable goods. By focusing on what human beings really need to be happy—a sense of belonging and purpose, good health, security for family and community—we can design ways of living that *increase* well-being while they reduce our ecological footprints.

Re-examining my impact on sustainability, with many millions of air miles during the course of my career, I have personally been responsible for a larger than average jet fuel footprint. While I cannot change the past, I can and do embrace new technologies, such as video conferencing, that will enable us all to connect and reduce our travel footprints in the future. Change is hard and takes time, but we all have to start somewhere. What is possible depends directly on our initial circumstances. Self-examination is the place to start, and we can be certain that even small acts, multiplied over many people, will add up to make big differences.

At present, however, we live by the tenets of a growth-based economic system that essentially compels us to use ever-larger quantities of resources. Under this system, well-being and prosperity—at least for some—are inseparable from economic growth, and so it is taken for granted that all growth is good and desirable. But growth depends primarily on increases in consumption and production, which rest on use of land, resources, and energy. Any attempt to reduce consumption and production significantly and rapidly to shrink a country's ecological footprint would cause the system to go into crisis, with significant negative impacts on everyone. We are stuck in a giant contradiction—in effect, a damned-if-we-do and damned-if-we-don't conundrum.

A sustainable future, therefore, depends on rejecting some of the assumptions of the economic system, predominant now, which puts us in this untenable bind. We need to replace it—slowly and intentionally—with a different model, one that is consistent with the Earth's finite supplies of land and resources and limited capacity to absorb our waste products. Such a system would include the re-defined notion of what it means to lead a good life that I just described, and it would recognize that human beings are motived to work hard and achieve their best for reasons other than pursuing narrowly defined economic self-interest. It must also distribute wealth far more evenly, avoiding the astonishingly lopsided situation we have now of the world's eight richest individuals possessing as much wealth as the poorest half of the entire human population.

None of these changes will be adequate if we don't also rein in population growth and reduce our numbers to reach a stable population that the world can support indefinitely. Given that the average per-capita ecological footprint must increase in the less-developed countries as a matter of humanitarian principle (as well as to reduce the likelihood of violent conflict and mass migration), continued growth in the total population as well will put the goal of reducing humanity's overall ecological footprint out of reach for a very long time, possibly forever. We do know, however, that raising people's standard of living, increasing access to education and health care, and empowering women and children—necessary goals in their own right—all have the effect of reducing the birth rate and lowering population growth. So, solving this problem is within our reach if birth control and family-planning information are made more widely and freely available, become more socially acceptable, and we make progress on the issues that lead people to have large families. The full empowerment of women and children everywhere is certainly a key element in achieving success in attaining a stable population.

The goals I have laid out will obviously take a great deal of effort and a long time to realize fully. Those are givens. But we will never get close to reaching these goals under our standard approach to long-term planning, which is to issue platitudes, make meaningless promises, and then return to the *status quo*. Only if we learn to love and respect one another do we have a chance. We need to build a global vision for a livable future 50 years from now, 100 years from now, even 400 years hence—and efforts to get there in one piece must be part of what we do every day. It's past time to face facts and get to work. Otherwise we will deserve Kurt Vonnegut's epitaph: "The good Earth—we could have saved it, but we were too damn cheap and lazy."[7]

The more we get done ourselves, the easier it will be for our children and their children to move the world back to sustainability. My granddaughter, Louisa Frances Raven, shown here the day after her fifteen-month birthday in September 2019, will reach my age in 2101. We all need to get to work!

Megacorax drawing by Alice Tangerini, illustrator for the Department of Botany, U.S. National Museum of Natural History, Smithsonian Institution.

— Appendix —

PLANTS NAMED FOR PETER H. RAVEN

One way that botanists have of recognizing people, usually other botanists or those who have been helpful to them, is to base the scientific names they provide for the plants they are describing for the first time, or to which they are giving new names for technical reasons, on the names of those they wish to honor. I have enumerated here the five genera and thirty-two species, subspecies, and varieties that have been named for me.

Of the three taxa illustrated here, *Megacorax* is a unique genus of the Onagraceae, to which I have devoted a lifetime of study, named for me by my former student Warren L. Wagner of the U.S. National Museum of Natural History, Smithsonian Institution, with a Mexican colleague. Its name means, embarrassingly, "great raven"! Another student, the late Ching-I Peng, commissioned the lovely image of *Begonia ravenii*, a Taiwanese species that he had named in my honor. His treasured gift is a wonderful memorial of our friendship. Finally, a Russian colleague, Leonid Ayeryanov, whom I have helped in various ways in his studies of the plants of Southeast Asia, with a Vietnamese colleague, named the lovely palm *Trachycarpus ravenii*, which he discovered in the mountains of Cambodia, for me.

Genera

Megacorax S. González & W. L. Wagner, Novon 12(3): 361. 2002. (Onagraceae) Mexico

Peteravenia R. M. King & H. Rob., Phytologia 21: 394. 1971. (Asteraceae) Mexico

Petroravenia Al-Shehbaz, Novon 4(3): 191. 1994. (Brassicaceae) Argentina

Phravenia Al-Shehbaz & Warwick, Taxon 60(4): 1161. 2011. (Brassicaceae) Argentina

Ravenochloa D. Z. Li & Y. X. Zhang, Plant Diversity 42: 132. 2020. (Poaceae) China

恩師窗文先生
學術成就舉世同飲
熱愛中華源遠流長
播導特傳嘉義彝裁培
之恩，雞言筆述，飯誌
謹成於一九八八年首發現並
以貝師奇特命名之
乙丑特用植揚岩生秋海
棠水墨糕繪二幅呈贈
睹物
崇致與感激
之至。

愛業
郭懿謹
誌二〇〇九圖胃四於
中央研究院

岩生秋海棠
己丑春游明熙高於中央研究院

Begonia ravenii C. I Peng & Y.K. Chen

Lovely plants of *Begonia ravenii* form colonies in wet places in the mountains of Taiwan.

Species, Subspecies, and Varieties

Allium ravenii F. O. Khass., Shomur. & Kadyrov, Stapfia 95: 173 (172–174; figs. 1–2). 2011. (Alliaceae) Kazakhstan

Amorphophallus ravenii V. D. Nguyen & Hett., Novon 26(1): 53. 2018. (Araceae) Cambodia

Anthurium ravenii Croat & R. A. Baker, Brenesia 16(suppl.) 1: 75(–76). 1979. (Araceae) Panama

Arctostaphylos hookeri subsp. *ravenii* P. V. Wells, Madroño 19: 200. 1968. (Ericaceae) U.S.A., California

Ardisia raveniana Lundell, Wrightia 5(3): 61. 1974. (Myrsinaceae) Mexico

Arracacia ravenii Constance & Affolter, Novon 5(1): 22. 1995. (Apiaceae) Mexico

Astragalus ravenii Barneby, Aliso 4: 131. 1958. (Fabaceae) U.S.A., California

Begonia ravenii C.I Peng & Y. K. Chen, Bot. Bull. Acad. Sin. n.s., 29(3): 217. 1988. (Begoniaceae) Taiwan

Berberis ravenii C. C. Yu & K. F. Chung, Phytotaxa 184(2): 88. 2014. (Berberidaceae) Taiwan

Calathea ravenii H. Kenn., J. Bot. Res. Inst. Texas 6(2): 379. 2012. (Marantaceae) Panama

Cardamine flaccida var. *ravenii* Rollins, Crucifer. Continental N. Amer. 273. 1993. (Brassicaceae) Mexico

Dendropanax ravenii M. J. Cannon & Cannon, Bull. Brit. Mus. (Nat. Hist.), Bot. 19: 11, fig. 4. 1989. (Araliaceae) Costa Rica

Dioscorea ravenii Ayala, Phytologia 55: 296. 1984. (Dioscoreaceae) Peru

Encelia ravenii Wiggins, Proc. Calif. Acad. Sci. ser. 4, 30: 251, pl. 16. 1965. (Asteraceae) Mexico

Eugenia ravenii Lundell, Wrightia 4: 94. 1968. (Myrtaceae) Mexico

Fuchsia ravenii Breedlove, Univ. Calif. Publ. Bot. 53: 57. 1969. (Onagraceae) Mexico

Haplopappus ravenii R. C. Jacks., Amer. J. Bot. 49: 123. 1962. (Asteraceae) U.S.A., Arizona

Koanophyllon ravenii R. M. King & H. Rob., Phytologia 22: 150. 1971. (Asteraceae) Mexico

Lomatium ravenii Mathias & Constance, Bull. Torrey Bot. Club 86: 379, fig. 3. 1959. (Apiaceae) U.S.A., California

Ludwigia ravenii C.I Peng, Syst. Bot. 9: 129. 1984. (Onagraceae) U.S.A., Florida

Mentzelia ravenii H. J. Thomps. &
    J. E. Roberts, Phytologia 21: 285.
    1971. (Loasaceae) U.S.A.,
    California

Miconia ravenii Wurdack, Phytologia
    14: 269. 1967. (Melastomataceae)
    Mexico

Microlepia ravenii S. J. Moore, Ann.
    Bot. Fenn. 48(3): 284(–286), fig. 1.
    2011. (Dennstaedtiaceae) Vietnam

Oenothera patriciae W. L. Wagner &
    Hoch, Syst. Bot. Monogr. 83: 213.
    2007. (Onagraceae) Southern U.S.A.

Oenothera ravenii W. Dietr., Ann.
    Missouri Bot. Gard. 64: 500(–502).
    1978. (Onagraceae) Argentina

Trachycarpus ravenii occurs farther
south than any other member of its
genus. Fan palms are commonly
cultivated.

Palicourea raveniana Borhidi, Acta
    Bot. Hung. 59(1–2): 47. 2017. (Rubiaceae) Venezuela

Phoradendron ravenii Kuijt, Syst. Bot. Monogr. 66: 381. 2003.
    (Loranthaceae) Costa Rica

Pogonia ravenii L. O. Williams, Fieldiana, Bot. 32: 200. 1970.
    (Orchidaceae) Costa Rica

Protium ravenii D. M. Porter, Ann. Missouri Bot. Gard. 58: 263, fig. 2.
    1971. (Burseraceae) Costa Rica

Prunus ravenii Zardini & Basualdo, Candollea 47: 252, fig. 1. 1992.
    (Rosaceae) Paraguay

Rudgea raveniana W. C. Burger Fieldiana, Bot. n.s., 33: 304. 1993.
    (Rubiaceae) Costa Rica

Salvia raveniana Ramamoorthy, Brittonia 36(3): 297. 1984.
    (Menthaceae) Mexico

Trachycarpus ravenii Aver. & K. S. Nguyen, Nordic J. Bot. 32(5): 563.
    2014. (Arecaceae) Cambodia

Tricyrtis ravenii C.I Peng & Tiang, Bot. Stud. (Taipei) 48: 358(–363),
    figs. 1–5. 2007. (Convallariaceae) Taiwan

Trixis ravenii Zardini & Soria, Ann. Missouri Bot. Gard. 78: 531, fig. 1.
    1991. (Asteraceae) Paraguay

Zapoteca ravenii H. M. Hern., Syst. Bot. 15: 226, figs. 1–5. 1990.
    (Fabaceae) Mexico

— Acknowledgments —

This project was initiated at the suggestion of the late Jonathan Klein-bard, assistant director of the Missouri Botanical Garden from 1997 to 2004. I am grateful to him for the inspiration, which has taken a number of years to reach fruition. It was at the suggestion of the late Bert Walker that I approached Kathy Evans of "Write for You" Life Stories, an out-standing memoirist and editor here in St. Louis. She later recruited Eric Engles of Editcraft, Inc., and Suzanne Fox of Bookstrategy, to all of whom I literally owe the completion of this work. Writing biographies is a very special skill that does not come naturally, and I have learned a great deal from each of these talented individuals.

Many people have contributed to assembling the information used in the preparation of this manuscript and to the images reproduced in it. In that connection, I specially would like to thank my daughter Alice Raven for sharing the abundant supply of genealogical material about our fam-ily that she has assembled.

Special thanks to Publisher Liz Fathman; her staff in Marketing and Communications; Lisa Pepper, for her unfailingly accurate and helpful editing; her colleagues at the Missouri Botanical Garden Press (Allison Brock and Amanda Koehler); and designer Katie Koschoff. Andrew Col-ligan, archivist at the Missouri Botanical Garden, has been unfailingly helpful in expediting searches of the Garden's Archives for bits of infor-mation that have made the final product more accurate than it could have been otherwise. In addition, Doug Holland and Linda Oestry of the Peter H. Raven Library, together with Roy Gereau, Garden curator, helped work out this book's title, with encouragement from the marketing team at University of Chicago Press. The many others who have helped are in-dividually acknowledged at appropriate places in the book, and I am most grateful to them all for being so generous with their time on so many occasions.

I deeply appreciate the care and attention provided by so many out-standing assistants during the course of over forty years at the Garden, with Florence Guth and Donna Rodgers the longest terms of these, each providing the many years of devoted and skillful assistance that made it possible to keep the institution moving forward.

For generously defraying the editorial expenses and thus helping greatly with the publication, I greatly appreciate contributions from the following individuals: Dr. James Aronson, Mary Randolph Ballinger, John and Penelope Biggs, Steven and Camilla Brauer, Dan and Connie Burkhardt, William H. T.† and Patricia Bush†, Dr. Christopher and Sharon Davidson, Jack and Laura Dangermond, Philippe de Spoelberch, Fairweather Foundation (Arthur Hall), Drs. Alan and Shirley Graham, David Kemper, Scott and Juliet† Schnuck, Dr. Suri and Edda Sehgal, Andrew C. and Barbara Taylor, Trudy Busch Valentine, George Herbert Walker III, and Ray† and Roma Witcoff.

Finally, I must acknowledge my wife, Patricia Duncan Raven, who gave up her career to support mine and makes my life so much nicer than it could possibly have been in any other way. I love her very much and appreciate her kindness every single day. She has helped a great deal in the preparation of this book and its meticulous editing. I delight in each of my children: Alice, whose intelligence and good humor have been a constant joy; Liz McQuinn, who lives near us and enriches our lives in many ways; Francis, who with his wife Carolyn Kousky is raising two wonderful sons, Noah and Nate, in a difficult but wonderful world; and Kate Raven, with her lovely daughter Louisa. All have read a draft of this book, but the opinions are mine—and hopefully theirs too. I love them all and hope what I have tried to do, and what I have expressed in this book, will help make their own lives better as the years go by.

— Notes —

1. FAMILY ROOTS

1 "CONVICTED BANKER WAS MYTH IN CHINA; F.J. Raven Rose Spectacularly in Shanghai to Become a Financial Power. BEGAN WORK AS SURVEYOR He Branched Off Into Realty, Banking and Trust Business—Then Came Collapse," *New York Times*, 2 February 1936, 36.

2 James Madison Lenhart is listed as one of the Pony Express riders in *The Pony Express Trail: Yesterday and Today* by William E. Hill (Caldwell, Idaho: Caxton Press, 2010), 271.

3 James Madison Lenhart entry in *The Morning Call*, San Francisco, 9 October 1891, 4, col. 3.

4 "Diary of Patrick Breen: About the Diary," The Bancroft Library, University of California at Berkeley, accessed on 29 September 2020, https://patrickbreen.word press.com/about-this-diary. Information on Patrick Breen's diary comes from the University of California's Bancroft Library's excellent blog on the document: https:// patrickbreen.wordpress.com/. The library also has images and transcriptions of each page digitized on the Online Archive of California: http://www.oac.cdlib.org/ark: /28722/bk0004b217j/?brand=oac4.

5 Patrick Breen Diary, [page 25], accessed on 29 September 2020, http://www.oac .cdlib.org/ark:/28722/bk0004b217j/?order=27&brand=oac4.

6 Charles Fayette McGlashan, *History of the Donner Party: A Tragedy of the Sierra* (Sacramento: H.S. Crocker Co., 1907), 197.

7 Ibid., 196–97.

8 "New Light on the Donner Party," Kristin Johnson, last modified 5 November 2006, accessed on 29 September 2020, https://user.xmission.com/~octa/Donner Party/Breen.htm.

9 Marjorie Pierce, *East of the Gabilans. The Ranches, the Towns, the People—Yesterday and Today* (Western Santa Cruz: Tanger Press, 1976).

10 See "Appendix I: Men of Shanghai," *Fortune* XI, 1 January 1935, 115.

11 See note 1 above.

3. BOTANICAL PRODIGY

1 The author of the manual, Dr. Willis Linn Jepson, is considered among the most distinguished of California's early botanists. He taught botany at the University of California at Berkeley for four decades and co-founded the Sierra Club with John Muir and Warren Olney and others in Olney's office in 1892. Jepson published his incomplete *A Flora of California* starting in 1909 and his *A Manual of the Flowering Plants of California* in 1925.

2 Barely hanging on in nature—the area where we found it is the only place it survives in the wild—*Amsinckia grandiflora* has fortunately been preserved living by the University of California Botanical Garden in Berkeley.

3 In generalized form, this important question has interested me all my life and remains a central question for thoughtful systematic botanists.

4 At the time we visited the Sierra together, Stebbins's magisterial book, *Variation and Evolution in Plants*, was about to appear, and he was planning a move that fall to the Davis campus of the University of California, where he would found a genetics department. In time, he would be recognized as the twentieth century's most renowned student of plant evolution.

5 As it turned out, Stebbins's instinct had been right. When he examined our samples subsequently, he found that their chromosome number was $2n = 90$, very different from $2n = 18$ in the northern populations. This indicated that they would not form fertile hybrids if they were brought together and that the southern populations probably had sets of chromosomes from other distinct species, and certainly should not be considered part of the same species as the northern ones. Tom Howell soon separated them as distinct species; they are still regarded as such to the present day, although both are now assigned to the genus *Tonestus*.

6 He visited Convict Creek with Jack Major, an ecologist friend of his from the University of California, Davis. There they discovered a treasure trove of plants, some of them characteristic of the Rocky Mountains and previously unknown in California or known only in limited local areas of the state.

4. FORMAL TRAINING

1 Published in the following year, the Lewises' monograph would turn out to be a classic of the biosystematic approach, sorting out the species of the genus in a useful and original way.

5. ENGLAND

1 Woodson and his colleagues had recorded about 4,500 species from Panama by that time (1961) and considered that they were nearing the completion of a thorough survey of all the plants in the republic. They were badly mistaken, however: just over half a century later, the species count has been increased to 10,100. The true number of plant species in Panama is estimated to exceed 11,000, nearly equal to the total flora of Europe in an area just over half the size of England.

6. STANFORD

1 In order to preserve the native wildlife and plants on San Clemente Island, it was deemed necessary to remove the goats. By 1991, forty years after my visits, the island was finally goat-free, and native plants—including a few that had been feared to be extinct—had sprung up exuberantly all over the island.

7. EXPANSION AND CHANGES

1 Now in its eighth edition, *The Biology of Plants* is the world's leading botany text. It has been translated into a number of different languages and has been useful for generations of students.

2 Soon, however, I went wrong when I tried to show in an article published in *Science* that these organelles had been incorporated multiple times into different branches of the tree of life during the course of evolution. Science proceeds by the proposal of hypotheses and their subsequent testing, and although mine proved to be

incorrect, it was useful in clarifying this important area in the years that followed. Each of these kinds of organelles seems now to have originated only once.

8. NEW ZEALAND

1 Two major earthquakes in 2011 so badly damaged the ChristChurch Cathedral that it became necessary to pull down its beautiful 207-foot spire, which had stood as a landmark and symbol of the city for so very long. The remainder of the iconic cathedral will be restored according to plans that are being developed.

2 The observations we made of *Fuchsia* and our extensive collections were later used by my former student, Paul Berry, in his detailed study of them and their relative on Tahiti, which he coauthored with Eric Godley.

3 Lawrie served as director of the Royal Botanic Garden Sydney, from 1972 to 1985, constructing a conservatory, enhancing the research program, and building up the living collections, particularly of Australian natives. He was a specialist on the large and very difficult complex of tree genera related to *Eucalyptus* and successfully collaborated with Barbara, a close, lifetime colleague, in detailed studies of several other groups of plants.

4 T. F. Cheeseman in his *Manual of the New Zealand Flora* (1925) had also recognized thirty-seven species, but their outlines and distinguishing features were often quite different from those of the ones we recognized following our detailed studies. Our book about them, *The Genus Epilobium in Australasia*, was published by the New Zealand Department of Scientific and Industrial Research in 1976.

5 Eventually, species listed under the Endangered Species Act would include Raven's manzanita and the Presidio clarkia, two local species that I rediscovered in the 1950s.

9. THE MISSOURI BOTANICAL GARDEN

1 Gyo was one of thirty Japanese students who had been allowed to transfer to Washington University from the University of California, Berkeley, early in World War II. His parents, however, were imprisoned in one of the Japanese internment camps.

2 In 1999, Howard Baer was named one of the twenty most influential St. Louis business leaders of the twentieth century by *St. Louis Commerce* magazine.

3 Securing the Butler Property was one of Bill Klein's many outstanding accomplishments. Bill left to become director of the Morris Arboretum near Philadelphia in 1977, after five very productive years at our Garden. I missed him greatly. The name Shaw Arboretum held until 2000, when, on further consideration of its nature, the facility was officially rechristened the Shaw Nature Reserve.

10. GLOBAL REACH

1 The poppies were of interest to the U.S.D.A. because the large oriental poppy, *Papaver bracteatum*, commonly grown in gardens, was found to produce molecules that could be made into morphine and codeine without going through the intermediate step of making opium. The idea of developing a crop from which those painkillers could be manufactured without increasing the availability of opium was naturally appealing, but in order to do so, the Department of Agriculture scientists had to understand the nature and limits of the species involved.

2 Among other things, Bob persuaded me to organize three collections of written papers advancing the knowledge of the coral tree genus *Erythrina*, which was of

great interest to him because of the wide array of chemicals that its members produce. After his death in 1983, I was made executor of Krukoff's will and charged with the task of scattering his ashes in the Pacific Ocean off Guatemala, which I accomplished a few months later.

3 In 2001, Doug and Olga Martha published the first three volumes of the *Flora de Nicaragua*, dealing with seed plants. The fourth and final volume, on ferns and lycophytes, with Amy Pool as coeditor, was published in 2009. Altogether, the work describes the nearly 6,000 species of vascular plants that were then known for the country. About 1,000 additional species have been discovered subsequently, and the number continues to grow.

4 The first volume of *Flora Mesoamericana* appeared in 1994, and the entire multi-volume series is scheduled to be completed within 30 years, by which time it will cover approximately 19,000 species, about a third of them found only in the region covered by the flora. When completed, this work will be the largest and most comprehensive work on botany ever published in the Spanish language. Survey work done in conjunction with the preparation of the flora turned up many new species, including two previously unrecognized families, Lacandoniaceae and Ticodendraceae.

5 Julian Steyermark could not manage to complete the flora himself, but when he died four years after coming to St. Louis, he left us the balance of his estate, about $800,000, to use for that purpose. With those funds I was able to hire my former student Paul Berry to complete the work; Berry fulfilled Steyermark's dream seven years after his death. Steyermark's *Flora of the Venezuelan Guayana* became the most extensive complete flora published for any area of northern South America.

6 With funding from the Missouri Department of Conservation, an exemplary state agency, I was able in 1987 to hire George Yatskievych, a recent graduate of Indiana University with extensive knowledge of Midwestern plants, to revise Steyermark's 1963 *Flora of Missouri*. George carried out extensive research on Missouri's plants and supervised the publication of a model three-volume flora from 1999 to 2013. Thanks to a generous gift from Anne L. Lehmann, we were able to illustrate almost every plant species that occurs in Missouri with new, original drawings made from fresh living material.

II. CONSERVATION AROUND THE WORLD

1 A flora published by the Muséum in Paris starting in 1938 estimated that 8,500 species occurred on the island. We now estimate the flora of Madagascar to number at least 14,000 species. All but a very few of these species are found only there.

2 As the years went by, the system added features such as images, geo-referencing, and sophisticated search and report tools. Tropicos has grown to contain information on more than 1.33 million plant names and 4.87 million specimens, with 685,000 images and over 150,000 references. It supports more than seventy projects individually and in the general database.

3 The Center for Plant Conservation was housed for twenty-one years at the Missouri Botanical Garden, growing to become a healthy network of more than forty institutions.

4 ATREE began as a small organization but has since grown to be regarded as one of the best environmental "think tanks" in Asia. With its growing staff and Kamal Bawa's skillful assembly of the resources that make its important programs in

education and the environment possible, ATREE continues to provide outstanding service to India and, in doing so, has also become a model for other countries as well.

12. CHAMPION OF SCIENCE, EDUCATION, AND INTERNATIONAL COOPERATION

1 My work on the National Museum Services Board opened up some interesting connections, leading to my service on the Executive Committee of the American Association of Museums (AAM; 1980–83) and as a member of the AAM's Commission on Museums for a New Century (1981–84). I learned a great deal about the nation's thousands of museums during that period.

2 Since 2007, the Doomsday Clock has represented all possible causes of global catastrophe, including climate change. Since late 2018, the clock has stood at 2 minutes to midnight, the only time since 1953 it has reached that point.

3 Of the many honorary degrees that I have received, the one awarded by the University of Missouri–Columbia in May 1992 felt the most deserved and well-earned because of the efforts I had made as a member of the board of curators.

14. NEW ADVENTURES AT CAREER'S END

1 Tragically, our dear friend Ebbe Nielsen, who had invited us to Australia in the first place, died of a heart attack only a month before our arrival. As a memorial to Ebbe, we followed the itinerary he had set out for us, and I delivered a memorial lecture for him in Canberra.

2 The Butterfly House is named in honor of Sophia M. Sachs, wife of the late Sam Sachs, who founded Sachs Electric Company and played a key role in the early development of the city of Chesterfield.

3 John Muir, *My First Summer in the Sierra* (Boston: Houghton Mifflin, 1911; Sierra Club, 1988 edition), 110.

EPILOGUE: TOWARD A GREEN AWAKENING

1 Edward O. Wilson, *Biophilia* (Cambridge: Harvard University Press, 1984), 182.

2 Deborah Hardoon, "An Economy for the 99%" Oxfam International, 16 January 2017, accessed 28 September 2020, https://policy-practice.oxfam.org.uk/publications /an-economy-for-the-99-its-time-to-build-a-human-economy-that-benefits -everyone-620170.

3 See Andrew P. Dobson et al., "Ecology and economics for pandemic prevention," *Science* 369 (2020): 379–381. doi: 10.1126/science.abc3189.

4 Stewart Brand, *The Whole Earth Catalog*, 1968 [+ subsequent issues], [front matter].

5 The idea that violence has declined over time may seem to be in conflict with the current state of the world, but Steven Pinker makes a good case in *The Better Angels of Our Nature: Why Violence Has Declined* (New York: Viking Press, 2011).

6 Adlai E. Stevenson, Speech to the Economic and Social Council of the United Nations, Geneva, Switzerland, 9 July 1965. See https://stevensoncenter.org/about /stevenson.php.

7 Kurt Vonnegut, *A Man Without a Country* (New York: Random House, 2005), 122.

— Illustration Sources and Credits —

[ii] From *Co-Evolution of Pollinators and Evening Primrose* © Bryan Haynes 2011. (See color image, #49.)

[xi] Photograph © Pat Raven 2011. Courtesy of Pat Raven.

2 Public domain photograph from National Archives and Records Administration (NARA); Washington, D.C.; NARA Series: Passport Applications, January 2, 1906–March 31, 1925; Roll #: 126; Volume #: Roll 0126 - Certificates: 42444-43365, 13 January 1911–31 January 1911.

5 Public domain image courtesy of Raven family collection.

11 Public domain image courtesy of UC Berkeley, Bancroft Library: http://www.oac .cdlib.org/ark:/28722/bk0004b217j/?order=30.

13 Photograph by Benjamin Bean, ca. 1850. Public domain image courtesy of California State Parks.

15 Courtesy of Raven family collection.

18 Photograph by Walter Francis Raven. Courtesy of Raven family collection.

21 Courtesy of Raven family collection.

26 Photograph by Pat Raven. Courtesy of Pat Raven.

30 "Exhibit," *The San Francisco Examiner*, 13 June 1948, 32. (With grateful acknowledgment to *The San Francisco Examiner*, the Bancroft Library, California Academy of Sciences Library & Archive, and Tom Daniel.)

33 Photograph by Rob Badger. Courtesy of WinterBadger Collection.

40 Photograph © California Academy of Sciences. Courtesy of California Academy of Sciences.

41 Courtesy of Missouri Botanical Garden.

43 Courtesy of Raven family collection.

44 Photograph © California Academy of Sciences. Courtesy of California Academy of Sciences.

50 Photograph by Walter Francis Raven. Courtesy of Raven family collection.

54 Photograph © California Academy of Sciences. Courtesy of California Academy of Sciences.

60 Courtesy of Raven family collection.

60 Courtesy of Raven family collection.

67 Courtesy of Raven family collection.

77 Reprinted with permission of the UCLA Herbarium.

84 Photograph © Christine Evers 2018. Courtesy of Christine Evers.

85 Courtesy of Raven family collection.

94 Courtesy of the Hunt Institute for Botanical Documentation, Carnegie Mellon University, Pittsburgh, PA.

95 Photograph by Ewen Roberts, 2008, from Flickr.com. Used under CC BY 2.0 License.

96 Courtesy of the Hunt Institute for Botanical Documentation, Carnegie Mellon University, Pittsburgh, PA.

97 Photograph by Ian Alexander, 2019, from Commons.wikimedia.org. Used under CC BY SA 4.0 License.

110 Courtesy of Paul R. Ehrlich.

114 Courtesy of Raven family collection.

118 Stock image courtesy of Shutterstock.com.

119 Photograph © University of California Botanical Garden at Berkeley. Courtesy of University of California Botanical Garden at Berkeley.

125 Photograph by Chuck Painter, Stanford News Service.

127 Photograph by Luther Goldman. Courtesy of Washington Biologists' Field Club.

138 Courtesy of Sally Curtis.

140 Courtesy of Mara Yamauchi.

147 Courtesy of Manaaki Whenua Landcare Research.

152 Drawing by Keith R. West. From Peter Raven and Tamra Engelhorn Raven, *The Genus Epilobium (Onagraceae) in Australasia: A Systematic and Evolutionary Study* (Christchurch, 1976).

155 Photograph by Tamra Engelhorn Raven. Courtesy of Raven family collection.

156 Courtesy of Barbara G. Briggs.

169 Courtesy of Missouri Botanical Garden.

174 Courtesy of Missouri Botanical Garden.

185 Courtesy of Missouri Botanical Garden.

188 Courtesy of Missouri Botanical Garden.

197 Photograph by Dan Levin. Courtesy of Dan Levin.

198 Courtesy of Missouri Botanical Garden.

199 Courtesy of Missouri Botanical Garden.

199 Courtesy of Missouri Botanical Garden.

201 Photograph by John Donaldson. Courtesy of John Donaldson.

203 Courtesy of the Hunt Institute for Botanical Documentation, Carnegie Mellon University, Pittsburgh, PA.

219 Courtesy of Raven family collection.

228 Photograph by William F. Laurance. Courtesy of William F. Laurance.

233 Photograph by Fortunat Rakotoarivony. Courtesy of Missouri Botanical Garden.

239 Courtesy of Missouri Botanical Garden.

241 Courtesy of Missouri Botanical Garden.

247 Photograph by Mark McDonald, University of Missouri Archives.

251 Courtesy of Missouri Botanical Garden.

253 Official White House photograph courtesy of National Academy of Science Archives.

259 Photograph by Brendan Smialowski. Courtesy of George Soros.

261 Photograph by Rocío Deanna. Courtesy of Rocío Deanna.

274 Photograph by Mary Lou Olson. Courtesy of Missouri Botanical Garden.

279 Photograph by Tim Parker. Courtesy of Missouri Botanical Garden.

279 Photograph by Matthew Garrett, 2014. Courtesy of Peter Crane.

280 Courtesy of Missouri Botanical Garden.

293 Courtesy of Enterprise Holdings.

297 Photograph by Pat Raven. Courtesy of Pat Raven.

304   Photo with permission by Sun Hang, Kunming Institute of Botany, Chinese Academy of Sciences.

333   Photograph © Kathryn Raven. Courtesy of Kathryn Raven.

[334]   Public domain image by Alice Tangerini, Smithsonian Botanical Illustrator. Courtesy of the Smithsonian Institution.

336   Courtesy of Raven family collection.

338   Photograph © Leonid V. Averyanov 2013. Courtesy of Leonid V. Averyanov.

COLOR IMAGES, #1–27

1. Photograph by Walter Francis Raven. Courtesy of Raven family collection.

2. Top left: Photograph © Michael R. Jeffords 2018. Courtesy of Michael R. Jeffords. Top right and bottom right/left: Photographs © Arthur V. Evans 2019. Courtesy of Arthur V. Evans.

3. Photograph © Michael R. Jeffords 2014. Originally published in Michael R. Jeffords, *Butterflies of Illinois* (Illinois Natural History Survey, 2014), 294. Courtesy of Michael R. Jeffords.

4. Photograph © Arthur V. Evans 2019. Courtesy of Arthur V. Evans.

5. Painting © Karolyn Darrow 2018. Courtesy of Karolyn Darrow.

6. Photograph © Guy Bruyea 2009. Courtesy of Guy Bruyea.

7. Photograph by Donald Hobern, 2018, from Flickr.com. Used under CC BY 2.0 License.

8. Photograph © Stephen Sharnoff 2016. Courtesy of Stephen Sharnoff.

9. Photograph © Zoya Akulova-Barlow 2020. Courtesy of Zoya Akulova-Barlow.

10. Photograph © Sylvia Sharnoff 1998. Courtesy of Stephen Sharnoff.

11. Photograph © Stephen Sharnoff 2010. Courtesy of Stephen Sharnoff.

12. Photograph by Curtis Clark from Wikimedia.org. Used under CC BY SA 2.5 Generic License.

13. Top: Photograph by Katja Schulz, 2017, from Flickr.com. Used under CC BY 2.0 License. Bottom: Photograph by Greg Hume, 2008, from Commons.wikimedia .org. Used under CC BY 3.0 License.

14. Photograph by Mara Koenig/USFWS, courtesy of Midwest Region of the United States Fish and Wildlife Service. From Flickr.com. Used under CC BY 2.0 Generic License.

15. Photograph © Rob Badger 2003. Courtesy of Rob Badger.

16. Photograph by Ansel Adams, United States, 1902–1984, *Dr. Ledyard Stebbins and Students on Field Trip*, May, 1966, Gelatin silver print. Courtesy of the Sweeney/ Rubin Ansel Adams Fiat Lux Collection, University of California, Riverside/ California Museum of Photography.

17. Photograph © Michael Chasse 2017. Courtesy of Michael Chasse.

18. Art by Joshua Ellingson, Presidio Trust, 2016. Courtesy of Presidio Trust.

19. Photograph by Tom Hilton, 2010, from Flickr.com. Used under CC BY 2.0 License.

20. Photograph by ilya_ktsn, 2008, from Flickr.com. Used under CC BY 2.0 License.

21. Photograph © Barry Breckling 2009. Courtesy of Barry Breckling.

22. Courtesy of Raven family collection.

23. Photograph © Stephen S. Matson 2018. Courtesy of Stephen S. Matson.

24. Courtesy of Raven family collection.

25. Photograph by Josh Hull/USFWS, courtesy of United States Fish and Wildlife Service Pacific Southwest Region. From Flickr.com. Used under CC BY 2.0 License.
26. Photograph by David Harelson, Presidio Trust. Public domain (Federal government) image courtesy of the Presidio Trust.
27. Photograph by Richard Mack. Courtesy of Richard Mack.

COLOR IMAGES, #28–50

28. Photograph © David Lyttle 2018. Courtesy of David Lyttle.
29. Photograph by Michal Klajban, 2016, from Commons.wikimedia.org. Used under CC BY SA 4.0 International License.
30. Photograph by John Barkla, https://inaturalist.nz/observations/9377649. Used under CC0 1.0 Universal (CC0 1.0) Public Domain Dedication License.
31. Photograph © Larry Emerson 2019. Courtesy of Larry Emerson.
32. Photograph © Larry Emerson 2019. Courtesy of Larry Emerson.
33. Photograph © Michael R. Jeffords. Courtesy of Michael R. Jeffords. Originally published in Michael R. Jeffords and Susan L. Post, *Curious Encounters with the Natural World* (University of Illinois Press, 2017), xvii.
34. Photograph © Henry Domke 2012. Courtesy of Henry Domke.
35. Photograph courtesy of Missouri Department of Conservation.
36. Courtesy of Missouri Botanical Garden.
37. Ink drawing by Koichi Kawana. Courtesy of Missouri Botanical Garden.
38. Photograph by Heather Marie Osborn. Courtesy of Missouri Botanical Garden.
39. Courtesy of Raven family collection.
40. Courtesy of Missouri Botanical Garden.
41. Photograph by Dr. Tsai-Wen Hu, Taiwan Endemic Species Research Institute. Courtesy of Raven family collection.
42. James P. Blair/Nat Geo Image Collection. Photograph by James P. Blair, published in *National Geographic*, "The Plant Hunters: A Portrait of the Missouri Botanical Garden" by Boyd Gibbons, August 1990. Courtesy of National Geographic Partners.
43. Courtesy of Missouri Botanical Garden.
44. Courtesy of Missouri Botanical Garden.
45. Photograph by Charles Schmidt, 2003. Courtesy of Missouri Botanical Garden.
46. Photograph by Nathan Wambold, 2016. Courtesy of Missouri Botanical Garden.
47. Photograph © Brian J. Huntley. Courtesy of Brian J. Huntley.
48. Photograph by Charles Schmidt, 2008. Courtesy of Missouri Botanical Garden.
49. *Co-Evolution of Pollinators and Evening Primrose* © Bryan Haynes 2011. Courtesy of Bryan Haynes.
50. Photograph © Adam Robinson Photography 2019. Courtesy of Adam Robinson.

— INDEX —